A. Oksche E.M. Rodríguez
P. Fernández-Llebrez (Eds.)

The Subcommissural Organ

An Ependymal Brain Gland

With 98 Figures, some in Color

Springer-Verlag
Berlin Heidelberg New York
London Paris Tokyo
Hong Kong Barcelona
Budapest

Professor Dr. Drs. h.c. Andreas Oksche
Institut für Anatomie und Zytobiologie, Justus-Liebig-Universität, Aulweg 123,
35392 Giessen, Germany

Professor Dr. Esteban M. Rodríguez
Instituto de Histología y Patología, Universidad Austral de Chile,
Valdivia, Chile

Professor Dr. Pedro Fernández-Llebrez
Departamento de Biología Animal, Facultad de Ciencias, 29071 Málaga, Spain

ISBN-13:978-3-642-78015-8 e-ISBN-13:978-3-642-78013-4
DOI: 10.1007/978-3-642-78013-4

Library of Congress Cataloging-in-Publication Data The Subcommissural organ: an ependymal brain gland/ A. Oksche, E.M. Rodríguez, P. Fernández-Llebrez (eds.). Includes bibliographical references and index. ISBN-13:978-3-642-78015-8 DM 198.00 1. Subcommissural organ.
I. Oksche, A. II. Rodríguez, E.M. (Esteban M.) III. Fernández-Llebrez, P. (Pedro), 1954[DNLM: 1. Subcommissural Organ – congresses. WL 102 S941 1993] QP378.S8 1993 591.1′88—dc20 DNLM/ DLC for Library of Congress

Softcover reprint of the hardcover 1st edition 1993

Cover design: Erich Kirchner, Heidelberg, with illustration by Karola Michael, Giessen
Reproduction of the figures: Gustav Dreher, Stuttgart
Typesetting: Best-set Typesetter Ltd., Hong Kong

25/3130-5 4 3 2 1 0 – Printed on acid-free paper

"Innerhalb des Centralcanales fand ich sehr häufig einen im Querschnitt kreisförmig begrenzt erscheinenden Strang, der, 0,0015‴ im Durchmesser haltend, einem Achsencylinder sehr ähnlich aussieht und höchstens etwas stärker lichtbrechend ist. "." Da dieser Strang, wenn ich ihn überhaupt zu Gesicht bekam, stets von derselben Gestalt war, und nicht einmal jene Formverschiedenheiten darbot, welche die Axencylinder an Chromsäurepräparaten so häufig zeigen, kann ich nicht annehmen, dass er gleichbedeutend sei mit den unregelmässigen Massen, welche den Centralcanal bisweilen vollständig oder zum Theil erfüllen und im Rückenmark anderer Thiere oder des Menschen von mehreren Forschern erwähnt worden sind."

"Within the central canal I frequently observed a cord, which in cross-sectional profile is circular-appearing, measuring 0.10015‴ in diameter, very much resembles an axon, and at best displays a somewhat stronger light-refractive capacity. "." Since this cord, if I was lucky enough to find it, always displayed constant features and did not exhibit even those variations in form which are characteristic of axons in chromic acid preparations, I cannot assume that it is identical with the irregular masses which occasionally fill the central canal totally or partly and have been mentioned by several investigators to occur in the spinal cord in other animals or in man."

E. Reissner (1860): Beiträge zur Kenntnis vom Bau des Rückenmarkes von *Petromyzon fluviatilis* L.
Archiv für Anatomie, Physiologie und wissenschaftliche Medicin, Jg. 1860: 545-588; *see p 552*

.... "we may state that we do not consider that the therm 'ependymal groove' is sufficiently distinctive for so remarkable and constant a feature of the vertebrate brain." "Inasmuch as it lies beneath the posterior commissure, we propose to speak of it in the future as the 'sub-commissural organ'".

"Reissner's fibre and the sub-commissural organ occur throughout the vertebrate series from the cyclostomes to the primates, and there can be little doubt that where they are (as in nearly all cases) well developed, they must have some important function."

A. Dendy and G.E. Nicholls (1910) On the Occurrence of a Mesocœlic Recess in the Human Brain, and its Relation to the Sub-commissural Organ of Lower Vertebrates; with special reference to the Distribution of Reissner's Fibre in the Vertebrate Series and its possible Function. Anat. Anz. 32:496–509; *see pp 497, 506*

"Verfolgen wir den historischen Gang der Erkenntnis der Funktion der verschiedensten Organe, so fällt uns auf, dass in vielen Fällen die Erforschung des anatomischen Aufbaues weit gediehen war, bevor die Funktion und der Zusammenhang eines Organes mit anderen Organen richtig erkannt wurde. In vielen dieser Fälle war aber die genaue anatomische Kenntnis die Voraussetzung für die Möglichkeit der physiologischen Forscherarbeit."

"If we follow the historical development of the insight into the function of different organs, we realize that in many cases the investigation of the anatomical structure had widely outpaced the clear-cut knowledge of function and organic interrelationships. In many of these cases, however, exact anatomical knowledge was the prerequisite for the start of physiological investigations."

W. Kolmer (1921) Das "Sagittalorgan" der Wirbeltiere. Z. Anat. Entwicklungsgesch. 60:652–717; *see p 652*

Preface

During the past two decades the progress in neuroendocrine research, stimulated by the increasing general interest in neurosciences, has been very impressive. Most of these efforts have concentrated on neuroendocrine nerve cells and their systems. Even if some aspects have remained open to discussion, the principal functional role of the neuroendocrine units capable of elaboration of biological active peptides (peptidergic neurons) is quite well understood. The same holds true for the central aminergic neurons and for such photoreceptor-derived paraneuronal elements as the pinealocytes. The primordium of the central nervous system possesses potencies for central sensory and secretory differentiations. Among the latter, a non-neuronal ependymal structure – the subcommissural organ – has remained enigmatic in terms of its biological significance. The subcommissural organ is a common, very constant, and conservative property of the vertebrate brain, from cyclostomes to mammals, and its appears early in ontogeny. The spectacular secretory activity of this brain gland, located in the diencephalic roof at the entrance to the mesencephalic aqueduct, results in the formation of an intraventricular secretory product – Reissner's fiber. This peculiar structural complex has attracted investigators to use a wide spectrum of modern cytological and, more recently, also molecular methods to investigate the secretory process and the secretory product, primarily glycoproteins, in greater detail. So far, however, the progress in structural insight has outpaced our knowledge of the function of the subcommissural organ.

In order to review the present state of the art in the field and to stimulate further conceptual discussion, the editors of this volume decided to organize a conference on the subcommissural organ which was then held in Vélez-Málaga, Spain, between October 7 and 10, 1991. The list of participants documents that most scientists active in this area attended the symposium. The present book is not a mere chronological record of events or collection of individual presentations delivered at the conference. It also contains general overviews linking the specialized analysis of the subcommissural organ to general aspects of the cerebrospinal fluid, ventricular lining, circumventricular organs, and glycoprotein chemistry. In this respect, the avenues of thought and the hypotheses arising from them are very diverse. Furthermore, the volume encompasses an edited, although complete, record of four discussion sections and a historical synopsis covering in essence the period from the discovery of Reissner's fiber in 1860 to the onset of the present epoch of the research. Moreover, it contains novel information on the ontogeny of the secretory activity of the subcommissural organ, the nature of a dissolved, microscopically unstructured fraction of the secretory material released into the cerebrospinal fluid, and the existence of an immunoreactive proteinaceous component of the secretion in the subcommissural organ of human fetuses. To date, the problem of function of the subcommissural organ has not been solved. We hope that our book will stimulate new ideas and lead to extended analytic and experimental activities which will result in the clarification of the function of a mysterious secretory complex in the brain.

We want to acknowledge the substantial financial support by the following Spanish institutions that made it possible to organize this first inclusive symposium on the subcommissural organ and to assemble most scientists working in this field: Ministerio de Educación y Ciencia de España, Consejería de Educación y Ciencia de la Junta de Andalucía, Universidad de Málaga, and Ayuntamiento de Velez-Málaga.

A.O. and E.M.R. are greatly indebted to the Volkswagen-Stiftung (Hannover, Germany) for the generous support of their investigations between 1984 and 1992, including a contribution towards color illustrations. They are grateful to Ms. Karola Michael (Giessen) for her artwork and to Ms. Inge Lyncker and Dr. Frank Nürnberger (Giessen) for their help in preparing this publication. Furthermore, they thank Dr. T. Thiekötter and his staff at Springer-Verlag in Heidelberg for their invaluable advice and active support in publishing the present volume.

March 1993

A. Oksche
E.M. Rodríguez
P. Fernández-Llebrez

Contents

I. Historical Reflections

II. Phylogeny and Ontogeny

III. Biochemistry

IV. Secretory Process, Including Model Systems

V. Relation to Other Circumventricular Organs: Conceptual Aspects

VI. Neural Inputs

VII. Connections with Other Brain Structures

VIII. Physiology of the Cerebrospinal Fluid

IX. Experimental Aspects

X. Discussion Sessions (With 3 Figures)

List of Participants

J.A. ANDRADES
Departamento de Biología Celular
Facultad de Ciencias
29071 Málaga, España

J. BECERRA
Departamento de Biología Celular
Facultad de Ciencias
29071 Málaga, España

C. BOUCHAUD
Université Pierre et Marie Curie
Faculté des Sciences,
Institut des Neurosciences
UA CNRS 1199, Département de Cytologie
75230 Paris Cedex 05, France

P. BROWN
University of Manchester
School of Biological Sciences
Department of Physiological Sciences
Stopford Building, Oxford Road
Manchester M13 9PT, England

A. CALLE
Departamento de Biología Celular
Facultad de Ciencias
29071 Málaga, España

E. CARMONA-CALERO
Departamento de Anatomía
Facultad de Medicina
Universidad de La Laguna
La Laguna, Tenerife, España

A. CASTAÑEYRA PERDOMO
Departamento de Anatomía
Facultad de Medicina
Universidad de La Laguna
La Laguna, Tenerife, España

M. CIFUENTES
Departamento de Biología Animal
Facultad de Ciencias
29071 Málaga, España

I. CREVEAUX
Laboratoire de Biochimie Médicale
Faculté de Médecine
28 place Henri Dunant
63001 Clermont-Ferrand Cédex, France

I. DE AOS
Departamento de Biología Celular
Facultad de Ciencias
29071 Málaga, España

N. DELHAYE-BOUCHAUD
Université Pierre et Marie Curie
Faculté des Sciences,
Institut des Neurosciences
UA CNRS 1199, Département de Cytologie
75230 Paris Cedex 05, France

M. DIDIER-BAZES
Laboratorie d'Anatomie Pathologie
INSERM U171,
Faculté de Médecine Alexis Carrel,
Rue Guillaume Paradin
69372 Lyon Cedex 08, France

J.H.B. Diederen
Dept. Experimental Zoology
University of Utrecht, Padualaan 8,
3508 TB Utrecht, The Netherlands

G. Estivill
Departamento de Biología Animal
Facultad de Ciencias
29071 Málaga, España

P. Fernández-Llebrez
Departamento de Biología Animal
Facultad de Ciencias
29071 Málaga, España

B. Ghiani
Istituto di Anatomia Comparata
Università degli studi di Genova
Viale Benedetto XV, 5
16132 Genova, Italy

J.M. Grondona
Departamento de Biología Animal
Facultad de Ciencias
29071 Málaga, España

M. Hansel
Sektion Biowissenschaften
Universität Leipzig, Talstrasse 33
04103 Leipzig, Germany

S. Hein
Instituto de Histología y Patología
Universidad Austral de Chile
Valdivia, Chile

C.B. Hirschberg
Dept. Biochemistry and Molecular Biology
University of Massachusetts,
Medical Center
55 Lake Avenue North
Worcester, MA 01655, USA

A. Jimenez
Departamento de Biología Celular
Facultad de Ciencias
29071 Málaga, España

H.C. Jones
University of Florida Health Science
Center, Department of
Pharmacology and Therapeutics
PO Box 100267
Gainesville, Florida 32610-0267
USA

K. Knigge
University of Rochester,
Medical Center
School of Medicine and Dentistry
Division of Neuroendocrinology
601 Elmwood Ave., Box 609
Rochester, New York 14642, USA

H.-W. Korf
Zentrum für Morphologie
Johann Wolfgang Goethe-Universität
Theodor-Stern-Kai 7,
60596 Frankfurt a. Main

B. Krisch
Anatomisches Institut
der Universität Kiel, Neue
Universität, Olshausenstr. 40–60
24118 Kiel, Germany

A.M. Lancha
Dept. Biología Celular
Facultad de Biología
Universidad de La Laguna
38071 La Laguna, Tenerife, España

W. Lehmann
Sektion Biowissenchaften
Universität Leipzig, Talstrasse 33
04103 Leipzig

H. Leonhardt
Anatomisches Institut der
Universität Kiel, Neue Universität,
Olshausenstr. 40–60,
24118 Kiel, Germany

M.D. López-Avalos
Departamento de Biología Celular
Facultad de Ciencias
29071 Málaga, España

J.M. Mancera
Departamento de Biología Celular
Facultad de Ciencias
29071 Málaga, España

F. Marín-Girón
Departamento de Biología Celular
Facultad de Ciencias
29071 Málaga, España

A. Meiniel
Laboratoire de Biochimie Medicale,
CJF INSERM 88.06
28 Place Henri-Dunant BP 38
63001 Clermont-Ferrand Cedex,
France

R. Meiniel
Laboratoire de Biologie Animale
CNRS 677, Université Clermont II
63170 Aubiere, France.

M.D. Mota
Departamento de Biología Animal
Facultad de Ciencias
29071 Málaga, España

W. Naumann
Abt. Zellbiologie und Regulation
Universität Leipzig, Talstr. 33
04103 Leipzig, Germany

F. Nualart
Instituto de Histología y Patología
Universidad Austral de Chile
Valdivia, Chile

F. Nürnberger
Institut fur Anatomie und
Zytobiologie, Justus-Liebig-Universität
Aulweg 123, 35392 Giessen,
Germany

A. Oksche
Institut fur Anatomie und
Zytobiologie, Justus-Liebig-Universität
Aulweg 123, 35392 Giessen,
Germany

R. Olsson
Department of Zoology,
University of Stockholm,
10691 Stockholm, Sweden

M. Pérez-Martín
Departmento de Biología Celular
Facultad de Ciencias
29071 Málaga, España

J. Pérez
Departamento de Biología Animal
Facultad de Ciencias
29071 Málaga, España

J.M. Pérez-Fígares
Departamento de Biología Celular
Facultad de Ciencias
29071 Málaga, España

P. Riera
Departamento de Biología Celular
Facultad de Medicina,
J Clavería S/n
33006 Oviedo, España

E.M. Rodríguez
Instituto de Histología y Patología
Universidad Austral de Chile
Valdivia, Chile

S. Rodríguez
Instituto de Histología y Patología
Universidad Austral de Chile
Valdivia, Chile

J.A. Santamaría
Departamento de Biología Celular
Facultad de Ciencias
29071 Málaga, España

L. Santos
Departamento de Biología Celular
Facultad de Ciencias
29071 Málaga, España

K. Schoebitz
Instituto de Histología y Patología
Universidad Austral de Chile
Valdivia, Chile

W.B. Severs
Department of Pharmacology,
The Milton S. Hershey
Medical Center
Pennsylvania State University,
PO Box 850,
Hershey, Pennsylvania 17033, USA

G. Sterba
Abt. Zellbiologie und Regulation,
Universität Leipzig, Talstr. 33
04103 Leipzig, Germany

R. Yulis
Instituto de Histología y Patología
Universidad Austral de Chile
Valdivia, Chile

I. Historical Reflections

Ernst Reissner
(1824–1878)
Anatomist of the "Membrane" and "Fiber"

Ernst Reissner was born in Riga on 12 September 1824. At the University of Dorpat (now Tartu), Estonia, he was a student of K.B. Reichert and apparently a very successful investigator, because already at the age of 23 years he was awarded the "*goldene Preis-Medaille*," with the promise that the university would cover the cost of printing the manuscript. In 1861, he received his doctorate (*doctor habilitatus*) with the thesis on the inner ear "*De auris internae formatione*," after having been appointed at Dorpat University since 1851. In 1857, he became Professor of Anatomy, a position which he held with great success until he retired in 1875, only 51 years old. After a number of years of distinguished work the honorary title of a Real State Councillor (*Wirklicher Staatsrath*) was conferred on him by the imperial Russian government. Reissner died on 4 September 1878. His name has been preserved for following generations through the eponyms "Reissners membrane" (*membrana vestibularis*) in the inner ear and "Reissners fiber" in the brain and spinal cord.

Ragnar Olsson

Recommended Reading

Album Academicum der Kaiserlichen Universität Dorpat (1867) Laakman, Dorpat, p 215
Hasselblatt A, Otto G (1889) Album Academicum der Kaiserlichen Universität Dorpat, Mattiesen, Dorpat, p 349
Piirimäe H (1982) Tartu Ülikooli Ajalugu II, 1798–1918. (History of Tartu University; in Estonian) Tallinn
Politzer A (1913) Geschichte der Ohrenheilkunde, vol II. Enke, Stuttgart, p 24

Helmut Hofer (1912–1989) and the Concept of the Circumventricular Organs

A. OKSCHE

Helmut Hofer: Vienna, 1985

If this conference on the subcommissural organ (SCO) had taken place only a few years earlier, Helmut Hofer would have been among the most inspiring discussants, surprising the audience with brillant reflections and new avenues of thought. When listing the SCO among the circumventricular organs we should be aware that this now widely accepted term was coined by Hofer in 1958 after thorough discussions with Wolfgang Bargmann. Since the publication of this fundamental concept in 1959, Hofer was fascinated by the SCO, a highly secretory component of the circumventricular system.

Helmut Hofer was a very skilled and dedicated comparative anatomist. His broad scientific background helps to understand his inventive approach to the problem of the circumventricular organs. Organlike structures lining the ventricular cavity had been observed and analyzed since the beginning of the century by a number of scientists, including such eminent neurobiologists as Studnička, Legait, Bargmann, and Wislocki (for details, see Leonhardt 1980). Hofer aimed to establish an overall concept for these partly neuronal, partly ependymo-glial structures endowed with sensory, secretory, and transporting properties. He noticed the peculiar position of these organs between the inner (ventricular) and the outer (subarachnoid) CSF compartments. He also emphasized the very rich and mostly specialized vascularization of these areas, the peculiarities of their blood–brain barrier, and, for a representative number of circumventricular organs, the median position of these differentiations. Hofer was careful enough not to postulate a unifying functional theory; he was not convinced that all circumventricular organs belong to one family and correlated them to the adjacent brain structures. Furthermore, he emphasized that some of the circumventricular organs are membranous, others occupy the entire thin ventricular wall, and still others are embedded in proliferated, well-organized divisions of the brain matter. Thus, Hofer's analysis and his conclusions were developmental, structural, and topographical. Some later authors have used the term "circumventricular organs" in a less precise way. Until the problems of the circumventricular organs can be solved on a functional basis, Hofer's definition appears to offer the most flexible framework for further investigations.

Helmut Hofer was born on 22 October 1912 in Weisskirchen, Moravia, in those days a part of the Austro-Hungarian empire. After graduation from a traditional Austrian *Gymnasium* (secondary school) in Vienna, he studied biology and medicine between 1932 and 1937 at the University of Vienna and received his Ph.D. (Dr. phil.) in Zoology. Among his professors were the comparative anatomist Versluys, who became his scientific mentor, the neuroanatomist and embryologist Hochstetter and the anatomist Pernkopf, scientists of supreme international reputation. The personality and the ideas of these great masters had a lasting influence on Hofer and his further scientific development. Hofer's first independent professional appointment at the Zoological Museum in Dresden, the capital of Saxony, was interrupted by military service (1940–1945) in the German army; Austria had lost its independence in 1938.

After the end of the war, between 1945 and 1953, Hofer was research associate at the Department of Zoology of the University of Vienna; there he obtained the *venia legendi* and became *Privatdozent* after the submission of his *Habilitationsschrift* (academic thesis). During this period he completed a number of thorough studies on the comparative anatomy of the skull with particular consideration of the avian cranium. In 1953, Hofer was invited by the distinguished neuroanatomist Hugo Spatz to join the Max Planck Institute for Brain Research that had been moved from Berlin to Giessen, until the definitive facilities of this department were established in Frankfurt. These 9 years in Giessen became a very stimulating and productive period in the scientific life of Hofer, culminating, in 1958, in his fundamental concept of circumventricular organs. Under the influence of Hugo Spatz, for whom he felt a deep admiration, Hofer materialized his genuine interest in primatology by investigating the relationships of the primate brain to the skull and the meningeal formations. Here he began his studies of the ventricular wall. At this time Hofer had the academic status of an Adjunct Associate Professor at the Department of Zoology of the University of Giessen, whose director Professor Emmo Ankel was a supporter of Hofer's comparative interests. In 1960, Thenius and Hofer published a monograph on the phylogeny of mammals (*Stammesgeschichte der Säugetiere*), a classic in the field.

When the Max Planck Institute for Brain Research was moved to Frankfurt in 1962, Hofer became the head of the Section of Primatology until he, in 1965, left Germany for the United States.

Between 1965 and 1975 Helmut Hofer was first Visiting Scientist and then Research Associate at the Delta Primate Research Center in Covington, Louisiana. During this period he also became Adjunct Associate Professor at Tulane University, New Orleans. The decade in Covington was characterized by a further potentiation of Hofer's primatological investigations. The opus magnum of this period is the book by Hofer and G. Altner "*Die Sonderstellung des Menschen*," a documentation of his profound knowledge of phylogeny and primatology. In addition, he published a number of research reports on specialized brain structures in Holostei, Marsupialia, and Xenarthra. He became a member of numerous scientific organizations and committees. In 1974, Helmut Hofer received the very prestigious Alexander von Humboldt Senior Scientist Award enabling him to work for a full year at a German university of his choice. He decided to join the newly founded University of Kassel that became the last station in his scientific activities. The end of his fellowship year coincided with his retirement from the Delta Primate Center, at which time Hofer was invited to accept a Visiting Professorship at the University of Kassel covering two semesters in 1975 and 1976. At the University of Kassel Hofer had a laboratory until his death on 26 July 1989. During this period his inspiring personality attracted a number of graduate and postgraduate students. This was his intellectual contribution to the scientific profile of the new university. In appreciation of his numerous scientific accomplishments, in 1985 the University of Kassel bestowed an honorary doctoral degree upon Helmut Hofer. During the Kassel period his investigations were mainly concerned with the SCO and the comparative anatomy of the oral cavity.

With reference to the SCO, Hofer published a number of remarkable observations. He was particularly fascinated by the caudalmost site of the release of the secretory material aggregated in Reissner's fiber into the systemic circulation. After studies in amphioxus, followed by very provocative ultrastructural observations on the vascular discharge of this material in lampreys, Hofer succeeded in demonstrating similar structures in the filum terminale and ampulla caudalis of monkeys, an area designated by him as "the terminal organ of the subcommissural complex." In the meantime his observations in the lamprey have been confirmed and extended by Rodríguez and collaborators. These scientific accomplishments deserve particular attention in context with the present conference.

Helmut Hofer was a great scholar and an investigator possessing an extraordinary measure of intuition. He was a fascinating lecturer and inspiring academic teacher. And last but not least, he was an amiable companion, well-educated, stimulating and open-minded, a master of spirited unconventional aperçu. We miss not only a distinguished member of the scientific community but also a dear personal friend.

References

Hofer H (1959) Zur Morphologie der circumventriculären Organe des Zwischenhirns der Säugetiere. Verh Dtsch Zool Ges 22:202–251

Hofer H (1986) Observations on the discharge of the secretion from the ependymal cells of the subcommissural organ (SCO) of some South American primates. Gegenbaurs morphol. Jahrb 132:205–230

Hofer H (1987) The terminal organ of the subcommissural complex of chordates: definition and perspectives. Gegenbaurs morphol. Jahrb 133:217–226

Leonhardt H (1980) Ependym und circumventriculäre Organe. In: Oksche A, Vollrath L (eds) Neuroglia I. Springer, Berlin Heidelberg New York, pp 177–665 (Handbuch der mikroskopischen Anatomie des Menschen, part IV, vol 10)

Historical Landmarks in the Investigation of the Subcommissural Organ and Reissner's Fiber

E.M. Rodríguez and A. Oksche

The aim of this review is to provide a precise account of the fundamental discoveries and the conceptual progress in investigating the subcommissural organ (SCO), under particular consideration of the sequence of the events and of the priorities. Our studies of classical papers have shown that copying of historical information from earlier overviews has led to the perpetuation of certain mistakes and even to a loss of sharp contours in basic statements.

In 1860, Reissner, anatomist at the University of Dorpat, published a comprehensive monograph on the light-microscopic structure of the spinal cord of *Petromyzon fluviatilis*. He called attention to a string, which in cross sections has a diameter of 1.5 μm and is characterized by its high refringence, its extremely regular shape and by lying free within the central canal. Reissner discriminated this thread from the amorphous mass that usually occupies the lumen of the central canal; the origin and ending of the fiber, however, remained unknown. In 1866, Kutschin, again in *P. fluviatilis*, confirmed Reissner's observation and named the threadlike structure within the central canal in honor of its discoverer "Reissner's fiber."

Stieda (1868) reported on the presence of a Reissner's fiber (RF) in all teleost specimens he examined for this purpose. He described RF as a cylindrical rod of homogeneous structure with a uniform diameter of 3.8 μm. However, Stieda considered RF to be an artifact caused by the fixation of the tissue. Sanders (1878) observed the occasional presence of a rodlike structure in the central canal of the teleost fish *Mugil* and agreed with Stieda (1868) in considering RF to be an artifactual coagulum of the cerebrospinal fluid. Although, in 1894, Sanders continued to interpret RF as an artifact, he stated it to be conspicuous in *Myxine*; furthermore, he noted that at the end of the central canal RF gains the appearance of a "coiled mass."

In 1895, Studnička classified the caudal end of the central canal in the spinal cord of *Petromyzon* and *Myxine* as a "terminal sinus" endowed with a thin wall and communicating with the surrounding tissue via a "wide pore." In 1899, Studnička made several important observations, most of which were later proven to be valid. Thus, he described RF in a series of species, although his detailed observations were made in *Petromyzon fluviatilis*. He confirmed Sanders' (1894) observation of a coiled mass of RF inside the terminal sinus, but considered this "coiled condition" to represent a normal state. He also recognized that at its caudal end RF escapes the terminal sinus via a terminal porus of the neural tube and extends to the neighboring connective tissue. Studnička regarded RF as a secretory product of the ependymal cells of the central canal.

It appears that the first (rather cursory) description of what later became known as "the subcommissural organ" was given by Edinger (1892) in sharks and followed by Studnička (1900), who called attention to uncommonly tall ependymal cells covering the posterior commissure of *Petromyzon fluviatilis*. He depicted these cells in instructive line drawings (Figs. XXXIX: 6, 7 in Studnička 1900). However, Studnička made no reference toward a relationship between this specialized ependyma and RF.

The publication by Sargent (1900) represents the first real leap in the knowledge of the SCO and RF and establishes the basis of what is presently regarded as the SCO–RF complex, illustrated by an amazingly accurate schematic representation of this assembly (Fig. 1). He was the first to perform a comparative study of RF, including 60 species of all classes. From his observations he concluded that RF is an organic structure formed in the living animal, being present in all classes of vertebrates. Sargent also performed the first histochemical study of RF by using several staining methods. He emphasized that RF displays an internal structure; when double stained, its medullary portion appears Congo-red positive, surrounded by a peripheral ring possessing affinity for hematoxylin. In this early investigation it was also established that the thickness of RF largely depends on the species, age of specimen, and anatomical location showing also a coarser RF during the embryonic stage.

The most amazing contribution of Sargent (1900), however, is his detailed anatomical description of RF: caudally RF extends to the very end of

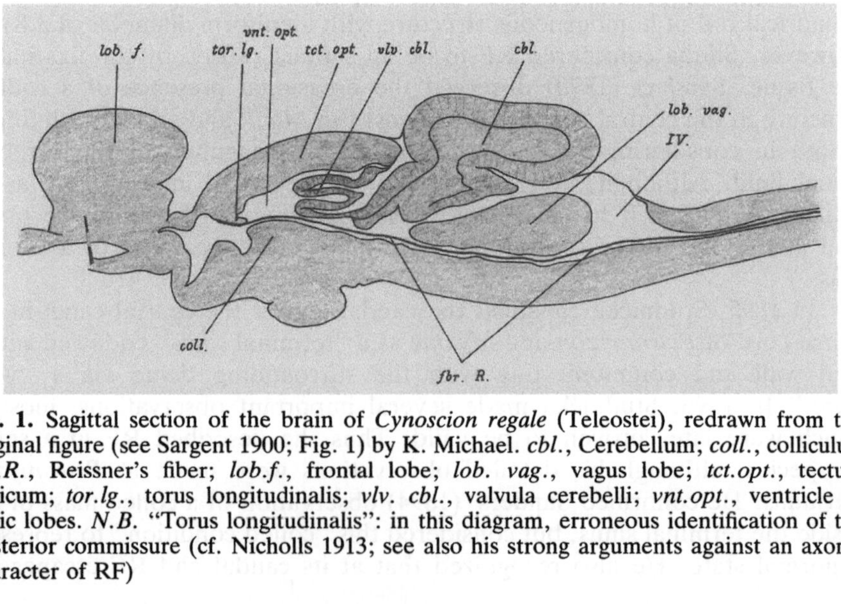

Fig. 1. Sagittal section of the brain of *Cynoscion regale* (Teleostei), redrawn from the original figure (see Sargent 1900; Fig. 1) by K. Michael. *cbl.*, Cerebellum; *coll.*, colliculus; *fbr.R.*, Reissner's fiber; *lob.f.*, frontal lobe; *lob. vag.*, vagus lobe; *tct.opt.*, tectum opticum; *tor.lg.*, torus longitudinalis; *vlv. cbl.*, valvula cerebelli; *vnt.opt.*, ventricle of optic lobes. *N.B.* "Torus longitudinalis": in this diagram, erroneous identification of the posterior commissure (cf. Nicholls 1913; see also his strong arguments against an axonal character of RF)

the central canal where "it disappears in the loose mass of cells forming the primitive terminal part of the cord." Furthermore, Sargent was able, for the first time, to trace RF "backward" to the fourth ventricle and Sylvian aqueduct. His statement concerning the rostral "termination" of RF is remarkable: "RF reaches the torus longitudinalis" (*post. comm.*), "which projects ventral and caudal from the anterior wall of the third ventricle (Fig. 1). Passing along the median fissure of the torus for two thirds its length, the fiber passes beneath the membrane which covers the torus; the fiber is deflected into the deeper parts of the torus. Just what becomes of it from this point on cannot be definitely stated at present. The evidence, however, so far as it goes, indicates that it branches repeatedly, and that the branches thus arising are intimately associated with the gelatinous tracts of the torus . . . When coursing along the central canal, one may see fine fibrillae coming off from RF and running outward toward the periphery of the canalis centralis." Sargent (1900) considered RF to be a nervous structure and firmly dissagreed with a secretory nature as suggested by Studnička (1899).

In 1902, Dendy described in the roof of the third ventricle of lamprey larvae a pair of longitudinal grooves formed by elongated and ciliated columnar cells covering the posterior commissure. He regarded these cells as specialized ependymal elements. In this publication Dendy proposed that RF "may serve by ciliary action to promote the circulation of the cerebrospinal fluid."

Sargent (1903) was the first to demonstrate unequivocally the existence of specialized cells covering the posterior commissure, named by him "ependymal groove," throughout a large series of vertebrates. He also noted that the ependymal groove is inconspicuous in mammals. In this report it was established for the first time that the ependymal cells of the ependymal groove and RF are intimately connected. Sargent regarded the "ependymal groove as an attachment plate for RF." In 1904, Sargent indicated that RF is an "ependymal groove"-dependent structure and persisted in his view that RF represents a mesencephalic axonal formation. In this latter report he mentioned experiments performed on elasmobranchs by means of surgical severance of RF. He concluded that "when RF is severed, the power to respond quickly to optic stimuli is lost" and postulated that the ependymal groove and RF may provide an apparatus for optic reflexes. In contrast, Kolmer (1905) regarded RF of ammocoetes to be a secretory product of ependymal cells, thus sharing the opinion of Studnička (1899).

In 1902, Boecke had produced the first description of the "infundibular organ" in amphioxus. This author (Boecke 1908) published a comprehensive and detailed description of the so-called infundibular organ, in the same species. Boecke concluded that the cells of the infundibular organ are of neuroepithelial nature, serving sensory functions. The presence of an RF-like structure in the central canal of amphioxus was first reported by Wolff (1907).

The existence of RF in primates was first indicated by Horsley (1908), who examined tailed monkeys (cynomolgus, rhesus). Horsley performed lesion experiments in macaque monkeys, stating that when the continuity of RF is destroyed, it exhibits none of those degenerative changes characteristic of severed nerves.

Nicholls (1908, 1909) postulated that RF is a highly elastic structure that is not connected to the walls of the central canal as assumed by Sargent (1900). Nicholls (1909), in *Gasterosteus aculeatus*, also noted that the filum terminale is reduced to an epithelial tube devoid of surrounding nervous matter. This tube, which contains RF, leaves the vertebral column and approaches the skin of the caudal fin. Here RF escapes from the tube through a terminal foramen and contacts the surrounding connective tissue, in accord with Studnička's (1899) previous observations in *Petromyzon fluviatilis*.

Dendy (1909) advanced the hypothesis that RF and the associated ependymal groove might be involved in regulating flexure of the body. He considered that in live animals RF is under considerable tension, and in consequence, it would be permanently subjected to varying strains with every deviation from the longitudinal axis of the body. He further suggested that such variations in RF tension might act as stimuli upon the columnar cells of the ependymal groove.

In 1910, Dendy and Nicholls published a joint paper containing further crucial information. They "did not consider that the term ependymal groove was sufficiently distinctive for so remarkable and constant a feature of the vertebral brain." They continued: "Inasmuch as it lies beneath the posterior commissure, we propose to speak of it in the future as the subcommissural organ." Furthermore, Dendy and Nicholls presented a detailed description of the SCO organ of the mouse, cat, and chimpanzee, including line drawings depicting the light-microscopic structure of the SCO almost to the extent as it is known today. They established firmly the connection between the SCO and RF, and indicated that SCO and RF are a constant structure of the brain from cyclostomes to primates. This report also included an investigation of the human SCO; it contains the first information indicating that in the human the SCO reaches its maximum development during embryonic life, whereas in adult humans it represents an entirely vestigial structure. [This observation was later confirmed by Puusepp and Voss (1924)].

In 1912, Nicholls published a further paper including some fine structural findings on the SCO and RF and the results of a series of experiments. Among the former worth mentioning is his description of the origin and termination of RF: "From the ventricular surfaces of the cells of the subcommissural organ a number of fine, long cilia-like processes grow out and coalesce to form a single thread which may be followed backwards through the brain ventricles . . . Reissner's fiber passes straight into the sinus terminalis and is there continued into the apex of a conical mass, the base of which is continuous with the hinder wall of the terminal sinus. For this mass

I have proposed the name 'terminal plug,' which apically seems structureless and whose constituent fibrillae may be supposed to be separating centrifugally to become attached to the hind wall of the terminal sinus."

Based on his previous findings (Nicholls 1909) in elasmobranchs, showing that in the filum RF is located within an ependymal tube devoid of nervous matter, Nicholls (1912) transected the fiber at this distal level. He arrived at two important conclusions: (1) breakage of RF produces a temporary loss of control of body position, leading him to the assumption that RF represents a part of a sense organ recording body movements; (2) a new terminal sinus is formed proximal to the lession, harboring RF that grows into this new sinus. This study was completed by extended observations on the development and the nature of RF and SCO (Nicholls 1913).

In a comprehensive record on the occurrence of central sense organs and their intraependymal neurons (= CSF-contacting neurons) in *Petromyzon*, Tretjakoff (1913) dealt also with his extensive observations on RF and SCO. He firmly denied the assumption that RF might represent a bundle of nerve fibers. Furthermore, he was unable to demonstrate sensory cells in the SCO groove, resulting in a disapproval of the theory of Dendy and Nicholls. In the opinion of Tretjakoff, the SCO groove serves only the formation of RF and the attachment of its proximal portion. He assumed that, as a consequence of lateral bending of the spinal cord, RF may exert pressure on the outer segments of the intraependymal sensory cells. In this respect, Tretjakoff tried to establish comparative parallels among the RF apparatus and the intracerebral static (otholitic) organs of ascidian larvae as well as ctenophores.

The controversial aspects of the problem were reapproached by Nicholls (1917) in a long article giving detailed account on his transection experiments. In addition to a new series of incisions, Nicholls provided a record of the histological postexamination of specimens from his previous experimental series published in 1912. The paper of Nicholls contains a detailed and precise review of the suggestions which had, to date, been made concerning the nature and function of RF and SCO. However, Nicholls' remark that Reissner (1860) believed that his "*Centralfaden*" was "simply a nerve fiber" goes too far and cannot be substantiated by a strong quotation from Reissner's paper. Also, Nicholls (1917), in an attempt to refute Tretjakoff's (1913) hypothesis, obviously misinterpreted the CSF-contacting neurons. He reemphasized the opinion that RF is "formed by the coalescence of cilia-like processes springing from cells which, while largely collected upon subcommissural organ, are not limited to that organ, other cells occurring scattered in the ependymal lining of the central canal contributing to this fiber." He concluded: "In my opinion the fiber is to be regarded as a thread of protoplasm . . . Perhaps the most remarkable characteristic of the fiber is its extreme elasticity." Furthermore, Nicholls' new experimental studies supported additional information concerning the spiral winding of the fiber as well as the duration of the reaction to the incision and the problem of the regeneration of RF. The time required for regeneration of Reissner's fiber was apparently not less than a week, in the case of the experimental material probably several weeks. The regeneration mechanism, however, remained obscure. From his new experiments Nicholls concluded, in accord with Dendy, that "the fibre serves to control automatically the flexure and pose of the body." Moreover, Nicholls assumed that the related sensory cells are "largely concentrated in the sub-commissural organ" although he was unable to exclude the existence of other scattered elements along the entire length of the canalis cerebralis.

A fine-histological study of the SCO was performed by Bauer-Jokl (1917). She called attention to the stratification of the epithelial tissue of the SCO and postulated a secretory role for this specialized area of the ventricular wall. Bauer-Jokl emphasized the spectacular development and differentiation of the SCO in the dog and bovine. However, she did not reveal a particular relationship between SCO and RF, thus remaining in a certain contradiction to the statements of Dendy and Nicholls. According to Bauer-Jokl, a well-developed SCO can be observed in human fetuses, but it is impossible to detect this organ in adults.

In 1921, Kolmer published an extensive comparative paper postulating the existence of a "sagittal organ," composed of SCO, RF, and sensory nerve cells located in the wall of the central canal and protruding a process into its lumen. He was not able to prove the existence of sensory cells in the SCO, although he did not exclude this possibility. At the same time Kolmer regarded RF as the product of a synthetic (secretory) activity of the SCO, released into the CSF and moved caudad by ciliary action. He emphasized that the formation of RF resembles elaboration of the cupula material in the labyrinth of the inner ear. Moreover, Kolmer commented on the rudimentary SCO of the hedgehog, the regressive ontogenetic development of the human SCO, the missing RF in human embryos and newborns, and the outgrowth of RF in regenerating tail segments of lizards. Since he used perfusion fixation for preservation of his specimens, he obtained histological material of supreme technical quality.

In 1922, Kolmer's paper was followed by Agduhr's (1922) monographic article dealing with the problem of a "central sensory organ in vertebrates." Agduhr's work based on observations in 206 vertebrate species had been conducted in parallel with Kolmer's investigations. This situation, overshadowed by aspects of priorities, offered Agduhr the opportunity of a critical analysis of Kolmer's findings. Agduhr was unable to find a correlation between the degree of formation of RF and the extent of the population of the intraependymal nerve cells. He emphasized that the latter are well developed also in cases with rudimentary or missing SCO and RF. In contrast, he indicated that in species with rudimentary or missing SCO, no RF could be found. Agduhr came to the conclusion that there is no evidence that RF and the intraependymal nerve cells are integral parts of one sensory organ. He suggested that the extended apparatus of intraependymal nerve cells may respond to changes in CSF circulation and pressure, and – in its entirety – serve the control of intracerebral pressure.

This is further substantiated by Agduhr's extensive experimental work with the teleost *Osmerus*, where he interrupted RF by surgical intervention. According to Agduhr, the interpretation of these experiments is difficult due to the fact that major pathways of the spinal cord had also been lesioned. In conclusion, Agduhr felt unable to support Sargent's (optic reflex apparatus) or Dendy's (control of body flexion) theories. In further experiments he fixed animals under different states of flexion of their bodies (and thus also spinal cords); no evidence was found that this procedure had led to a closer contact between RF and the outer segments of intraependymal sensory cells. In Agduhr's opinion, the "ependymal sensory organ" is an apparatus independent from the SCO–RF complex. Thus, he refuses to accept the term "sagittal organ" as proposed by Kolmer and suggests using the term "median organ" exclusively for the SCO–RF complex. Very apparently Agduhr was able to observe a more intimate connection of RF material with the SCO cells as had been documented by Kolmer. Without doubt, Agduhr's attention was mainly focused on the system of intraependymal (sensory) nerve cells (now usually named CSF-contacting neurons). He indicated that such sensory neurons may occur also in the wall of lateral ventricles.

Krabbe (1925, 1933) distinguished two layers in the SCO, i.e., the ependymal and the subependymal division; he introduced the term "hypendyma" (1933) to depict a tissue layer located under the ependymal cells of the SCO, containing numerous blood vessels and loosely arranged glial and parenchymalike cells.

Jordan (1925) by using a series of different staining methods concluded that SCO and RF share the same staining properties.

A series of papers were published almost simultaneously on the ontogenetic development of the SCO and RF in birds (Schumacher 1928; Kux 1929; Mathis 1929). Kux (1929) reported that in canary embryos RF is first seen in the central canal at embryonic (E) day 10, reaching the caudal neuropore at E day 13. Kux (1929) and Mathis (1929) showed that RF grows out through the distal neuropore of avian embryos and terminates freely in the amniotic fluid.

In 1931, Kolmer extended his previous observations on the system he had named the "sagittal organ" and reemphasized the close relationship of RF to intraependymal receptor neurons in the central canal. In his opinion, this central sensory organ exists in all (or nearly all) vertebrates. From thorough observations in several ophidian and lacertilian species, namely, animals curving their body frequently and extensively, Kolmer assumed with high certainty that RF is capable of contacting the intraependymal sensory cells and declined Agduhr's (1922) criticism. Since in the adult human, SCO and RF are missing and the central canal appears to be obliterated, Kolmer investigated in detail the situation in well-perfused rhesus monkeys and showed that they possess all components of the "sagittal organ" complex. In addition, he studied in these monkeys the "attachment" of RF to the SCO and its termination in the ampulla of the central canal. From the course of the "regeneration process" of RF in the regenerating tail of lizards, Kolmer was convinced that RF material is continuously shifted caudad by means of forces generated by ciliary beat. This may also explain the fact that in giant snakes (python, boa) RF is several meters long. Kolmer denies that the free cells associated with RF (previously observed by Sargent, Kolmer, and Agduhr) can be correlated to the formation and outgrowth of RF (Sargent); he regards them as "*Wanderzellen*", i.e., phagocytotic elements.

Credit should be given to Turkewitsch (1936) for his careful examination of the development of the bovine SCO, a frequently used model in recent analytic work. This study was a part of systematic investigations of the entire epithalamic region. Turkewitsch paid particular attention to the hypendymal cells of Krabbe. He observed similarities in the development between the hypendyma of the bovine SCO and the posterior lobe of the human pineal organ. In his opinion, the bovine pineal organ is devoid of differentiations resembling the posterior lobe of the human pineal. The excessive development of the hypendymal cells in the bovine SCO is viewed in the light of this evidence.

Pesonen (1940b) concentrated on the study of the human SCO, and confirming the findings by Dendy and Nicholls (1910) and Puusepp and Voss (1924) concluded that in the human the SCO attains its maximum development during embryonic life and undergoes regression after the birth. He called attention to the rich vascularization of the SCO (see also Pesonen 1940a).

By performing a careful light-microscopic study of the SCO of rats and lizards, Reichold (1942) provided strong evidence that the SCO is a secretory structure. In a number of cases he could observe that fine threads apparently arising from the tips of cilia were joining together to form RF; due to the limitations of resolving power of the light microscope it was impossible to identify the exact source of this material.

In 1942, E. Legait published a comprehensive monograph dealing with several circumventricular structures, including the SCO. With respect to the human SCO he arrived at conclusions similar to those of Dendy and Nicholls (1910), Puusepp and Voss (1924), and Pesonen (1940b). His main contribution was, however, the cytological analysis of the frog SCO under various experimental conditions. He reported that in this species the secretory activity of the SCO shows alterations in parallel to color change and in response to osmotic stimuli. In a later publication, Legait (1949) called attention to the close association between cells of the SCO and blood vessels. An inspiring comparative analysis of the SCO can be found in Bargmann's (1943) outstanding monograph on the pineal organ.

Eberl-Rothe (1951) appears to be the first to succeed in collecting fresh RF from some mammals.

In 1950, Stutinsky published a very important paper demonstrating for the first time (in anurans) the occurrence of a selectively stainable, 'Gomori-positive', secretory material in the SCO cells. After this report a new epoch in SCO research began. Now it became possible to visualize the site of secretion and the pathways of the release of the secretory material and to compare the activity of secretory ependymal elements with that of secretory nerve cells (neurosecretion). The events and accomplishments of this new epoch are reflected by numerous references and comments in individual articles of this volume as well as in the recordings of the discussion sessions. Therefore, we resist the temptation to extend our review beyond the period of Stutinsky's discovery. For a detailed account of more recent developments, see Rodríguez et al. (1992).

Evaluating the proceedings of this conference we conclude that the secretory products of the SCO can be released in different ways. In addition to RF material, a CSF-soluble form of the secretion does exist. Furthermore, there is an apparent discharge of secretory material to the perivascular space and to the leptomeningeal compartment of the CSF. Thus, we regard the SCO parenchyma as the central element in this complicated arrangement and avoid speaking of the "SCO–RF complex" when considering basic functional aspects.

Finally, a comment should be made with respect to the perspectives of future SCO research. Despite the valuable information collected during the last four decades, two crucial aspects of the SCO remain unsolved: (1) the exact chemical structure of the secretory products and of their encoding genes and (2) the functional role of the SCO. Investigations aiming to solve the first question are in progress in several laboratories; the use of microsequencing techniques, recombinant DNA technology, and cDNA probes is either in progress or in sight. The major challenge, however, is to discover the biological function of the SCO. Although the same statement was made already by Nicholls (1912), today we seem to have a better chance of meeting this challenge, since appropriate experimental models and methodological tools are now available.

References

Agduhr E (1922) Über ein zentrales Sinnesorgan (?) bei den Vertebraten. Zeitschr Ges Anat 66:223–360

Bargmann W (1943) Die Epiphysis cerebri. In: Möllendorff W von (ed) Handbuch der mikroskopischen Anatomie des Menschen, VI/4. Springer, Berlin, pp 309–502

Bauer-Jokl M (1917) Über das sogenannte Subcommissuralorgan. Arb Neurol Inst Univ Wien 22:41–79

Boeke J (1902) Über das Homologon des Infundibularorganes bei *Amphioxus lanceolatus*. Anat Anz 21:411–414

Boeke J (1908) Das Infundibularorgan im Gehirne des Amphioxus. Anat Anz 32:473–488

Dendy A (1902) On a pair of ciliated grooves in the brain of the ammocoete, apparently serving to promote the circulation of the fluid in the brain-cavity. Proc R Soc Lond 69:485–494

Dendy A (1909) The function of Reissner's fibre and the ependymal groove. Nature 82:217
Dendy A, Nicholls GE (1910) On the occurrence of a mesocoelic recess in the human brain, and its relation to the sub-commissural organ of lower vertebrates; with special reference to the distribution of Reissner's fibre in the vertebrates series and its possible function. Anat Anz 32:496–509
Eberl-Rothe G (1951) Über den Reissnerschen Faden der Wirbeltiere. Z Mikrosk Anat Forsch 57:137–180
Edinger L (1892) Untersuchungen über die vergleichende Anatomie des Gehirnes. 2, Das Zwischenhirn. Abhandl Senckenb Naturforsch Gesellsch 18:3–55
Horsley V (1908) Note on the existence of Reissner's fibre in higher vertebrates. Brain 31:147–159
Jordan H (1925) The structure and staining reactions of the Reissner's fibre apparatus, particularly the subcommissural organ. Amer J Anat 34:427–444
Kolmer W (1905) Zur Kenntnis des Rückenmarks von Ammocoetes. Anat Hefte 29: 165–214
Kolmer W (1921) Das "Sagittalorgan" der Wirbeltiere. Z Anat Entwicklungsgesch 60: 652–717
Kolmer W (1931) Über das Sagittalorgan, ein zentrales Sinnesorgan der Wirbeltiere, insbesondere beim Affen. Z Zellforsch 13:236–248
Krabbe KH (1925) L'organe sous-commissural du cerveau chez les Mammifères. Biol Medd K Dan Vidensk Selsk 5:1–83
Krabbe KH (1933) Anatomy of the subcommissural organ of the brain; review of literature. Nordd Med Tidskr 6:1030–1035
Kutschin K (1866) Über den Bau des Rückenmarkes des Neunaugen. Arch Mikrosk Anat 2:525–530
Kux E (1929) Der sekundäre Neuroporus und der Reissnersche Faden bei Entenembryonen. Z Mikrosk Anat Forsch 16:141–174
Legait E (1942) Les organes épendymaires du troisième ventricule. Doctoral thesis, University of Nancy, France
Legait E (1949) Le rôle de l'épendyme dans les phénomènes endocrines du diencéphale. Bull Soc Sci Nancy 1:11–12
Mathis J (1929) Über Bildungs-und Rückbildungserscheinungen am Schwanzende des Rückenmarksrohres bei älteren Ganskeimlingen (sekundärer hinterer Neuroporus, kaudale Rückenmarksreste, Reissnerscher Faden). Z Mikrosk Anat Forsch 16: 331–382
Nicholls GE (1908) Reissner's fibre in the frog. Nature 77:344
Nicholls GE (1909) The function of Reissner's fibre and the ependymal groove. Nature 82:217–218
Nicholls GE (1912) An experimental investigation on the function of Reissner's fibre. Anat Anz 40:409–432
Nicholls GE (1913) The structure and development of Reissner's fibre and the subcommissural organ. Part 1. Q J Microsc Sci 58:1–116
Nicholls GE (1917) Some experiments on the nature and function of Reissner's fibre. J Comp Neurol 27:117–199
Pesonen N (1940a) Über das Subkommissuralorgan beim Meerschweinchen. Acta Soc Med Fenn Duodecim Ser A 22:53–78
Pesonen N (1940b) Über das Subkommissuralorgan beim Menschen. Acta Soc Med Fenn Duodecim Ser A 22:79–114
Puusepp L, Voss HEV (1924) Studien über das Subkommissuralorgan. I. Das Subkommissuralorgan beim Menschen. Fol Neuropathol 2:13–21
Reichold S (1942) Untersuchungen über die Morphologie des subfornikalen und des subkommissuralen Organs bei Säugetieren und Sauropsiden. Z Mikrosk Anat Forsch 52:455–479
Reissner E (1860) Beiträge zur Kenntnis vom Bau des Rückenmarks von *Petromyzon fluviatilis* L. Arch Anat Physiol 77:545–588

Rodríguez EM, Oksche A, Hein S, Yulis CR (1992) Cell biology of the subcommissural organ. Int Rev Cytol 135:39–121
Sanders A (1878) Contributions to the anatomy of the central nervous system of teleosts. Phil Trans Roy Soc 169:735–776
Sanders A (1894) Researches in the nervous system of *Myxine glutinosa*. London, pp 1–44
Sargent PE (1900) Reissner's fibre in the canalis centralis of vertebrates. Anat Anz 17:33–44
Sargent PE (1903) The ependymal grooves in the roof of the diencephalon of vertebrates. Science 17:487
Sargent PE (1904) The optic reflex apparatus of vertebrates for short-circuit transmission of motor reflexes through Reissner's fibre; its morphology, ontogeny, phylogeny, and function. The fish-like vertebrates. Bull Mus Comp Zool 45:129–258
Schumacher S (1928) Über Bildungs- und Rückbildungsvorgänge am Schwanzende des Medullarrohres bei älteren Hühnerembryonen mit besonderer Berücksichtigung des Auftretens eines "sekundären hinteren Neuroporus". Z Mikrosk Anat Forsch 13: 269–327
Stieda L (1868) Studien über das centrale Nervensystem der Knochenfische. Z Wiss Zool 18:1–70
Studnička FK (1895) Über die terminale Partie des Rückenmarkes. Sitzungsber Böhm Ges Wiss Math Naturwiss Kl 37:1–8
Studnička FK (1899) Der "Reissnersche Faden" aus dem Central-Kanal des Rückenmarkes und sein Verhalten in dem Ventriculus (Sinus) terminalis. Sitzungsber Böhm Ges Wiss Math Naturwiss Kl 36:1–10
Studnička FK (1900) Untersuchungen über den Bau des Ependyms der nervösen Centralorgane. Anat Hefte Abt 2 15:303–430
Stutinsky F (1950) Colloïde, corps de Herring et substance Gomori positive de la neurohypophyse. C R Soc Biol 144:1357–1360
Tretjakoff D (1913) Die zentralen Sinnesorgane bei *Petromyzon*. Arch Mikrosk Anat 83:68–117
Turkewitsch N (1936) Die Entwicklung des subkommissuralen Organs beim Rind (*Bos taurus* L.) Gegenbaurs Morphol Jahrb 77:573–586
Wolff M (1907) Bemerkungen zur Morphologie und zur Genese des Amphioxus-Rückenmarkes. Biol Zbl 27:186–192

II. Phylogeny and Ontogeny

Phylogenetic and Conceptual Aspects of the Subcommissural Organ

A. Oksche

General Aspects

The subcommissural organ (SCO) is a unique secretory structure of the vertebrate brain. It consists of specialized ependymal and ependyma-derived cells of neuroepithelial origin – elements of glial lineage. Because of its extensive secretory activity and the chemical properties of its secretion, this array of cells assumes, among plain ependymal and glial elements, a position similar to that of neurosecretory nerve cells among ordinary neurons. In contrast to the extended apparatus of neuroendocrine, especially peptidergic neurons, the phenomenon of ependyma- or glia-related secretion (see Knowles 1969) appears to be restricted to a few specialized formations. In a number of ependymal complexes, formerly claimed to be secretory (see Studnička 1900; Legait 1942), the elements possessing a secretory capacity are actually closely adjacent neurons (see Oksche 1969). Furthermore, granular inclusions of glial cells supposed to indicate an extended interstitial gland of the brain (as suggested by Nageotte 1910) were more recently proven to be regular cell organelles.

The SCO occupies the diencephalic roof caudal to the pineal organ, covers the prominent posterior commissure, and marks the entrance into the cerebral (sylvian) aqueduct (Fig. 1A). In this area a borderline between the diencephalon (epithalamus) and the mesencephalon (pretectal area) cannot be drawn with certainty (Fig. 1B). Due to its topographical position between the ventricular cavity and a well-developed local vascular system, the SCO qualifies as a member of the group of circumventricular organs (Hofer 1959; Leonhardt 1980). Its secretory parenchyma establishes spatial contact (a) with two compartments containing circulating CSF, i.e., the ventricular cavity and the subarachnoid space, and (b) with the bloodstream of the general brain circulation. The subarachnoid and vascular contacts of the SCO are constituted by ependymal processes and columnar arrangements of ependyma-derived parenchymal cells. A peculiar, integral structure of the SCO is Reissner's fiber (RF; Reissner 1860). In the literature, the term SCO–RF complex has been frequently used. This peculiar threadlike element is an extracellular fraction of secretory material of the SCO discharged into the circulating cerebrospinal fluid (CSF).

A detailed review on the early period of investigation of the SCO was given by Bargmann (1943) in a comprehensive monograph on the pineal

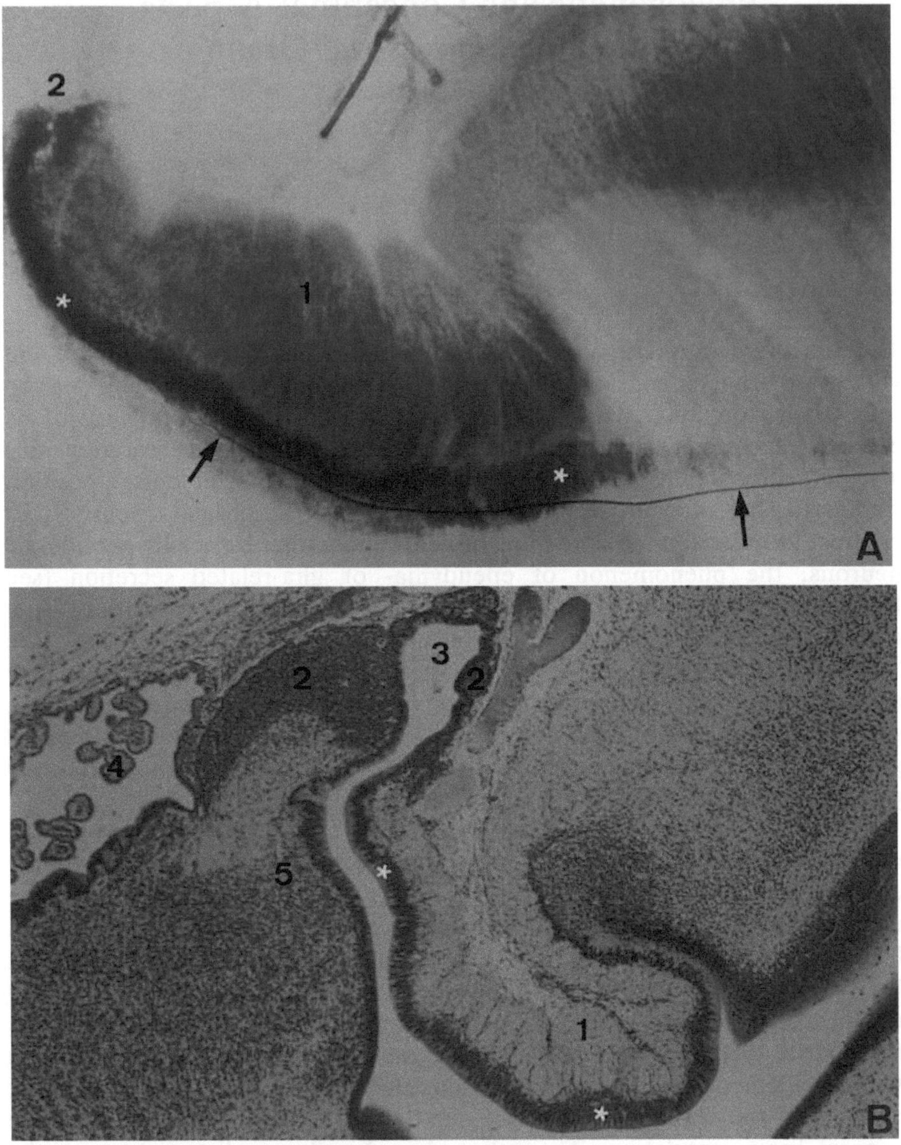

Fig. 1. A Subcommissural organ (*asterisks*), adult cat. Whole mount. Reissner's fiber (*arrows*) extending from the third ventricle into the cerebral aqueduct. *1*, Posterior commissure; *2*, pineal area. (Oksche 1969) Paraldehyde fuchsin, ×50. **B** Subcommissural organ (*asterisks*), 4-month-old human fetus. *1*, Posterior commissure; *2*, pineal primordium displaying an anterior and a posterior lobe separated by the pineal recess (*3*); *4*, choroid plexus of the third ventricle; *5*, habenular region. (Oksche 1961) Chrome-alum hematoxylin-phloxine (Gomori), ×50

organ. This report summarizes certain ideas concerning the development and possible functional significance of the SCO. Indeed, in the borderline area of the primordia, it is hardly possible to delimit either of these epithalamic organs, thus providing arguments for speculations. In ontogeny, the SCO is an initially differentiating structure of the brain, displaying an early onset of its secretory activity and formation of RF.

Phylogenetic Considerations

In phylogeny, the SCO is an ancient and highly constant structure of the vertebrate brain (see Oksche 1961, 1969; Rodríguez et al. 1984, 1987b). In spite of interspecific variations in the shape and fine-structural organization of its ependymal and hypendymal formations, there is a high degree of conformity concerning the location and general morphologic pattern of the SCO. Thus, SCO is present in the most archaic vertebrate brains. It is very prominent and active as a secretory structure already in both orders of cyclostomes, i.e., lampreys and hagfish (Adam 1957, 1963; Sterba 1962; Sterba et al. 1967).

Obviously, intrinsic cavities or at least wide spaces containing circulating fluid (CSF or its precursors) are a fundamental prerequisite for the condensation of the secretion into an RF-like structure. In this context, the rudimentary ventricular compartment (vesicle) to which the SCO of *Myxine* is exposed still provides the basic conditions required for the formation of a RF (Adam 1963). Concerning similar threadlike secretory structures, which occur in other sites of the central nervous system of acranian chordates (*Branchiostoma lanceolatum*) and early developmental stages of teleosts, the expertise of Ragnar Olsson has been called upon (this volume). To my knowledge, a RF-like structure has never been observed within the compact ganglia of invertebrates.

Very definitely, the presence of a pineal organ is not a necessary condition for the existence of the SCO. In the hagfish, *Myxine glutinosa*, and also in all crocodiles so far examined, where the pineal organ is reported to be missing, the SCO is well developed and active as a secretory structure. This holds true also for mammalian species devoid of a typical pineal body. Thus, in the armadillo, *Dasypus novemcinctus*, the SCO is abundant and highly active as a secretory structure (see Rodríguez et al. 1984; Hofer et al. 1976).

Only very few vertebrate species possess a poorly differentiated and hardly secretory active SCO, e.g., the European hedgehog (*Erinaceus europaeus*). This is, however, not a characteristic feature of the insectivore brain; for example, the SCO of the European mole (*Talpa europaea*) is conspicuous and rich in secretory material (see Hofer et al. 1976). Again, in these cases, a particular correlation between the pineal and the SCO is not evident.

The situation in primates deserves special attention. The human SCO shows its highest development between the third and fourth month of fetal life (Figs. 1B, 2, 3a,b), followed by a gradual regression during the second half of pregnancy and a further decline after the birth. Thus, in 6-year-old children only rudiments of SCO parenchyma can be found (Oksche 1961). In contrast, in a representative number of species of Old- and New-World monkeys (e.g., rhesus monkey, *Cebus*, *Ateles*, *Aotes*) the SCO is well developed and active as a secretory structure even in mature animals. Of eminent interest, however, is the situation in anthropoid apes. According to our observations, in very old chimpanzees the SCO shows a reduction in height and regression of its secretory parenchyma. So far, we have no evidence of sex-dependent developments.

Development of Concepts

Since the discovery of the SCO, a number of contradictory hypotheses have been advanced considering not only secretory but, under involvement of RF, also sensory functions (for details, see Bargmann 1943; Rodríguez et al. 1987b). Most of these hypotheses were not based on substantial evidence. From the introduction of new, selective staining methods for visualizing hypothalamic neurosecretion (starting with Bargmann 1949; see Diepen 1962), a strong impetus arose also for SCO research. Stutinsky (1950) was the first to show that dyes of the Gomori type also stain the secretion of the SCO, an advancement that stimulated further experimental and histochemical work.

In this line, parallel studies were conducted in the laboratories of Ragnar Olsson, Günter Sterba, and in my laboratory, followed by the work of Rodríguez and associates (1984, 1986, 1987a,b, 1990), and A. and R. Meiniel (Meiniel et al., this volume; Meiniel and Meiniel 1985; Meiniel et al. 1991). (For details, see this volume.)

Ragnar Olsson succeeded in transecting RF within the spinal cord and in demonstrating other ependymal areas capable of forming RF-like structures (see Olsson 1955, 1956, 1958a,b). Oksche (1954, 1956, 1961, 1969) placed emphasis on possible vascular and leptomeningeal routes of SCO secretion, showing striking interspecific variations and displaying seasonal and experimentally induced changes. Sterba and his associates introduced new staining methods (Sterba 1962), analytical procedures in vitro (see Sterba 1972; Sterba and Bargmann 1977) and, for the first time, immunocytochemical methods (Sterba et al. 1981, 1982). All these groups have contributed to the elucidation of the basic chemical properties of the SCO

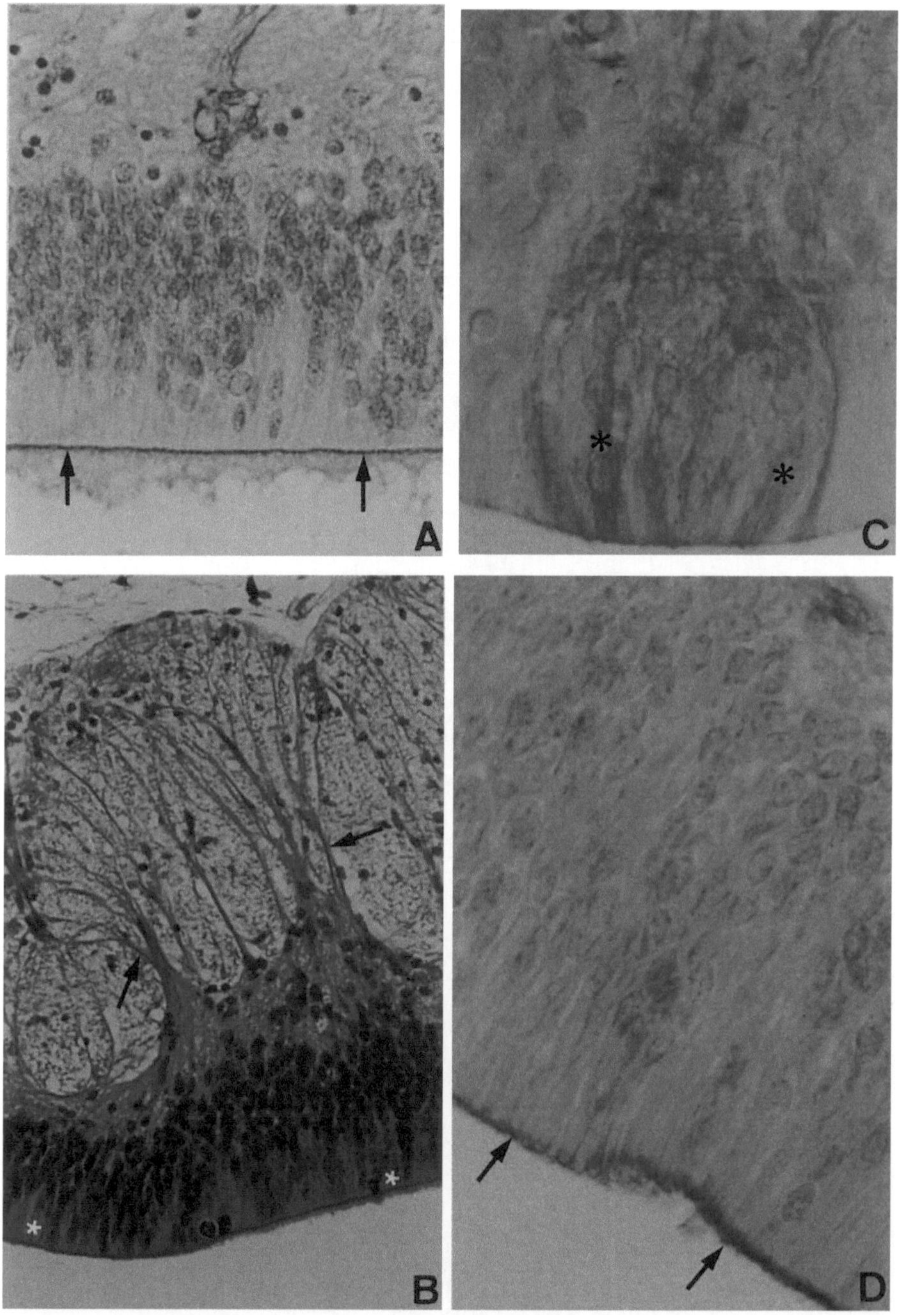

Fig. 2A–D. Human fetal subcommissural organ (SCO). **A** A 3-month-old fetus. Note selectively stained apical material (*arrows*). For the first report on signs of secretory activity in human fetal SCO, see Oksche (1956); compare also Oksche (1969). Chrome-alum hematoxylin-phloxine (Gomori), ×400. **B** A 4-month-old fetus. General overview showing the polar organization of the SCO. Ventricular surface (*asterisks*). Note ependymal-type processes (*arrows*) arising from the basal, less compact portion of the organ. Iron hematoxylin (Heidenhain)-phloxine, ×150. **C** A 4-month-old-fetus. Circumscribed aggregation of reactive, apparently secretory cells (*asterisks*) extending to the apical surface of the SCO. PAS-hemalum, ×560. **D** A 4-month-old fetus. Selectively stainable material (*arrows*) at the apical surface of the SCO. Paraldehyde fuchsin (Halmi-Dawson), ×560

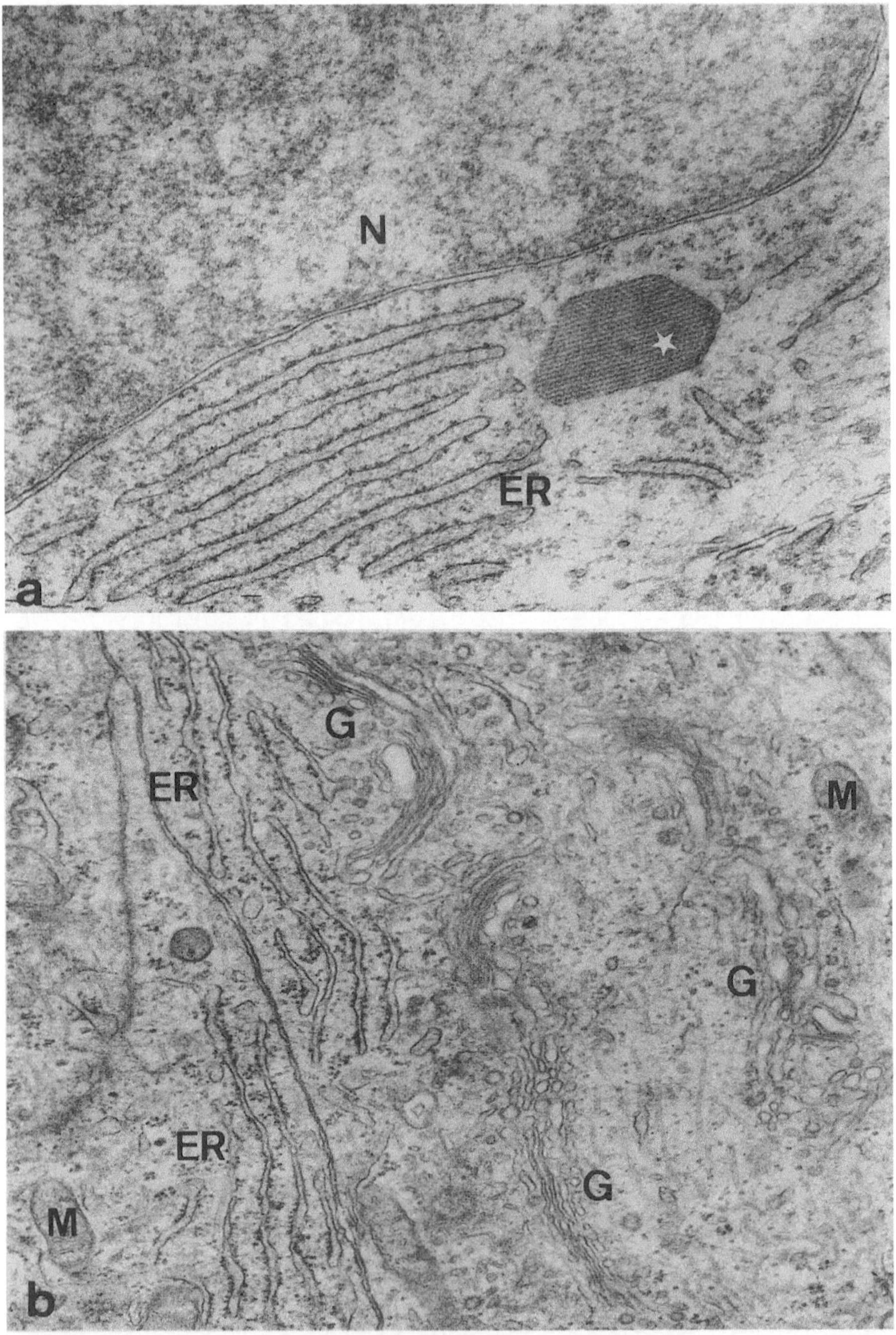

Fig. 3a,b. Ultrastructure of the human fetal SCO (estimated age: 109 days) indicating synthetic activity of the ependymal cells. Note parallel arrays of cisternae of the granular endoplasmic reticulum (*ER*) filled with flocculent material and extended Golgi complexes, (*G*). *M*, mitochondria; *star*, crystalloid inclusion; *N*, cell nucleus. **a** ×50 000; **b** ×35 000 secretion; this was extended by Diederen (1972), who introduced autoradiographic methods for study of dynamics of the secretory process. (For detailed references, see Leonhardt 1980.)

These developments have been reviewed and analyzed by Leonhardt (1980) in a monograph on circumventricular organs. Special credit should be given to Hofer (1959) for his definition of circumventricular organs and for his substantial contribution to our knowledge of the SCO, under particular reference to primitive vertebrates and monkeys (see Leonhardt 1980).

A new era of SCO research began when Meiniel and Meiniel (1985) and Rodríguez and associates (1986) started to use light- und electron-microscopic immunocytochemistry in combination with lectin histochemistry and in vitro chemistry, with the strong aim to approach the field of molecular biology. This progress will play a central role in our discussions. Furthermore, Didier and colleagues (see Chouaf et al. 1991) succeeded in demonstrating the ependymo-glial, nonneuronal character of the SCO by using a series of molecular markers. For a long time progress in fine-structural research was more advanced than was insight into basic chemical aspects of the SCO. At the ultrastructural level, the elementary secretory process in the SCO cell has now been quite precisely elucidated. In contrast, the functional significance of the SCO remains enigmatic.

Several core-glycosylated glycoproteins have now been identified in the SCO (Rodríguez et al. 1987a; Herrera and Rodríguez 1990; R. Meiniel et al. 1991; see also Hein et al. and A. Meiniel et al., this volume). While stored in the rough endoplasmic reticulum, the precursors are N-linked, high mannose-type glycoproteins. The processed (post-Golgian) forms stored in secretory granules are N-linked, complex-type glycoproteins with sialic acid marking the end of the terminal chain. The principal pathway of discharge of the SCO secretion is directed into the ventricular CSF. In addition, pathways to the CSF-containing subarachnoid space and to the perivascular space must be considered. It is open to discussion whether identical products are released into each of these compartments.

Conclusions

In our opinion, two unsolved aspects of the SCO–RF complex appear to be crucial: (1) the exact chemical structure of the secretory glycoproteins and their encoding genes, and (2) the functional role of the SCO.

With respect to the analytical aspect, developments appear to be encouraging. The secretory material of the bovine SCO has been purified and is being used for microsequencing studies. Polyclonal and monoclonal antibodies against different components of the SCO secretion have been raised. Recombinant DNA technology using cDNA probes, in situ hybridization, and sensitive enzyme immunoassays offer promising avenues for further investigations. Tissue culture work will promote chemical analysis of the secretory products and allow study of the secretory activity of SCO cells deprived of their serotoninergic innervation.

Furthermore, an even more important task is to determine the biological function of the SCO–RF complex. Whether the SCO produces a hormonelike compound or some kind of trophic agent is open to discussion. An appropriate set of bioassays is greatly needed. So far, classical procedures as used in experimental endocrinology (hormone research) have not led to convincing, unequivocal results. Thus, we are now in great need of ideas leading to a new approach, too. An immunologically based selective blockade or even complete deletion of the SCO offer new avenues of experimental research (Rodríguez et al. 1990). Surgical lesions in the region are never free of disturbing side effects, such as damage to the fiber systems of the posterior commissure or escape of CSF via artificial openings in the ventricular wall (roof).

On the other hand, the possibility of the occurrence of one or several trophic agents interacting with cellular, especially neuronal, membranes should also be discussed. A number of recognition, adhesion, and growth-promoting molecules belong to the class of glycoproteins. Also, a role in ion-exchange mechanisms or as a signal in synchronization of biological events might be a matter of further debate. Moreover, the close proximity of the SCO to the pineal organ should not be overlooked. There is no indication of a portal vascular system between these two circumventricular organs, but their existing capillary links and the communicative role of the circulating CSF are to be considered.

Bioactive molecules of the type mentioned above released into the ventricular and possibly also into the subarachnoid CSF or temporarily stored in the depot structure of RF would finally be distributed over the entire central nervous system and could interact with cellular units possessing receptors for these molecules. If these receptors could be demonstrated using appropriate antibodies, it would be possible to identify the target sites within the central nervous system. The latter goal can be more easily achieved by using radioactively labeled secretory glycoproteins and searching for their binding sites by means of radioautography.

A further puzzling, unsolved problem is the fact that the secretion of the SCO gains access to the vascular system at different levels: (a) to the capillaries proper of the SCO, especially along its basal aspect and the leptomeningeal parenchymal route; (b) to the capillaries in the adjacent leptomeningeal formations; and (c) to the capillaries in the region of the ampulla caudalis (terminal ventricle) of the central canal in the spinal cord. The fact that these vessels possess a blood–brain barrier does not exclude a possible passage of small molecules into the circulation. But then again one would have to search for possible target sites of these molecules. New selective experimental procedures and new working hypotheses may help to answer these crucial questions.

References

Adam H (1957) Beitrag zur Kenntnis der Hirnventrikel und des Ependyms bei den Cyclostomen. Anat Anz 103:173–188

Adam H (1963) Brain ventricles, ependyma and related structures. In: Brodal A, Fänge R (eds) The biology of *Myxine*. Universitetsforlaget, Oslo, pp 137–149

Bargmann W (1943) Die Epiphysis cerebri. In: Von Möllendorff W (ed) Handbuch der mikroskopischen Anatomie des Menschen, vol 6, part 4. Springer, Berlin, pp 309–502

Bargmann W (1949) Über die neurosekretorische Verknüpfung von Hypothalamus und Neurohypophyse. Z Zellforsch 34:610–634

Chouaf L, Didier-Bazes M, Hardin H, Aguera M, Fevre-Montange M, Voutsinos B, Belin MF (1991) Developmental expression of glial markers in ependymocytes of the rat subcommissural organ: role of the environment. Cell Tissue Res 266:553–561

Diederen JHB (1972) Influence of light and darkness on the subcommissural organ of *Rana temporaria* L. A cytological and autoradiographical study. Z Zellforsch 129: 237–255

Diepen R (1962) Der Hypothalamus. In: Bargmann W (ed) Handbuch der mikroskopischen Anatomie des Menschen, vol 4, part 7. Springer, Berlin Göttingen Heidelberg

Galeotti G (1897) Studio morphologico e citologico della volta del diencefalo in alcuni vertebrati. Riv Pat Nerv 2:481–517

Herrera H, Rodríguez EM (1990) Secretory glycoproteins of the rat subcommissural organ are N-linked complex-type glycoproteins. Demonstration by combined use of lectins and specific glycosidases, and by the administration of Tunicamycin. Histochemistry 93:607–615

Hofer H (1959) Zur Morphologie der circumventriculären Organe des Zwischenhirns der Säugetiere. Verh Dtsch Zool Ges Frankfurt 1958, Zool Anz 22:202–251

Hofer H, Merker G, Oksche A (1976) Atypische Formen des Pinealorgans der Säugetiere. Anat Anz 140 [Suppl]:97–102

Kwowles F (1969) Ependymal secretion, especially in the hypothalamic region. J Neuro Visc Relat [Suppl] IX:97–110

Legait E (1942) Les organes épendymaires du troisième ventricle. L'organe sous-commissural. L'organe sub-fornical. L'organe para-ventriculaire. Med Thesis, University of Nancy

Leonhardt H (1980) Ependym and circumventriculäre Organe. In: Oksche A, Vollrath L (eds) Neuroglia I. Handbuch der mikroskopischen Anatomie des Menschen, vol IV, part 10, Springer, Berlin Heidelberg New York, pp 117–665

Meiniel R, Meiniel A (1985) Analysis of the secretions of the subcommissural organ of several vertebrate species by use of fluorescent lectins. Cell Tissue Res 239:359–364

Meiniel R, Duchier-Liris N, Molat J-L, Meiniel A (1991) The complex-type glycoprotein secreted by the bovine subcommissural organ: an immunological study using C_1 B_8 A_8 monoclonal antibody. Cell Tissue Res 266:483–490

Nageotte J (1910) Phénomènes de sécrétion dans le protoplasma des cellules névrogliques de la substance grise. C R Soc Biol (Paris) 68:1068–1069

Oksche A (1954) Über die Art und Bedeutung sekretorischer Zelltätigkeit in der Zirbel und im Subcommissuralorgan. Anat Anz 101 [Suppl]:88–96

Oksche A (1956) Funktionelle histologische Untersuchungen über die Organe des Zwischenhirndaches der Chordaten. Anat Anz 102:204–419

Oksche A (1961) Vergleichende Untersuchungen über die sekretorische Aktivität des Subcommissuralorgans und den Gliacharakter seiner Zellen. Z Zellforsch 54:549–612

Oksche A (1969) The subcommissural organ. J Neuro Visc Relat [Suppl] IX:111–139

Olsson R (1955) Structure and development of Reissner's fibre in the caudal end of *Amphioxus* and some lower vertebrates. Acta Zool (Stockh) 36:167–198

Olsson R (1956) The development of Reissner's fibre in the brain of the salmon. Acta Zool (Stockh) 37:235–250

Olsson R (1958a) Studies on the subcommissural organ. Acta Zool (Stockh) 39:71–102

Olsson R (1958b) The subcommissural organ. Handström, Stockholm

Reissner E (1860) Beiträge zur Kenntnis vom Bau des Rückenmarks von *Petromyzon fluviatilis* L. Arch Anat Physiol 1960:545–588

Rodríguez EM, Oksche A, Hein S, Rodríguez S, Yulis R (1984) Comparative immunocytochemical study of the subcommissural organ. Cell Tissue Res 237:427–441

Rodríguez EM, Herrera H, Peruzzo B, Rodríguez S, Hein S, Oksche A (1986) Light- and electron-microscopic immunocytochemistry and lectin histochemistry of the subcommissural organ. Evidence for processing of the secretory material. Cell Tissue Res 243:545–559

Rodríguez EM, Hein S, Rodríguez S, Herrera H, Peruzzo B, Nualart F, Oksche A (1987a) Analysis of the secretory products of the subcommissural organ. In: Scharrer B, Korf HW, Hartwig HG (eds) Functional morphology of neuroendocrine systems. Springer, Berlin Heidelberg New York, pp 189–202

Rodríguez EM, Oksche A, Rodríguez S, Hein S, Peruzzo B, Schoebitz K, Herrera H (1987b) The subcommissural organ. In: Gross EM (ed) Circumventricular organs and body fluids, vol II. CRC Press, Boca Raton, pp 1–41

Rodríguez S, Rodríguez EM, Jara P, Peruzzo B, Oksche A (1990) Single injection into the cerebrospinal fluid of antibodies against the secretory material of the subcommissural organ reversibly blocks formation of Reissner's fiber: immunocytochemical investigations in the rat. Exp Brain Res 81:113–124

Sterba G (1962) Das Subcommissuralorgan von *Lampetra planeri* (Bloch). Zool Jahrb Anat 80:135–158

Sterba G (ed) (1969) Zirkumventrikuläre Organe und Liquor. Fischer, Jena

Sterba G (1972) Subcommissuralorgan und Liquorregulation. Biol Rundschau 10:309–324

Sterba G, Bargmann W (eds) (1977) Circumventriculäre Organe. Nova Acta Leopoldina, Suppl 9. Deutsche Akademie der Naturforscher, Halle

Sterba G, Müller H, Naumann (1967) Fluoreszenz- und elektronenmikroskopische Untersuchungen über die Bildung des Reissnerschen Fadens bei *Lampetra planeri* (Bloch). Z Zellforsch 76:355–376

Sterba G, Kleim I, Naumann W, Petter H (1981) Immunocytochemical investigation of the subcommissural organ in the rat. Cell Tissue Res 218:659–662

Sterba G, Kiessig C, Naumann W, Petter H, Kleim I (1982) The secretion of the subcommissural organ. A comparative immuncytochemical investigation. Cell Tissue Res 226:427–439

Studnička FK (1900) Untersuchungen über den Bau des Ependyms der nervösen Centralorgane. Anat, Hefte Wiesbaden 15:303–430

Stutinsky F (1950) Colloïde, corps de Herring et substance Gomori positive de la neurohypophse. C R Soc Biol (Paris) 144:1357–1360

Reissner's Fiber Mechanisms: Some Common Denominators

R. Olsson

In 1860, Professor Ernst Reissner published an anatomical study of the lamprey spinal cord, in which he described a fiberlike structure consistently occurring in the central canal (Fig. 1). This structure, which was later given the name "Reissner's fiber", was found to be a regular component of the cavities of the vertebrate central nervous system, extending caudad from the third brain ventricle. The fiber is a secretory product (cf. Studnička 1899), which in vertebrates is produced by the subcommissural organ at the diencephalic roof (Kolmer 1921; Mazzi 1952; Wingstrand 1953). It stains intensely with chrome-alum hematoxylin (Mazzi 1952); it is acellular and not surrounded by cytomembranes (Afzelius and Olsson 1957). It was early concluded that a fiber mechanism of this general design is very conservative, having the same fundamental pattern in all members of the subphylum Craniata (vertebrates).

However, fibers with similar fine structure and with the same affinity for the Gomori stain were also described in the spinal cord equivalents of subcraniate animals: the acranian *Branchiostoma* (Wolff 1907; Olsson and Wingstrand 1954), the appendicularian *Oikopleura* (Olsson 1969), and even in some ascidian tadpoles (Numakunai et al. 1965; Olsson 1972). It is not usual to homologize secretions, but it seemed reasonable to designate all fibers of this type as "Reissner's fibers" and to consider them as homologous structures.

If we consider the different sources of these fibers, though, the situation appears rather confused. The craniate fiber (*5* in Fig. 2) is produced by the dorsal *subcommissural organ* (SCO) (Kolmer 1921), but at ontogenetically early stages (*3* in Fig. 2) it may be produced by a ventral *flexural organ* (FO) (Olsson 1956) situated at the embryonal brain flexure (*plica ventralis encephali*). The fiber source in the lancelet *Branchiostoma* (*2* in Fig. 2) is also ventral, i.e., the *infundibular organ* (IO; Olsson and Wingstrand 1954), named by Boeke (1902), who believed that it was the homologue of the craniate infundibulum. The appendicularian (*Oikopleura*) fiber originates at a *fibrogen cell* (FC) (*6* in Fig. 2), which is apparently a specialized ependymal cell (Holmberg and Olsson 1984). Finally, the few reports of fibers in ascidian tadpoles (Numakunai et al. 1965; Olsson 1972) give no clue to the site of origin, but a reasonable guess must be that one of the four ependymal cells (most likely the ventral one) lining the central canal in the visceral ganglion (Katz 1983) is the equivalent of the appendicularian fibrogen cell.

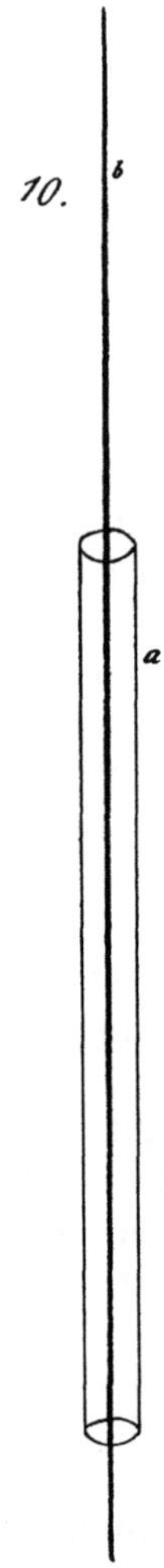

Fig. 1. Central canal in the spinal cord of *Petromyzon fluviatilis* L. *a* Lumen. *b* An intraluminar thread. (From E. Reissner 1860; Fig. 10)

However, in spite of all differences in areas of origin, there is one important common denominator of the ventral fiber producers: they are all descendants from *lamina basalis*, the floor-plate matrix in the embryonic neural tube. The flexural organ is a gland composed of the rostralmost floor-plate cells at the brain flexure. However, it is not clearly delimited caudally (Olsson 1956); ventromedian cells with the same cytology and reaction for conventional stains were described in fish embryos as far caudally as in the urophysis area (Fridberg 1962). Also these cells seemed to release a secretion similar to that of the FO into the cerebrospinal fluid, although a connection with the fiber could not be observed.

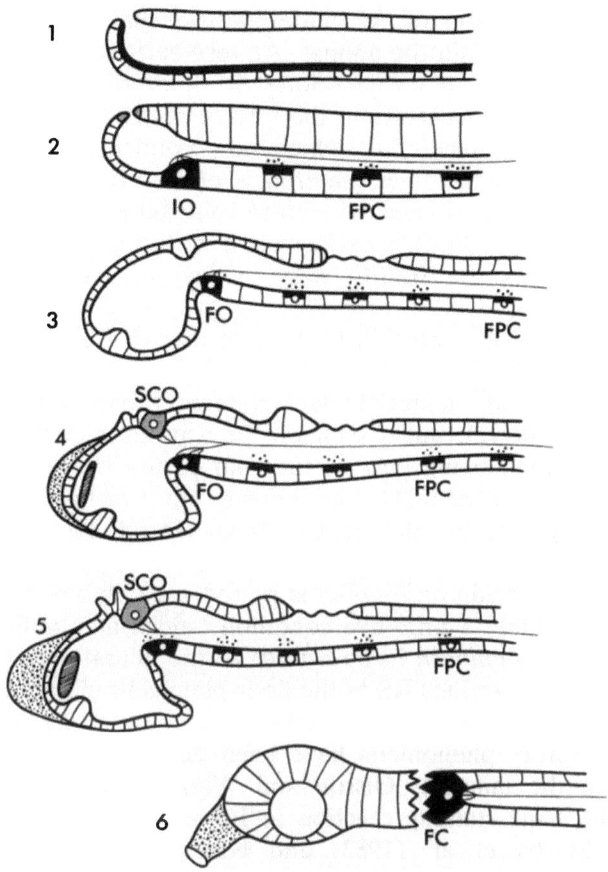

Fig. 2. Schematic sagittal sections showing cells and glands producing RF secretion (*black* or *dotted*) in various chordates. *1*, Embryonic neural tube with open neuropore (*upper left*) and floor-plate cells. *2*, Adult lancelet *Branchiostoma*, with fiber-forming infundibular organ (*IO*) and floor-plate cells (*FPC*). *3*, Embryonic craniate (*Salmo*) with juvenile flexural organ (*FO*). *4*, Slightly older stage than the previous one during the short time when both flexural and subcommissural organs (*SCO*) contribute to a Reissner's fiber. *5*, Adult craniate with a SCO as the sole fiber producer. *6*, An appendicularian (*Oikopleura*) with almost compact brain and one fibrogen cell (*FC*) in the neural cord of the tail. (Artwork by B. Mayrhofer)

A close relationship between the FO and the SCO is also apparent during the ontogeny. Both possess not only the capacity to produce Reissner's fibers (RF), they may even contribute to the production of one and the same fiber during a short embryonic period (*4* in Fig. 2; Olsson 1956; Naumann 1986; Olsson et al., unpublished). With an antiserum raised against bovine RF Naumann (1986) could demonstrate immunoreactivity at the apical border of the floor cells in very early developmental stages of the salamandrid *Pleurodeles* (*1* in Fig. 2). This was an important study which demonstrated a succession of RF material-producing sources, both temporally and spatially, within the animal. At very early stages the material is visible as a reactive apical border along the floor cells. At later stages, immunoreactive cells concentrate in the flexure region where they form an FO, the reactive substance(s) of which now condenses in the ventricle forming a RF. Eventually, fiber material is produced also from the SCO and finally the FO can no longer be seen to contribute to the fiber.

Phylogenetically the SCO is exclusively a vertebrate organ. The protochordates never pass the IO/FO stage, which is purely embryonic in vertebrates.

The SCO is the one exception to the rule that all Reissner's substance-producing cells are floor-plate descendants ("Reissner's substance," RS, is here used to indicate any material which is immunoreactive with vertebrate RF antisera). It is dorsal, but it seems to be a product of the dorsolateral alar plate rather than being part of the roof plate. The very early ontogenetic development of the SCO is not known, but it seems to have a paired origin. In petromyzonts, two dorsolateral SCOs (Dendy 1902; Nicholls 1912; Fig. 3) produce one fiber each, which converge to form a single RF. The SCO of the salamandrid *Ambystoma* is paired rostrally (Wenk and Pfister 1970). This is probably a primitive condition, which may indicate that the SCO originates from *anlagen* in both folds of the neural groove possessing the same capacity to produce RS as the floor-plate cells in the midline of the neural tube.

RS-like secretory phenomena have been described also in the neuroporus area. In the lancelet, Olsson and Wingstrand (1954) found cells which gave the same staining reaction as RS with Gomori's chrome-alum hematoxylin. Sterba et al. (1983) and Naumann (1986) demonstrated

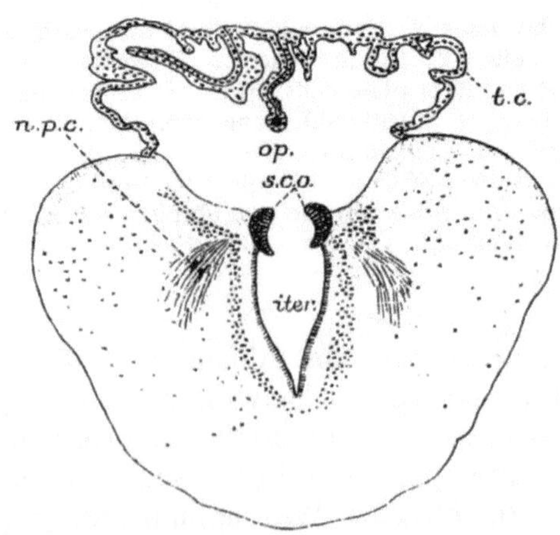

Fig. 3. Paired subcommissural organs (*S.C.O.*) in the midbrain of *Petromyzon* (*Lampetra*) *fluviatilis*. From Nicholls (1912); Fig. 36: *t.c.*, tela choroidea (choroid plexus); *n.p.c.*, posterior commissure; *o.p.*, ventricular cavity (recess); *iter*, aqueduct

immunorectivity in the neuroporus region using RF antisera in such diverse animals as the protochordate *Branchiostoma* and the vertebrate (urodele) *Pelobates*. All these results seem to support the concept that all RS producers are decendants of a single large glandular area in the brain of a primitive ancestor (Olsson 1956; Naumann 1986).

The discovery that the lancelet IO reacts with antibodies against bovine fiber (RF) material but not with antibodies against the organ which produces the fiber (SCO) (R. Olsson et al., unpublished) may indicate that, in vertebrates, the fiber contains material that has been delivered not only from the SCO but also from another source that might be the equivalent of the lancelet IO. A reasonable guess seems to be that this source is to be found in floor-plate cells which release their secretory product freely into the cerebrospinal fluid, from which it may be added to the SCO material in the fiber.

Although there have been a great number of publications dealing with different aspects of SCOs and RFs from different craniates, there are astonishingly few contributions which consider the whole RS mechanism from biological and phylogenetic standpoints. The very simplified diagram in Fig. 2 is an attempt to sketch the great variety of this mechanism within the chordate phylum. At the same time it also brings many questions up for discussion, most of which, however, still remain unanswered.

1. What is the function of the floor-plate cells?

At least some of these cells produce factors which are essential for normal differentiation and cell patterning of neurons in the developing brain (Hirano et al. 1991; Yamada et al. 1991). However, it is not known whether RS is involved in these effects.

2. Why is RS in the cerebrospinal fluid eventually condensed to fiber form?

Not all RS condenses to a fiber. There are apparently three different generations of RS producers: (1) Floor-plate cells delivering their product into the central canal; it is akin to that of the FO (Naumann 1986) but it does not form a fiber; (2) FO (IO) cells as glands at the most anterior point in the brain where a fiber is needed or useful; (3) SCO cells at a new location and with somewhat different cytology, but likewise producing a RF.

As to the strange form of the secretion as a fiber which slowly (Ermisch 1973) passes the central nervous system cavities, there still seems to be no better explanation than the detoxication hypothesis, which means that harmful molecules in the cerebrospinal fluid may bind to fiber material and be removed from the nervous system (Olsson 1956, 1958; Sterba 1969; Ermisch 1973). There are indications that RS-producing cells scattered in the ependyma coexist with fiber-producing cells (Naumann 1986; Rodríguez et al. 1985). It is therefore possible that RS has different functions when delivered into the central canal as amorphous secretion or when it has formed a fiber.

3. Why is the ventral fiber producer replaced by an SCO in older vertebrate embryos?

The SCO is the third generation of RS-producers and it is exclusively a vertebrate structure. Immunocytochemical studies (R. Olsson et al., unpublished) seem to indicate that the fiber produced by the SCO has at least one more protein than the secretion within the cells. It is also possible that the SCO has acquired a new function in addition to the ancient one to form a fiber in the cerebrospinal fluid.

4. Where does a central nervous system need a RF?

No RF can be seen in the, in terms of evolution, youngest parts of the chordate brain, namely, the prosencephalon in vertebrates and its counterpart in the acranian lancelets, the brain vesicle. The appendicularian *Oikopleura* has lost all ependyma-lined cavities in the central nervous system except in the tail, where a central canal with a tiny RF is retained. This condition was surmised (Holmberg and Olsson 1984) to show a relationship between the fiber and myomotor nerve roots.

Finally, a complete RF mechanism is missing in a few vertebrates. This may be an indication that this probably most ancient structure of the chordate central nervous system becomes less important when the nervous mass is very large in relation to the cavities where the RF is effective, or the cavities are very large in relation to the fiber.

References

Afzelius BA, Olsson R (1957) The fine structure of the subcommissural cells and of Reissner's fibre in *Myxine*. Z Zellforsch 46:672–685

Boeke J (1902) Über das Homologon des Infundibularorgans by *Amphioxus lanceolatus*. Anat Anz 15:411–414

Dendy A (1902) On a pair of ciliated grooves in the brain of the ammocoete, apparently serving to promote the circulation of the fluid in the brain-cavity. Proc R Soc Lond 69:485–494

Ermisch A (1973) Zur Charakterisierung des Komplexes Subcommissuralorgan – Reissnerscher Faden und seiner Beziehungen zum Liquor unter besonderer Berücksichtigung autoradiographischer Untersuchungen sowie funktioneller Aspekte. Wiss Z Karl-Marx-Univ, Leipzig, Math-Naturwiss R 22:297–336

Fridberg G (1962) Studies on the caudal neurosecretory system in teleosts. Acta Zool 43:1–77

Hirano S, Fuse S, Sohal GS (1991) The effect of the floor plate on pattern and polarity in the developing central nervous system. Science 251:310–313

Holmberg K, Olsson R (1984) The origin of Reissner's fibre in an appendicularian, *Oikopleura dioica*. Vidensk. Medd Dansk Naturhist Foren 145:43–52

Katz MJ (1983) Comparative anatomy of the tunicate tadpole, *Ciona intestinalis*. Biol Bull 164:1–27

Kolmer W (1921) Das "Sagittalorgan" der Wirbeltiere. Z Anat Entw Gesch 60:652–717

Mazzi V (1952) Caratteri secretori nelle cellule dell'organo sottocommissurale dei vertebrati inferiori. Arch Zool Ital 37:445–464

Naumann W (1986) Immunhistochemische Untersuchungen zur Ontogenese des Subcommissuralorgans. Acta Histochem [Suppl] 33:265–272

Nicholls GE (1912) The structure and development of Reissner's fibre and the subcommissural organ. Q J Micr Sci 58:1–111
Numakunai T, Ishikawa M, Hirai E (1965) Changes of the structures stainable with modified Gomori's aldehyde-fuchsin method in the tadpole larvae of the ascidian, *Halocyntia roretzi*, relating to tail resorption. Bull Mar Biol Stat Asamushi, Tohôko Univ 12:161–172
Olsson R (1956) The development of Reissner's fibre in the brain of the salmon. Acta Zool (Stockh) 37:235–250
Olsson R (1958) Studies on the subcommissural organ. Acta Zool (Stockh) 39:71–102
Olsson R (1969) Phylogeny of the ventricle system. In: Sterba G (ed) Zirkumventrikuläre Organe und Liquor. Fischer, Jena, pp 291–305
Olsson R (1972) Reissner's fiber in ascidian tadpole larvae. Acta Zool (Stockh) 53:17–21
Olsson R, Wingstrand KG (1954) Reissner's fibre and the infundibular organ in amphioxus – results obtained with Gomori's chrome alum haematoxylin. Univ Bergen Årbok (Publ biol stat) 14:1–14
Olsson R, Yulis CR, Rodríguez EM (unpublished results)
Reissner E (1860) Beiträge zur Kenntnis vom Bau des Rückenmarkes von *Petromyzon fluviatilis* L. Arch Anat Physiol Wiss Med, Leipzig, pp 545–588
Rodríguez S, Hein S, Yulis CR, Delannoy L, Siegmund I, Rodríguez E (1985) Reissner's fiber and the wall of the central canal in the lumbo-sacral region of the bovine spinal cord. Comparative immunocytochemical and ultrastructural study. Cell Tissue Res 240:649–662
Sterba G (1969) Morphologie und Funktion des Subcommissuralorgans. In: Sterba G (ed) Zirkumventrikuläre Organe und Liquor. Symposium Schloss Reinhardsbrunn 1968. Fischer, Jena, pp 17–32
Sterba G, Fredriksson G, Olsson R (1983) Immunocytochemical investigations of the infundibular organ in amphioxus (*Branchiostoma lanceolatum*; Cephalochordata). Acta Zool (Stockh) 64:149–153
Studnička FK (1899) Der Reissnersche Faden im Zentralkanal des Rückenmarks und sein Verhalten in dem Ventriculus terminalis. Sitz Ber Kgl Böhm Gesellsch Wiss Prag 36:1–10
Wenk K, Pfister C (1970) Über die Fermentreifung im Subcommissuralorgan von *Ambystoma mexicanum* in Beziehung zur morphologischen Differenzierung. Z Mikr Anat Forsch 81:313–328
Wingstrand KG (1953) Neurosecretion and antidiuretic activity in chick embryos with remarks on the subcommissural organ. Arkiv Zool 6:41–67
Wolff M (1907) Bemerkungen zur Morphologie und zur Genese des Amphioxus-Rückenmarkes. Biol Centralbl 27:186–192, 196–212, 225–233
Yamada, T, Placzek M, Tanaka H, Dodd J, Jessell TM (1991) Control of cell pattern in the developing nervous system: polarizing activity of the floor plate and notochord. Cell 64:635–64

Ontogenetic Development of the Subcommissural Organ with Reference to the Flexural Organ

K. Schoebitz, E.M. Rodríguez, O. Garrido, and M.A. Del Brió-Leon

Introduction

The ontogeny of the chick and rat subcommisural organ (SCO) has been the subject of several reports. Some of these studies were performed on the rat SCO (Kohl and Linderer 1973; Marcinkiewicz and Bouchaud 1983, 1986; Rakic and Sidman 1968; Naumann et al. 1987; Castañeyra-Perdomo et al. 1983). The SCO of chicken embryos represents the nonmammalian organ model whose ontogenetic development has been most thoroughly investigated (Wingstrand 1953; Ziegels 1977; Schoebitz et al. 1986; Bruel et al. 1987; Naumann et al. 1987; Karoumi et al. 1990). Using classical stains for neurosecretion, Wingstrand (1953) and Ziegels (1977) have reported that the secretory material of the SCO first appears in 7-day-old chick embryos.

We have reinvestigated certain developmental aspects of the SCO in the chick and rat, using a sensitive immunocytochemical procedure (Sternberger et al. 1970) and antisera reacting specifically with the secretory material of the SCO and Reissner's fiber (RF) (Rodríguez et al. 1984).

These two species were chosen because they may become useful animal models for experimental purposes. Special attention has been paid to the time sequence of certain events, namely, (a) onset of the secretory activity of the SCO cells; (b) intracellular distribution of the secretory material, primarily the presence of secretory granules at the ventricular cell pole; (c) appearance of released secretory material on the ventricular surface of the organ (pre-RF); (d) time of formation of RF. This information appears essential in order to design an experimental protocol for blocking immunologically the formation of the "first" RF appearing during ontogeny. Thus, whereas in the chicken RF first appears at day 11 of embryonic life, in the rat the first RF is visualized during the first two postnatal days. Therefore, in order to block the formation of the first RF, the antibodies should be made available around embryonic day 11 in the chicken and after birth in the rat.

Materials and Methods

Chick embryos were fixed every day from the first to the 21st day of incubation. For further details, see (Schoebitz et al. 1986). Rat embryos from day 10 to 22 (E10 to E22) were fixed by immersion in Bouin's fluid for 2 days, after exposing the brain cavities with a cut along each side of the head. Postnatally, rats 1–6, 10, 14, 18, 22, 26, 30, 34, 38, 42, and 46 days old were fixed by vascular perfusion with Bouin's fluid via a cannula inserted in the left ventricle of the heart. Embedding was in Paraplast. Serial sagittal and frontal sections through the brain and serial transverse sections through the spinal cord were obtained.

Adjacent serial sections of the brain and spinal cord were processed: (a) for wheat-germ agglutinin (WGA) binding using peroxidase-labeled lectin at a concentration of 2 µg/ml (see Herrera and Rodríguez 1990); (b) according to the immunoperoxidase method of Sternberger et al. (1970). An antiserum against Reissner's fiber (anti-RF serum) was used as the primary antibody (AFRU: A = antiserum, FR = fiber of Reissner, U = urea; see Rodríguez et al. 1984) at dilutions of 1:2000, 1:4000 and 1:100000. High dilutions of AFRU were used for the demonstration of mature secretory granules of the SCO (Rodríguez et al. 1986). In addition, sections of the spinal cord were stained with paraldehyde fuchsin according to Gabe (1968). This method was used to visualize RF in the central canal because of the low immunoreactivity of the rat RF to the anti-RF serum.

Results

Rat Embryos

SCO–RF Complex

First of all, the results obtained with immunostaining using the anti-RF serum dilution of 1:2000 will be described. During day E-10 and E-11 (Carnegie stage 11–13), no immunoreactive material was found in the developing brain. At day E-15, the immunoreactive material first appeared in some cells of the developing SCO. The cytoplasm of these cells was evenly immunostained, with no signs of polarization (Fig. 1a).

At day E-16, the ependymal cells of the SCO developed basal processes projecting to the leptomeninges. By day E-17, all cells of the SCO were strongly immunoreactive. When AFRU was used at a dilution of 1:100000, immunoreactive secretory granules were not seen in the apical portion of the ependymal cells. At day E-18, the secretory ependyma was formed by tall cylindrical cells. These cells displayed apical secretory granules, which were WGA-positive and immunoreactive with AFRU at a dilution of 1 : 100000. They also contained numerous large droplets (RER) located in the perinuclear region; these inclusions were highly immunoreactive also with high dilutions of AFRU. Up to E-21, no signs of ventricular release of secretory material were observed. In rat embryos fixed shortly before delivery day E-22), a few fine, immunostained fibrils were seen in the sylvian aqueduct

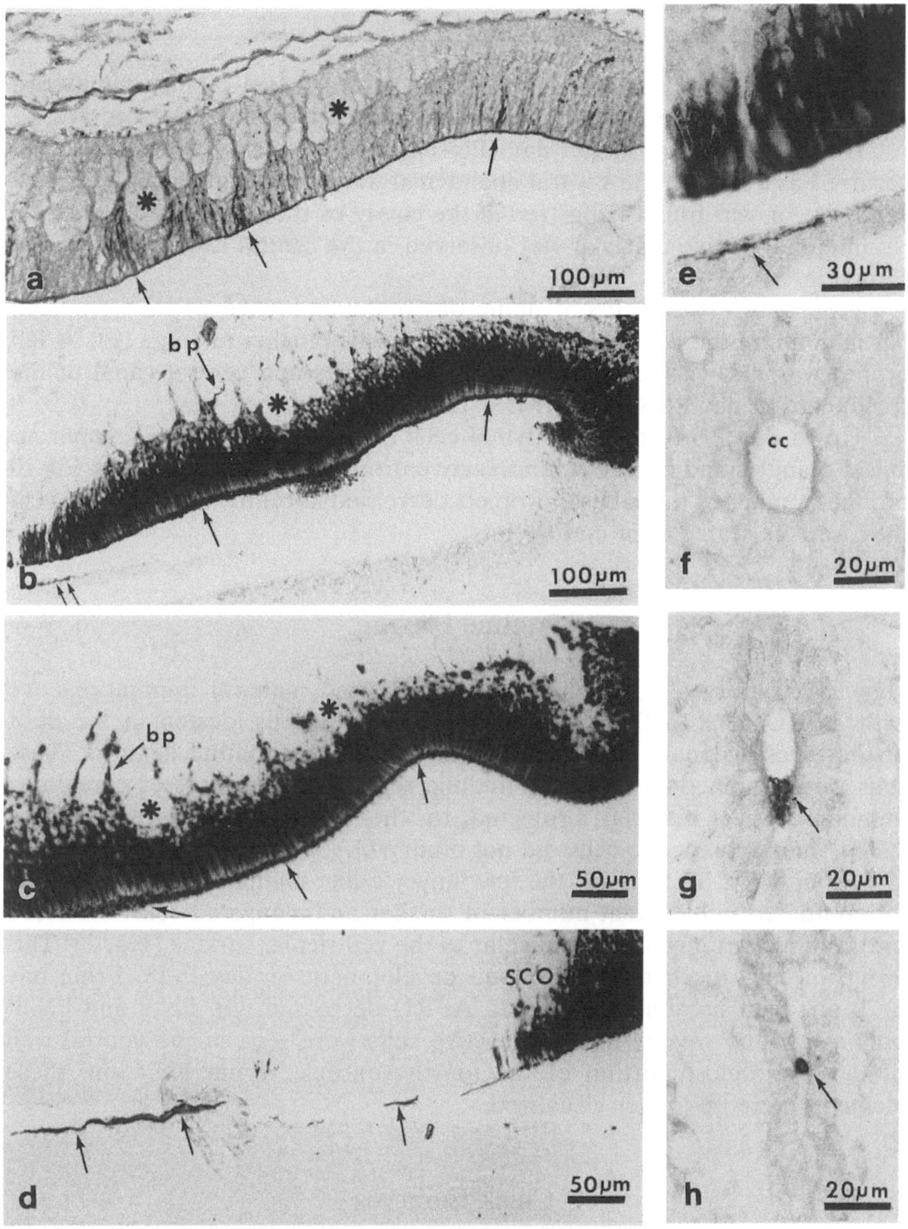

Fig. 1a–h. Sections through the rat SCO and central canal at different stages of development, immunostained with AFRU diluted 1:2000. In the 15-day-old rat embryo, AFRU-immunoreactive material appears in scattered SCO cells (**a**; *arrows*). During the last day of prenatal life, the SCO cells display strong AFRU immunoreactivity (**b**; *arrows*), and immunoreactive fibrous material appears in the ventricle (*double arrows*). At higher magnification (**e**; *arrow*), it can be shown that this material is not organized in the form of a typical RF. One day after birth, there are abundant secretory granules in the apical cytoplasm of the SCO cells (**c**; *arrows*); fibrous immunoreactive structures start to grow from the caudal end of the SCO (**d**; *arrows*), but do not reach the central canal (**f**; *cc*). At day PN-2, immunoreactive fibrous material is found in cervical segments of the central canal (**g**; *arrow*). RF proper appears for the first time in the lumen of the central canal at day PN-6 (**h**; *arrow*). *Asterisks* Posterior commissure; *bp* basal processes

(Fig. 1b,e). The fourth ventricle and the central canal of the spinal cord were, however, completely devoid of immunoreactive material.

During the first postnatal day (PN-1), immunoreactive fibrous material occurred as a film on the ventral ependymal wall of the aqueduct and fourth ventricle, or was found lying free in the cavity of the aqueduct (Fig. 1c,d). No immunoreactive material was observed in the central canal of the spinal cord (Fig. 1f).

From day PN-2 to PN-5, numerous immunoreactive fibrils were seen partially filling the central canal of the cervical spinal cord (Fig. 1g). A RF proper was first visible in the ventricular cavities and central canal of the spinal cord at day PN-6 (Fig. 1h).

Up to day PN-6, the ependymal cells of the SCO displayed numerous apical granules and basal immunoreactive droplets (RER). From day PN-10 on, the number of these basal droplets decreased abruptly, so that by day 14 they were as scarce as in mature rats.

Pontine Flexure

Day E-12 embryos (Carnegie stage 15) showed material immunoreactive with AFRU at a 1:2000 dilution in a cluster of cells located in the floor of the myelencephalic ventricle, in the area of the pontine flexure. These cells were confined to a narrow midline region; basal processes containing immunoreactive material projected to the outer brain surface. These AFRU-immunoreactive cells did not bind WGA.

From day E-13 to E-18, the few tanycyte-like immunoreactive elements differentiated into a great number of smaller and elongated cells, clustered into columns arranged perpendicular to the ventricular surface (Fig. 2). This group of cells reached its maximum development on day E-18. From day E-20 on, their immunoreactivity to AFRU decreased. On PN-2 and PN-3, only a few and weakly immunoreactive cells were seen in the ventral wall of the rostralmost portion of the fourth ventricle. From PN-4 on, these elements were no longer visualized.

Chick Embryos

During the first 2 days of incubation (Hamilton stages 6 and 12), no immunoreactive material was found in the developing brain. The embryos collected during the third day of incubation (Hamilton stages 17) showed a group of immunoreactive cells located in the dorsal wall of the posterior diencephalon. These immunoreactive cells were elongated with the immunoreactive material distributed in the apical cytoplasm. These elements did not exhibit a basal process.

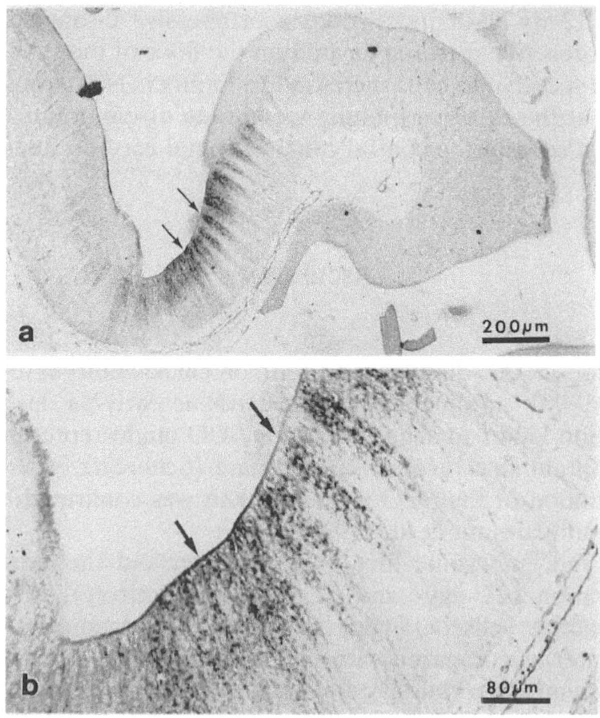

Fig. 2a,b. AFRU-immunostaining of a 15-day-old rat embryo revealed groups of immunoreactive cells in the region of the pontine flexure (**a**; *arrows*). These cells are aligned in columns perpendicular to the ventricular surface (**b**; *arrows*)

On day E-4 of incubation, the number of immunoreactive cells increased. At day E-5, the first nerve bundles of the posterior commissure started to appear, and the immunoreactive cells of the SCO developed basal processes containing immunoreactive material. In day E-6 embryos, the SCO was formed by a stratified layer of strongly immunoreactive cells. The ventricular surface of the SCO did not display immunoreactive material.

At day E-7, a thin layer (pre-RF) of immunoreactive material appeared on the free surface of the ependymal cells and on the floor of the sylvian aqueduct. SCO cells showed a few apical granules immunostained by AFRU at a 1:100000 dilution.

After day E-10, a thick layer of immunoreactive pre-RF material was observed on the free surface of the SCO. Numerous immunoreactive granules were now visible in the apical cytoplasm; they were readily immunostained with AFRU 1:100000.

Day E-11 embryos were characterized by the appearance of a RF. At this stage, RF coexisted with a fibrous immunoreactive structure located near the ventral wall of the aqueduct.

From day 12 to 21 of incubation, a progressive disappearance of the fibrous immunoreactive material located on the floor of the sylvian aqueduct was evident. Hypendymal cells increased in number. Numerous ependymal endings ended at the external limiting membrane of the brain.

RF was first seen in the central canal of spinal cord on day E-12.

Discussion

The appearance in the diencephalic roof of chick embryos of a material immunostained with an antiserum against RF as early as during the third day of incubation led us to suggest that the SCO might represent one of the first secretory brain structures to differentiate (Schoebitz et al. 1986). This early differentiation of the SCO in the chicken was confirmed by Naumann et al. (1987) and Karoumi et al. (1990).

Although the embryonic life of the chicken and the rat is almost of the same duration (21 days and 22 days, respectively), in rat embryos RF-immunoreactive cells belonging to the SCO first appear at day E-15, i.e., 12 days later as compared with the situation in the chicken. However, in the rat, the central nervous system starts to develop on day E-8, and so in this species the SCO differentiates at an early stage of the central nervous development, too.

Still, there are essential differences in the development of the SCO of chicken and rat. At birth the chick SCO appears to be fully developed; the development of the rat SCO continues postnatally with completion of differentiation by the fourth postnatal week (Kohl and Linderer 1973; Marcinkiewicz and Bouchaud 1983, 1986; Naumann et al. 1987).

However, the most apparent difference in the ontogeny of the SCO of chicken and rat concerns the time course of events leading to RF formation. Thus, in the chicken, ventricular release of immunocytochemically visible material in the form of fibrous structures occurs at day E-7; in the rat this event takes place during the last day of embryonic life (E-22). The chicken displays a RF proper at day E-11; however, the rat shows "Reissner's fibrils" in the central canal of the spinal cord only 2 days after birth and a RF proper at postnatal day 6.

That, in the rat, a RF is first found only after birth is a striking observation which leads to several important considerations: (1) In the rat, RF apparently does not play a role during embryonic life, since it is missing; (2) during the last 6 days of embryonic life the rat SCO displays high secretory activity and is endowed with numerous apical granules containing "mature" secretory material, indicated by its affinity to WGA and high dilutions of AFRU. It seems highly probable that during this period of embryonic life the SCO does release material into the ventricle which is not packed into fibrous structures but remains soluble in the cerebrospinal fluid. Based on the previous considerations, it may be postulated that whatever the function of the SCO during embryonic life might be, in the rat this role is mediated by a cerebrospinal fluid-soluble secretion. This is in agreement with recent findings concerning cerebrospinal fluid-soluble secretions of the SCO (Rodríguez et al., this volume).

The completion of the development of the rat SCO several weeks after birth, as indicated by several authors (see above), has been further substantiated by the present investigation. The high number and large size of the perinuclear RER cisternae containing RF-immunoreactive material during the last prenatal week and the first postnatal week could be taken as an indication of a high rate of synthesis of secretory material during this period. The sharp decrease in the number of these "droplets" (RER) starting during the second postnatal week would imply a decline in the secretory activity of the SCO. This is in full agreement with the fact that serotonin-immunoreactive fibers arrive at the SCO during the first postnatal week and that full development of the neural input to the SCO is accomplished by the end of the first postnatal month (Wiklund 1974; Wiklund et al. 1977; Marcienkiewicz and Bouchaud 1986; Matsuura et al. 1989). Our own studies have shown that immunoreactive serotonin fibers start to innervate the SCO during the second postnatal week. It may be suggested that the initiation of the decrease in the number of perinuclear immunoreactive droplets (RER) is the morphological correlate of the establishment of the inhibitory neural input.

A relevant finding of the present investigation is the presence in the pontine flexure of cells immunostained with anti-RF sera. It seems likely that this structure is homologous to the flexural organ described by Olsson (1956) in embryos of *Salmo salar*, *Esox lucius*, and *Xenopus laevis*. A recent immunocytochemical study performed in our laboratory using anti-RF sera has shown that the SCO and flexural organ of salmon embryos are strongly immunoreactive to the same antisera (R. Olsson et al. 1993, unpublished results). According to Olsson (1956), in salmon embryos RF material is first secreted by the flexural organ. As the SCO develops, the fiber-forming function is taken over by this latter organ, and the flexural organ undergoes regression. Thus, the flexural organ appears to be a secretory structure of the brain exclusively active during a short period of the embryonic life. The present investigation represents the first study providing evidence for the existence in a mammalian species of a secretory structure similar to the flexural organ described by Olsson (1956). The rat flexural organ is also a transient secretory structure that appears earlier in ontogeny (E-12) as compared to the SCO (E-15), and of which only a few cells persist during the first three postnatal days. The lack of a RF during the embryonic life of the rat indicates that in this species the flexural organ is not involved in the formation of a RF; instead, its secretion might be cerebrospinal-fluid soluble. The secretion, although AFRU positive, may be different from that in the SCO, since it does not bind WGA; this latter finding strongly suggests that the secretion of the flexural organ is not glycosylated via N-linkages. The columnar arrangement of the rat flexural cells and their poor vascularization may indicate a local effect of this secretion. Interestingly, between these columns the serotoninergic neurons of the brain stem start to differentiate (unpublished observations).

Acknowledgements. Supported by Grant 1/36 476 from the Volkswagen-Stiftung, Federal Republic of Germany; Grant 91/0956 from FONDECYT, Chile, and Grant S-89-01 from the Dirección de Investigaciones, Universidad Austral de Chile.

References

Bruel MY, Meiniel A, David D (1987) Ontogenetical study of the chick embryo subcommissural organ by lectin-histofluorescence and electron microscopy. J Neural Transm 70:145–178

Castañeÿra-Perdomo A, Meyer G, Ferres-Torres R (1983) Development of the subcommissural organ in the albino mouse. J Hirnforsch 24:368–370

Gabe M (1968) Techniques histologiques. Masson, Paris

Herrera H, Rodríguez EM (1990) Secretory glycoproteins of the rat subcommissural organ are N-linked complex-type glycoproteins. Demonstration by combined use of lectins and specific glycosidases, and by the administration of Tunicamycin. Histochemistry 93:607–615

Karoumi A, Meiniel R, Croisille Y, Belin MF, Meiniel A (1990) Glycoprotein synthesis in the subcommissural organ of the chick embryo. I. An ontogenical study using specific antibodies. J Neural Transm (Gen Sect) 79:141–153

Kohl W, Linderer T (1973) Zur Entwicklung des Subcommissuralorgans der Ratte. Morphologische und histochemische Untersuchungen. Histochemie 33:349–368

Marcinkiewicz M, Bouchaud C (1983) The ependymal secretion of the fetal and adult rat subcommissural organ. Morphological aspects linked to the synthesis, storage and release of the secretory products. Biol Cell 48:47–52

Marcinkiewicz M, Bouchaud C (1986) Formation and maturation of axoglandular synapses and concomitant changes in the target cells of the rat subcommissural organ. Biol Cell 56:57–65

Matsuura T, Kumamoto K, Ebara S, Sano Y (1989) Serotonergic innervation of the mammalian subcommissural organ: an immunohistochemical study. Biomed Res 10 [Suppl 3]:177–186

Naumann W, Müller G, Kloss P (1987) Immunoreactive glycoprotein of the subcommissural organ in the embryonic stages of the vertebrate brain. Wiss Z Karl Marx Univ Leipzig Math Naturwiss R 36:17–20

Olsson R (1956) The development of Reissner's fibre in the brain of the salmon. Acta Zool (Stockh) 37:235–250

Rakic P, Sidman RL (1968) Subcommissural organ and adjacent ependyma: autoradiographic study of their origin in the mouse brain. Am J Anat 122:317–336

Rodríguez EM, Oksche A, Hein S, Rodríguez S, Yulis R (1984) Comparative immunocytochemical study of the subcommissural organ. Cell Tissue Res 237:427–441

Rodríguez EM, Herrera A, Peruzzo B, Rodríguez S, Hein S, Oksche A (1986) Light- and electron-microscopic lectin histochemistry and immunohistochemistry of the subcommissural organ: evidence for processing of the secretory material. Cell Tissue Res 243:545–559

Schoebitz K, Garrido O, Heinrichs M, Speer L, Rodríguez EM (1986) Ontogenetical development of the chick and duck subcommissural organ and immunocytochemical study. Histochemistry 84:31–40
Sterba G, Kiessig C, Naumann W, Petter H, Klein I (1982) The secretion of the subcommissural organ. A comparative immunocytochemical investigation. Cell Tissue Res 226:427–439
Sternberger LA, Hardy PH Jr, Cuculis JJ, Meyer HG (1970) The unlabeled antibody enzyme method of immunohistochemistry; preparation of properties of soluble antigen-antibody complex (horseradish peroxidase-antiperoxidase) and its use in identification of spirochetes. J Histochem Cytochem 18:315–333
Wiklund L (1974) Development of serotonin-containing cells and the sympathetic innervation of the habenular region in the rat brain. A fluorescence histochemical study. Cell Tissue Res 155:231–243
Wiklund L, Lundberg JJ, Møllgård K (1977) Species differences in serotonergic innervation and secretory activity of rat, gerbil, mouse and rabbit subcommissural organ. Acta Physiol Scand [Suppl] 452:27–30
Wingstrand KG (1953) Neurosecretion and antidiuretic activity in chick embryos with remarks on the subcommissural organ. Arch Zool (Stockh) 6:41–67
Ziegels J (1977) Etude histochimique de l'organe sous-commissural du poulet au cours du development. CR Soc Biol 171:1306–1308

Developmental Aspects of the Subcommissural Organ: An Approach Using Lectins and Monoclonal Antibodies

R. MEINIEL, R. DIDIER, J.L. MOLAT, and A. MEINIEL

Introduction

The SCO ependymocytes are representatives of a particular differentiation process; although of ependymal lineage, they only partly express typical glial markers (Didier et al. 1986) and show also certain molecular features that can be compared to those of neurons (Gamrani et al. 1981). The secretory activity of the SCO is now well established (Hess and Sterba 1973; Sterba et al. 1981, 1982; Rodríguez et al. 1984a,b); complex-type glycoproteins released permanently into the ventricular cavity (Rodríguez et al. 1986; Meiniel et al. 1988a) contribute to the formation of Reissner's fiber (RF). The latter is a very constant structure of the chordate nervous system (Olsson 1986). In addition, a basal secretory pathway is also suspected (Oksche 1969; Kimble and Møllgaard 1973; Lösecke et al. 1984; Rodríguez et al. 1984b; Diederen et al. 1987). In fact, SCO secretory ependymocytes may be regarded as a type of neuroendocrine cells. To investigate some molecular features of this organ, we generated monoclonal antibodies (Mab). In addition to the study of the spatiotemporal differentiation of the secretory activity, these probes can be most useful for an experimental approach to the role of the secretory products in the course of development. The aim of this paper is to present results obtained during embryogenesis in two species of amniotes: the bovine and chick.

Results and Discussion

Features of the Secretory Products in the Adult Bovine SCO as Revealed by Use of Lectins and Mab $C_1B_8A_8$

The SCO has long been known to contain glycoproteins, which were visualized by classical histochemical staining procedures (see Oksche 1969; Leonhardt 1980). Recently, the application of lectins has helped us to learn more about the secretory compounds of the SCO. Concanavalin A (ConA) links to secretory material in the cytoplasm of ependymal and hypendymal cells (Meiniel and Meiniel 1985). This material corresponds to high mannose-type glycoproteins stored in the rough endoplasmic reticulum (RER) (Rodríguez et al. 1986; Meiniel et al. 1988a). After transformation in the Golgi apparatus, the complex-type glycoproteins released into the cerebrospinal fluid (CSF) strongly react with various lectins (wheat-germ agglutinin, WGA; *Lens culinaris* agglutinin, LCA; *Phaseolus vulgaris* erythroagglutinin, E-PHA; *Phaseolus vulgaris* leukoagglutinin, L-PHA). The epitope recognized by Mab $C_1B_8A_8$ is present in the high mannose-type glycoprotein (precursor form, ConA positive) and in the complex-type glycoprotein (secreted form, WGA positive) (Meiniel et al. 1988b; see Fig. 1a,b).

Development of the Secretory Process in the Bovine SCO

In the 2-month-old bovine fetus (59 days), the differentiation of the specialized ependymal cells of the SCO could only be revealed using specific ligands. At this stage, both ConA and $C_1B_8A_8$ (Fig. 1c) showed the presence of the precursor form of the secretory products: the high mannose-type glycoprotein. Observation of glycoproteins in the entire embryonic SCO as early as in 2-month-old embryos suggests that this specific antigen is expressed already somewhat earlier. In addition, absence of reaction with the other lectins used (WGA, LCA, E-PHA, L-PHA) points to an incomplete maturation of the carbohydrate moieties at this stage of development.

In the 3½-month-old fetus (103 days) the SCO is constituted by two distinct layers corresponding to the future ependymal and hypendymal formations (Olsson 1956). Using Mab $C_1B_8A_8$, strong immunofluorescence was detected in both formations (Meiniel et al. 1990). This diffuse fluorescence was superimposed on the reaction product observed after exposure to ConA, while the strong fluorescence visible at the borderline to the ventricular cavity and in granular structures located in the hypendymal formation was similar to that detected after WGA, LCA, E-PHA, and L-PHA exposure. The presence of secretory material in the borderline region of the SCO close to the ventricular system indicates that maturation of the complex-type glycoprotein in the Golgi complex occurs at this stage. Nevertheless, absence of RF at the level of the spinal cord suggests either that the release of secretory products in not yet accomplished or that other factors are required for condensation of these secretory products in the CSF. Long processes containing immunoreactive material (IRM) extend deeply into the posterior commissure towards the leptomeningeal spaces, suggesting a basal secretory pathway.

Further chronological events related to the secretory process of the SCO are marked by the appearance of RF all along the spinal cord in the 4-month-old fetus (120 days). Considering the high reactivity of the SCO epithelium using Mab $C_1B_8A_8$ at this stage of development we can postulate that, when the formation of the RF takes place, SCO secretory activity is enhanced.

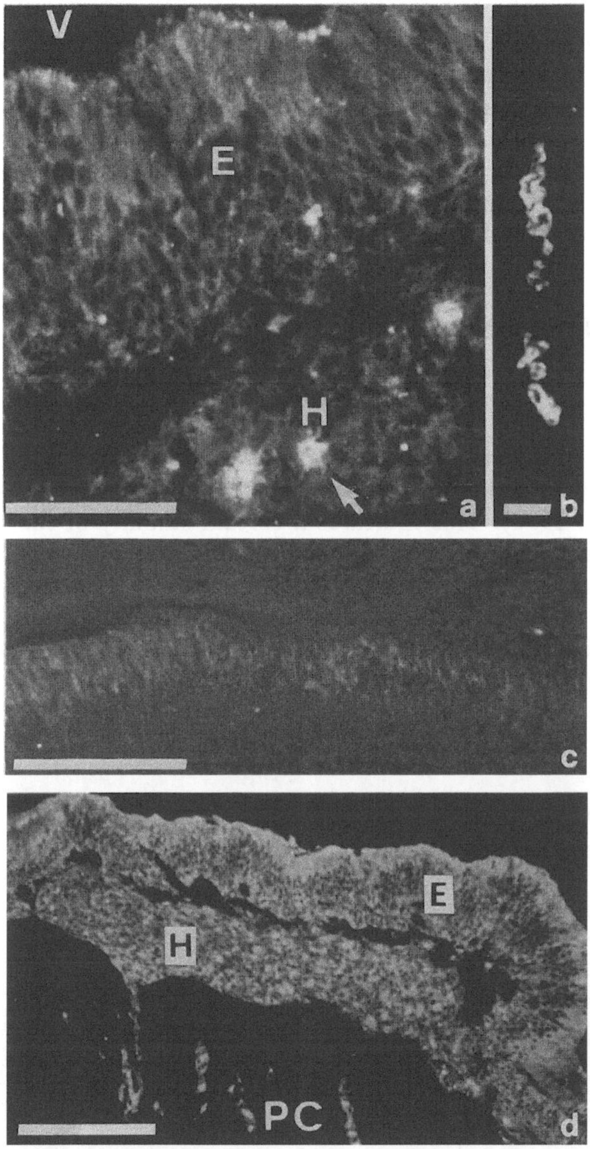

Fig. 1a–d. Immunoreactive material in the bovine SCO and RF after exposure to $C_1B_8A_8$ Mab. **a** Diffuse fluorescent material in the cytoplasm of ependymal (*E*) and hypendymal (*H*) cells in the adult. Strong fluorescence close to the ventricular cavity (*V*) and in rosettes (*arrow*). *Bar* = 100 µm. **b** Oblique section through the sacral part of the spinal cord in a bovine embryo (120 days). IRM is detected in RF. *Bar* = 20 µm. **c** Embryonic SCO at day 59; parasagittal section of the diencephalic roof. The SCO-anlage is labeled. *Bar* = 130 µm. **d** Embryonic SCO at day 259. The cytoplasm of ependymal (*E*) and hypendymal (*H*) cells is positive. Note the presence of reactive cell processes in the posterior commissure (*PC*). *Bar* = 130 µm

In the 8½-month-old fetus (259 days), the diencephalic roof has aquired a configuration closely resembling that of the adult and newborn specimens. Nevertheless, hypendymal cells are not yet arranged in rosettes (Fig. 1d), except in the most rostral and lateral parts of the SCO. Such hypendymal "rosettes" are specific for the SCO of normal bovines (Olsson 1956). Proof of the identity in cell lineage of the cells constituting the ependymal layer and of the cells constituting the future "rosettes" in the hypendymal layer could now be obtained using Mab $C_1B_8A_8$.

Development of the SCO in the Chick Embryo

Specificity of the Mabs

The secretory products in the 19-day-old chick embryo exhibit the same reactivity to the lectins that those of the adult bovine (Fig. 2a–c; see Karoumi et al. 1990). This indicates a certain stability of the carbohydrate moities, probably conferring a high degree of homology of the complex-type glycoprotein in the two species. Four Mabs, i.e., $H_2E_9G_8$, $D_{14}F_1D_{12}$, $G_1E_2B_{11}$ and $G_7H_4B_8$, were selected on the basis of their affinity for the material present in the SCO and RF of 19-day-old embryos (Fig. 2d,e) (Didier et al. 1992). The Mabs $H_2E_9G_8$ and $D_{14}F_1D_{12}$ were shown to have a

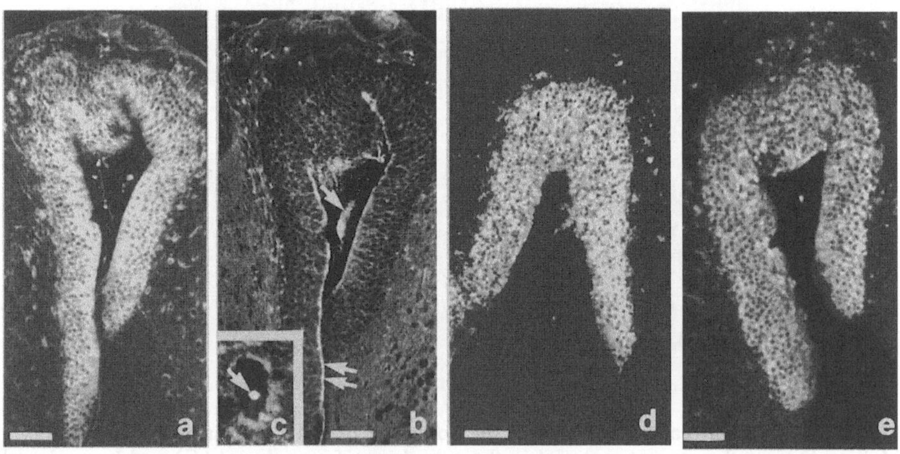

Fig. 2a–e. Transverse sections in 19-day-old chick embryos. **a** ConA. The entire SCO epithelium and a few cells scattered in the surrounding brain tissue are labeled. *Bar* = 50 µm. **b, c** WGA. **b** The labeling is located at the apical lining of the epithelium (*double arrow*). RF material is also positive (*arrow*). *Bar* = 50 µm. **c** Section through the central canal of the spinal cord; *arrow*, RF. **d** Mab $H_2E_9G_8$. Note the strong and specific immunoreactivity of the SCO. *Bar* = 50 µm. **e** Mab $G_7H_4B_8$. SCO is strongly labeled. Weak fluorescence in the adjacent nervous tissue, particularly in the perikarya of neurons. *Bar* = 50 µm

high specificity for the secretory products of the SCO. Using the Mabs $G_1E_2B_{11}$ and $G_7H_4B_8$, a complex labeling pattern was observed, suggesting that certain epitopes recognized in the SCO may also be present in other cells of the nervous system, thus leading to a cross-reactivity (see Didier et al. 1992).

These Mabs have been explored in order to learn more about certain developmental events in the chick SCO. They were tested starting from stage 14 of Hamburger and Hamilton (1951).

Spatiotemporal Differentiation of the SCO Using Mabs

With Mab $H_2E_9G_8$ no labeling could be detected before stage 17 (52–64 h). At this stage, some cells in the subcommissural *anlage*, displayed a strong fluorescence, indicating the presence of IRM in their cytoplasm (Fig. 3a). At further stages, the number of immunoreactive cells increased progressively in a caudal direction (Fig. 3b). These results are similar to those obtained in 1986 by Schoebitz et al. (with the use of an anti-RF serum), showing that the synthetic activity in the chick embryo SCO starts on the third day of incubation.

Using $D_{14}F_1D_{12}$ and $G_1E_2B_{11}$ antibodies, the first immunoreactive cells were observed later than with $H_2E_9G_8$, i.e., from stage 19 on (68–72 h). With Mab $G_7H_4B_8$ the whole neuroectoderm exhibited a strong immunoreactivity until day 4 of incubation (Fig. 3c). From day 5 on, the immunoreactivity progressively decreased in the neuroepithelial cells, but the reaction remained intense in the SCO epithelium. A very early reactivity in the entire neuroepithelium was also detected with classical stains (Olsson 1956; Oksche 1969), ConA (Bruel et al. 1987), and an anti-RF serum (Naumann et al. 1987). Although the nature of this transient coexpression remains unclear, we can postulate that the SCO cells represent a special cell lineage retaining embryonic features in spite of their particular differentiation.

Up to day 10 of incubation no IRM could be detected in transverse sections of the medulla oblongata whatever the immunoglobulin used. A

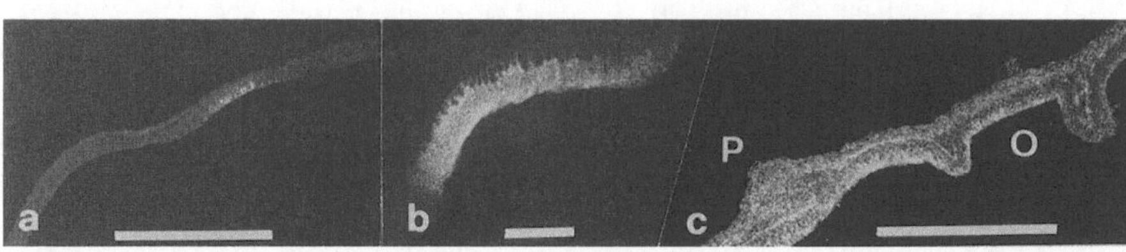

Fig. 3a–c. Sagittal sections at different stages of ontogeny. **a**, **b** Mab $H_2E_9G_8$. SCO of chick embryo at stage 17 (52–64 h; **b**). The SCO anlage is the only positive structure. *Bar* = 400 µm. At day 5 (**b**), the immunoreactive cells are increased in number. *Bar* = 130µm. **c** A 4-day-old chick embryo. Immunostaining with Mab $G_7H_4B_8$. Immunoreactive material is revealed in the entire neuroepithelium; *P* anlage of the pineal organ; *O*, anlage of the SCO. *Bar* = 450 µm

positive reaction was observed in the central canal of the thoracic spinal cord in 10-day embryos using Mab $H_2E_9G_8$. Thus, this probe enables to detect this structure at an earlier stage than the use of other procedures [day 11 using an anti-SCO antibody (Karoumi et al. 1990), and day 12 using histochemical staining (Schoebitz et al. 1986)].

Injection of Mab $H_2E_9G_8$ in the Chick Embryo

In 9-day-old chick embryos, Mab $H_2E_9G_8$ was injected into either the cerebroventricular cavity or the vascular system via a major extraembryonic vessel. The SCO and spinal cord were prepared for cryostat sectioning 24, 48, and 72 h after the injection. The sites of binding of $H_2E_9G_8$ (IgG1) were revealed using an anti-Y1 immunoglobulin labeled with fluorescein. Whatever the type of injection or how long the time period between injection and sacrifice, the Mab could be detected at the apex of the SCO epithelium (Fig. 4a,b). This indicates that the antibody, even when injected into the vascular system, can reach the ventricular cavity and bind to the secreted material. In addition, labeling was observed at the base of the SCO epithelium, suggesting a release of secretory products at this level. RF was present in 12-day-old embryos and a strong fluorescence of this structure indicated the binding of the Mab (Fig. 4c). Blockage of one epitope of the secretory product was thus ineffective to alter the condensation in RF. Recently, Rodríguez et al. (1990), using an anti-RF serum, were able to block reversibly the formation of RF in the adult rat.

Extention of such experimental procedures will probably be useful in providing additional knowledge concerning the secretory activity of the SCO in the course of the development and help to elucidate its role in the differentiation process.

In conclusion, the early expression of the secretory process (Fig. 5) and its intensity at certain stages of ontogeny suggest an involvement of the SCO in developmental processes (see Rühle 1971; Takeuchi et al. 1987). In

Fig. 4a–c. Injection of Mab $H_2E_9G_8$ in a 9-day-old embryo. At day 11, the Mab is detected at the apex of SCO (*arrow*) and in basal processes (*double arrow*) (**a** *bar* = 85 μm; **b** *bar* = 80 μm.) **c** RF material is also strongly labeled in a sagittal section through the medulla oblongata. *Bar* = 50 μm

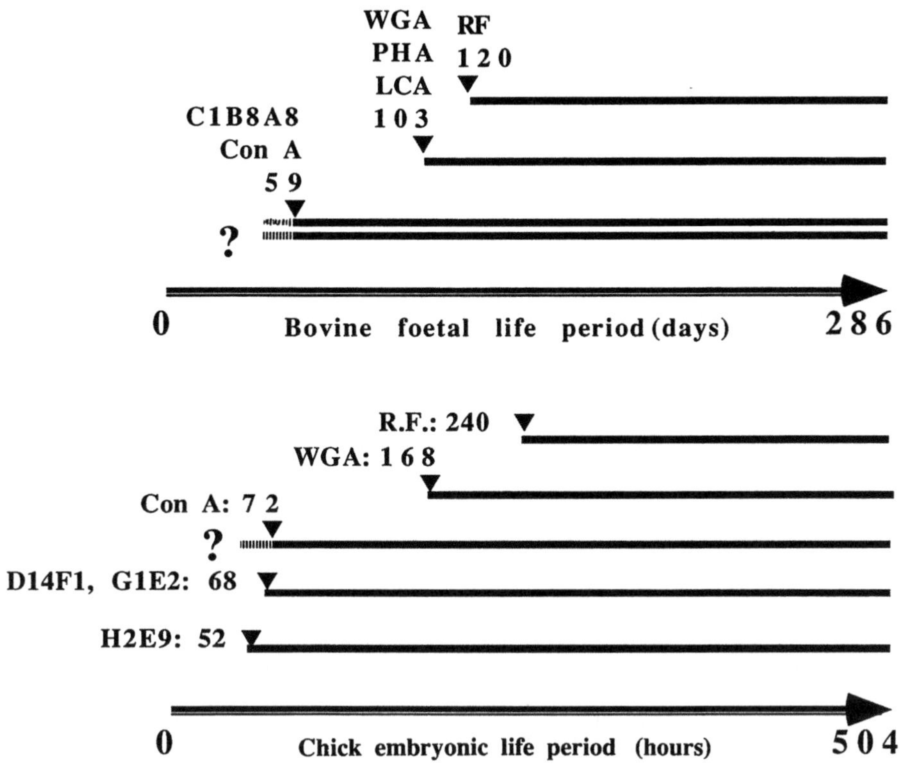

Fig. 5. Comparison of developmental aspects of SCO and Reissner's fiber in the bovine and chick

addition, regarding the particular and specific phenotype of the secretory ependymocytes of the SCO, these elements could constitute an interesting model for the study of the commitment of specialized cells to the nervous system.

Acknowledgements. This research was partly supported by a grant from the Association Française contre les Myopathies (AFM).

References

Bruel MT, Meiniel R, Meiniel A, David D (1987) Ontogenetical study of the chick embryo subcommissural organ by lectin histofluorescence and electronmicroscopy. J Neural Transm 70:145–168

Didier M, Harandi M, Aguera M, Bancel B, Tardy M, Fages C, Calas A, Stagaard M, Möllgård K, Belin MF (1986) Differential immunocytochemical staining for glial fibrillary acidic (GFA) protein, S-100 protein and glutamine synthetase in the rat subcommissural organ and non-specialized ventricular ependyma and adjacent neuropil. Cell Tissue Res 245:343–351

Didier R, Meiniel A, Meiniel R (1992) Monoclonal antibodies as probes for the analysis of the secretory ependymal differentiation in the subcommissural organ of the chick embryo. Dev Neurosci 14:44–52

Diederen JHB, Vullings HGB, Legerstee-Oostveen TG (1987) Autoradiographic study of the production of secretory material by the subcommissural organ of frogs (*Rana temporaria*) after injection of several radioactive precursors, with special reference to the glycosylation and turnover rate of the secretory material. Cell Tissue Res 248:215–222

Gamrani H, Belin MF, Aguera M, Calas A, Pujol JF (1981) Radioautographic evidence for an innervation of the subcommissural organ by GABA containing nerve fibres. J Neurocytol 10:411–424

Hamburger V, Hamilton HL (1951) A series of normal stages in the development of the chick embryo. J Morphol 88:49–92

Hess J, Sterba G (1973) Studies concerning the function of the complex subcommissural organ-liquor fibre: the binding ability of the liquor fibre to pyrocatechin derivatives and its functional aspects. Brain Res 58:303–312

Karoumi A, Meiniel R, Croisille Y, Belin MF, Meiniel A (1990) Glycoprotein synthesis in the SCO of the chick embryo. I: An ontogenetical study using specific antibodies. J Neural Transm 79:141–153

Kimble JE, Møllgaard K (1973) Evidence for basal secretion in the subcommissural organ of the adult rabbit. Z Zellforsch 142:223–239

Leonhardt H (1980) Ependym und circumventriculäre Organe. In: Oksche A, Vollrath L (eds) Handbuch der mikroskopischen Anatomie des Menschen, vol IV, part 10: Neuroglia I. Springer, Berlin Heidelberg New York, pp 117–665

Lösecke W, Naumann W, Sterba G (1984) Preparation and discharge of secretion in the subcommissural organ of the rat. An electron-microscopic immunocytochemical study. Cell Tissue Res 235:201–206

Meiniel A, Molat JL, Meiniel R (1988a) Complex-type glycoproteins synthesized in the subcommissural organ of mammals. Light-and electron-microscopic investigations by use of lectins. Cell Tissue Res 253:383–395

Meiniel R, Meiniel A (1985) Analysis of the secretions of the subcommissural organs of several vertebrate species by use of fluorescent lectins. Cell Tissue Res 239:359–364

Meiniel R, Duchier N, Meiniel A (1988b) Monoclonal antibody $C_1B_8A_8$ recognizes a ventricular secretory product elaborated in the bovine subcommissural organ. Cell Tissue Res 254:611–615

Meiniel R, Molat JL, Duchier-Liris N, Meiniel A (1990) Ontogenesis of the secretory epithelium of the bovine subcommissural organ. A histofluorescence study using lectins and monoclonal antibodies. Dev Brain Res 55:171–180

Naumann W, Müller G, Kloss P (1987) Immunoreactive glycoprotein of the subcommissural organ in the embryonic stages of the vertebrate brain. Wiss Z Karl Marx Univ Leipzig Math Naturwiss R 36:17–20

Oksche A (1969) The subcommissural organ. J Neuro Visc Rel [Suppl] 9:111–139

Olsson R (1956) Studies on the subcommissural organ. Acta Zool 39:71–102

Olsson R (1986) Basic design of the chordate brain. In: Arai T, Taniuchi A, Matsuura K, Fishes T, Uyeno R (eds) Proceedings of the 2nd international conference on Indo-Pacific. Ichthyological Society of Japan, pp 86–93

Rodríguez EM, Oksche A, Hein S, Rodríguez S, Yulis R (1984a) Comparative immunocytochemical study of the subcommissural organ. Cell Tissue Res 237:427–441

Rodríguez EM, Oksche A, Hein S, Rodríguez S, Yulis R (1984b) Spatial and structural interrelationships between secretory cells of the subcommissural organ and blood vessels. An immunocytochemical study. Cell Tissue Res 237:443–449

Rodríguez EM, Herrera H, Peruzzo B, Rodríguez S, Hein S, Oksche A (1986) Light- and electron-microscopic immunohistochemistry and lectin histochemistry of the subcommissural organ: evidence for processing of the secretory material. Cell Tissue Res 243:545–559

Rodríguez S, Rodríguez EM, Jara P, Peruzzo B, Oksche A (1990) Single injection into the cerebrospinal fluid of antibodies against the secretory material of the subcommissural organ reversibly blocks formation of Reissner's fiber. Immunocytochemical investigation in the rat. Exp Brain Res 81:113–124

Rühle HJ (1971) Anomalien im Wachstum der Achsenorgane nach experimenteller Ausschaltung des Komplexes Subcommissuralorgan-Reissnerscher Faden. Untersuchungen am Rippenmolch (*Pleurodeles waltlii* Michah 1830). Acta Zool (Stockh) 52:23–68

Schoebitz K, Garrido O, Heinrichs M, Speer L, Rodríguez EM (1986) Ontogenetical development of the chick and duck subcommissural organ. An immunocytochemical study. Histochemistry 84:31–40

Sterba G, Kleim I, Naumann W, Petter H (1981) Immunocytochemical investigation of the subcommissural organ in the rat. Cell Tissue Res 218:659–662

Sterba G, Kiessig C, Naumann W, Petter H, Kleim I (1982) The secretion of the subcommissural organ. A comparative immunocytochemical investigation. Cell Tissue Res 226:427–439

Takeuchi IK, Kimura R, Matsuda M, Shoji R (1987) Absence of subcommissural organ in the cerebral aqueduct of congenital hydrocephalus spontaneously occurring in MT/Hokldr mice. Acta Neuropathol 73:320–322

The Subcommissural Organ and Ontogenetic Development of the Brain

W.W. NAUMANN, W. LEHMANN, and P. DEBBAGE

The Subcommissural Organ – A Secretory Glial Cell Group

The subcommissural organ (SCO) and Reissner's fiber (RF) may be regarded as a functional complex displaying the following main features: (a) The SCO is a conservative, phylogenetically ancient structure; (b) its morphological differentiation begins early in ontogeny; and (c) it develops the capacity to release carbohydrate-rich glycoproteinaceous secretory products, among others the RF.

Polyclonal antisera raised against bovine RF as an antigen suspended in nondenaturing aqueous media enabled us to carry out immunocytochemical studies concerned with the differentiation of the cell types of the SCO and especially the distribution of their secretory material. Immunoelectron-microscopic studies of the SCO of adult rats and rabbits revealed details of its secretory capacity. The release of secretory material from ependymocytes was found to be directed via the ventricular route into the cerebrospinal fluid to form RF. Secretion by ependymal and hypendymal cells into the intercellular space of the subjacent tissue (Lösecke et al. 1986) and into subpial space is regarded as the basal route.

Ontogeny

Immunocytochemical studies indicate that the differentiation of the SCO begins early in ontogeny (Schoebitz et al. 1986) and reveal that material immunoreactive with antisera against RF can be demonstrated over a long period before any histological differentiation of the SCO becomes evident (Naumann 1986; Naumann et al. 1987). In the earliest stages of vertebrate ontogeny examined, most of the neuroepithelium of the brain displayed a positive immunoreaction (Fig. 1a). During the course of ontogeny characteristic changes in the pattern of immunostaining of the ventricular epithelium were observed. In contrast to the decrease in the anti-RF immunoreactivity in the lateral regions of the brain, the midline portion of the vertebrate brain started to develop this property or enhanced it at least in circumscribed regions.

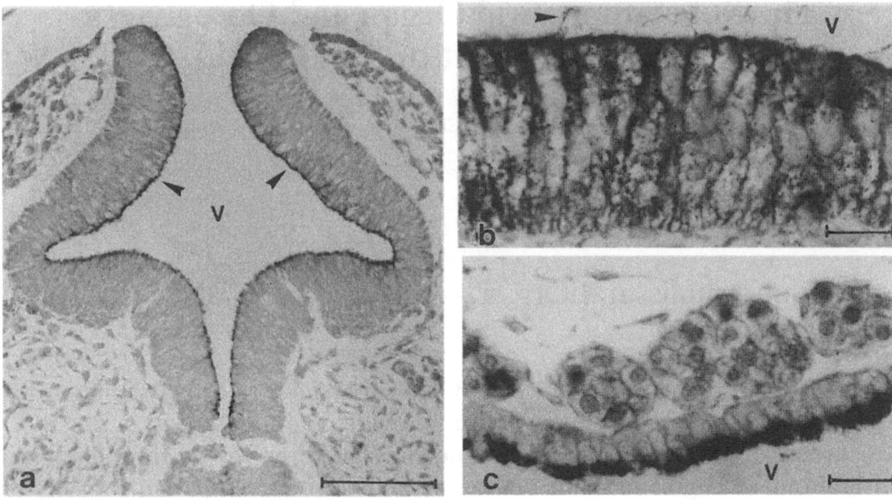

Fig. 1a–c. Distribution of anti-RF immunoreactive material in the embryonic rat brain studied prior to SCO differentiation using peroxidase-anti-peroxidase technique and an antiserum against RF. **a** Immunoreactive material covers completely the neuroepithelium (*arrows*) in an early rat embryo (ed 11). *Bar*, 1000 μm. **b** The neuroepithelial cells of the flexural organ (*Pleurodeles waltli*) displaying a radial orientation are strongly immunolabeled. Note apically released secretory material (*arrow*). *Bar*, 20 μm. **c** Early rhombencephalic neuroepithelium showing immunoreactive apical material (ed 14) *Bar*, 20 μm. *V*, ventricle

In principle, this is also valid for the embryonic development of anamniotes. However, anamniotes show certain specific features in anti-RF immunostaining before SCO differentiation, namely, strong immunoreactivity of the cells in the area of the basal deuterencephalon (Fig. 1b) and spinal cord. In this region cells of the ventricular zone release immunoreactive secretory material into the cerebrospinal fluid, where it aggregates to form an RF-like structure. Using classical paraldehyde-fuchsin staining, Olsson (1956) described a smaller number of such secretory cells in the region of the plica ventralis encephali in the brain of the salmon and named them the "flexural organ." In later stages of brain development this property of the flexural organ disappears and RF production is restricted only to the cells of the SCO.

In amniotes, the infundibular and preoptic recess as well as the subfornical organ were found to be immunoreactive at all early ontogenetic stages examined, whereas the presumptive cells of the rat rhombencephalic plexus were not selectively stained before embryonic day (ed) 14 (Fig. 1c). During just this developmental period of the rat brain, the flat cells became more cuboidal as characteristic of choroid epithelia in adults. Interestingly, the anti-RF immunoreactivity disappeared as soon as the infolding of the early plexus epithelium was completed. Finally anti-RF immunoreactivity persisted exclusively in the SCO.

The localization of anti-RF immunoreactivity during early developmental stages of the embryonic brain, first in the cells covering the ventricular cavity and then in the circumscribed area of the SCO, and its correlation in time with the development of the choroid plexus suggests that this material may play a role in the ontogenetic development of the vertebrate brain.

In addition to our light-microscopic results and the observations of Fujita (1969) indicating the secretory nature of the neuroepithelium, we used electron-microscopic analysis for further clarification of this problem. This approach enabled us to demonstrate numerous vesicles in the cytoplasm of the cells in the ventricular zone, the content of which could be immunolabeled by using an anti-RF serum. Furthermore, immunoreactive material was also localized in the cavity of the brain ventricles and in the extracellular space of the subjacent tissue. The distribution of anti-RF immunoreactive material within both embryonic and adult brains permits the conclusion that SCO secretory products are likely to occur in the extracellular matrix of the brain.

The time-dependent decrease in anti-RF immunoreactivity of the ventricular neuroepithelium raises the question of whether the SCO might be regarded as a relic of an original glial cell population as supposed by Leonhardt et al. (1987) for the development of the dorsal-midline differentiations of the brain. Regarding the origin of SCO cells in ontogeny, we should keep in mind that characteristic elements of the embryonic vertebrate brain are radial glial cells, which express the intermediate filaments GFAP and vimentin. In the prospective SCO area, radial glial cells can be stained immunocytochemically with an anti-vimentin serum (Fig. 2a), but a population of these cells is also immunoreactive with an anti-RF serum (Fig. 2b). We therefore suppose there is a direct development lineage leading from the radial glial cells to the ependymal and hypendymal cells of the SCO.

Functional Aspects – Analysis of the Secretory Product (RF)

The ideas of Puusepp and Voss (1924), the first authors to speculate about the role of the SCO in ontogeny, were confirmed by experimental studies of brain development in the absence of the SCO secretory product. It was suggested that axial regeneration and normogenesis in fishes and amphibians (Hauser 1969; Rühle 1971; Hauser and Murbach 1979), as well as experimental and congenital hydrocephalus of rats and mice (Takeuchi and Takeuchi 1986), were likely to be related to the state of the SCO or the presence of RF in the ventricular cavities.

Our considerations concerning RF function are based on RF solubility in aqueous media. UV-spectrophotometric measurements (Fig. 3a) and

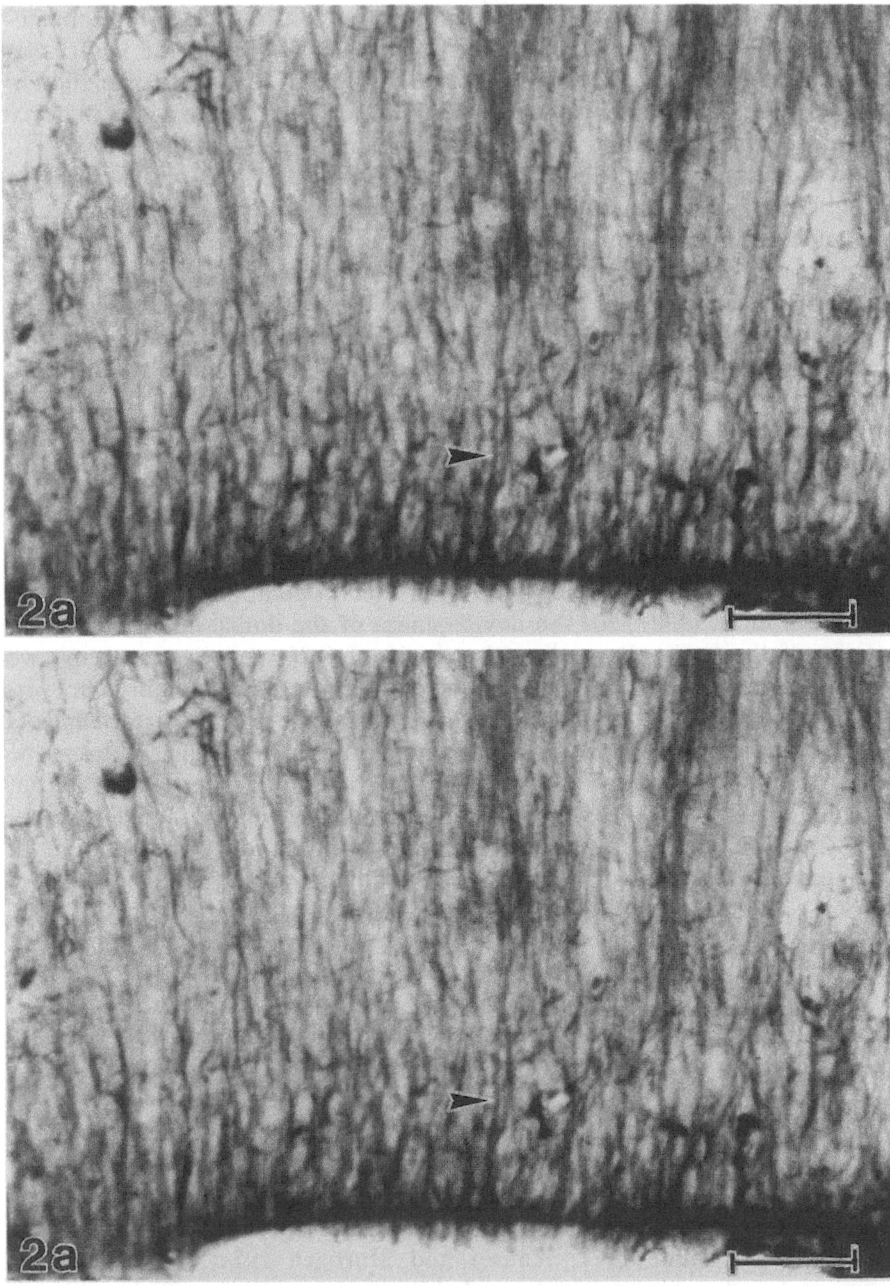

Fig. 2a,b. Immunocytochemical staining of radial glial cells (*arrows*) in the caudalmost part of the rat SCO (postnatal day 1) using anti-vimentin (**a**) and anti-RF (**b**) sera. *Bar*, 20 μm

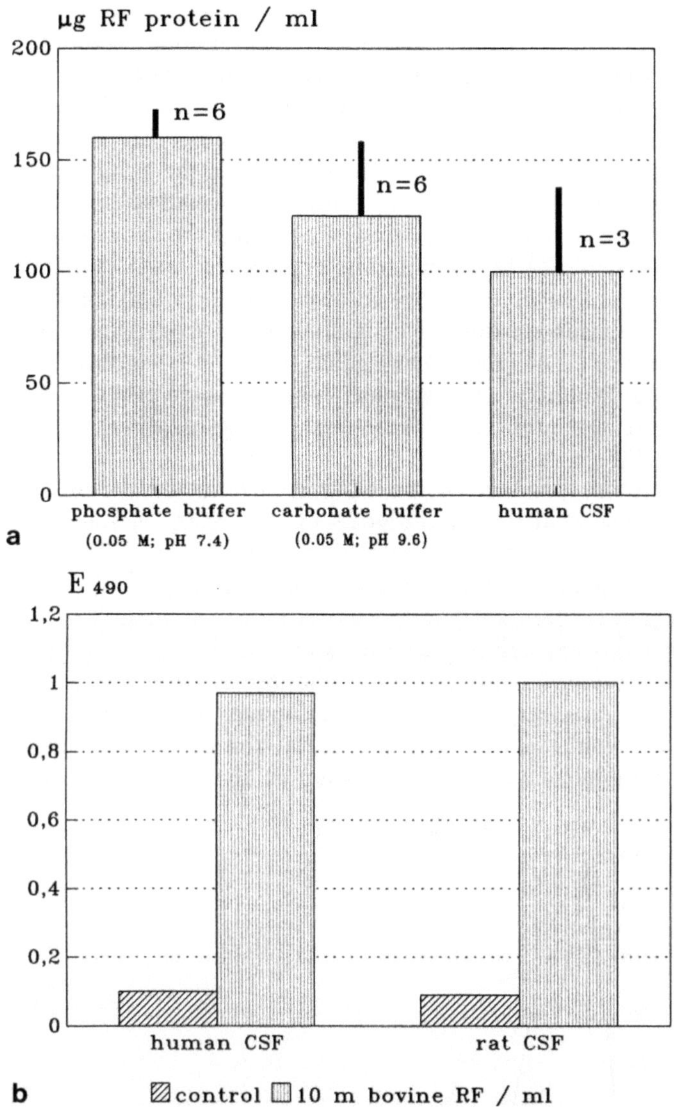

Fig. 3a,b. Solubility of bovine RF examined by use of UV-spectrophotometry at λ = 280 nm (**a**). Results obtained by studies with indirect ELISA using immobilized CSF samples, some of which contained solubilized RF proteins, and subsequent application of anti-RF serum (**b**). To solubilize RF, approximately 10 m bovine RF/ml of the sample were incubated overnight at 4°C. The supernatant was cleared from the particles before using. *E490*, extinction 490 nm immunological studies with indirect ELISA using an anti-RF serum (Fig. 3b) showed that bovine RF can be solubilized in commonly used buffer solutions or even in samples of human and rat cerebrospinal fluid.

In consequence, the SCO secretory product is not only a constituent of the extracellular matrix of the tissue surrounding the SCO; it is also a soluble component of the extracellular fluid of the adult and embryonic brain and is distributed within the central nervous system to an unknown extent. With this assumption in mind, we studied the capacity of an aqueous extract of bovine RF in aggregating embryonic brain cells (P. Debbage et al., manuscript in preparation) to search for a possible target of soluble RF material. In suspensions of cells dissociated from whole brains of embryonic chicken (ed 4) (Fig. 4) and mice (ed 12, 14, 16; data not shown), a two- to tenfold number of aggregates were counted when RF protein was added to the preparation; a maximum of aggregation was observed at 0.5–5 µg/ml RF protein.

The secretory product of the SCO is rich in glycoproteins containing N-linked oligosaccharides (Rodríguez et al. 1986; Meiniel et al. 1988), comparable to the substances mediating certain cell–cell and cell–matrix interactions. In the case of such interactions the partner molecule recognizing the sugar chain is often a sugar-specific protein – a "lectin" or "agglutinin." Keeping in mind the cell-aggregating effect of the RF solution, we studied its possible structural and functional similarities with lectins, namely, the immunological cross-reactivities and the capacity for carbohydrate binding.

Clear-cut results in analyzing structural homology were obtained in dot tests with immobilized bovine RF extract as an antigen and using immuno-

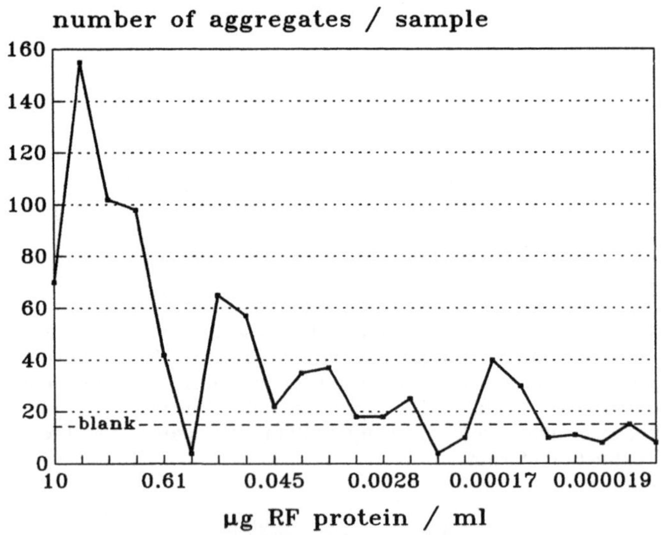

Fig. 4. Aggregation of dissociated chicken brain cells (ed 4) in the presence of different concentrations of aqueous bovine RF proteins. The aggregates per sample with diameters larger than 25 µm were counted

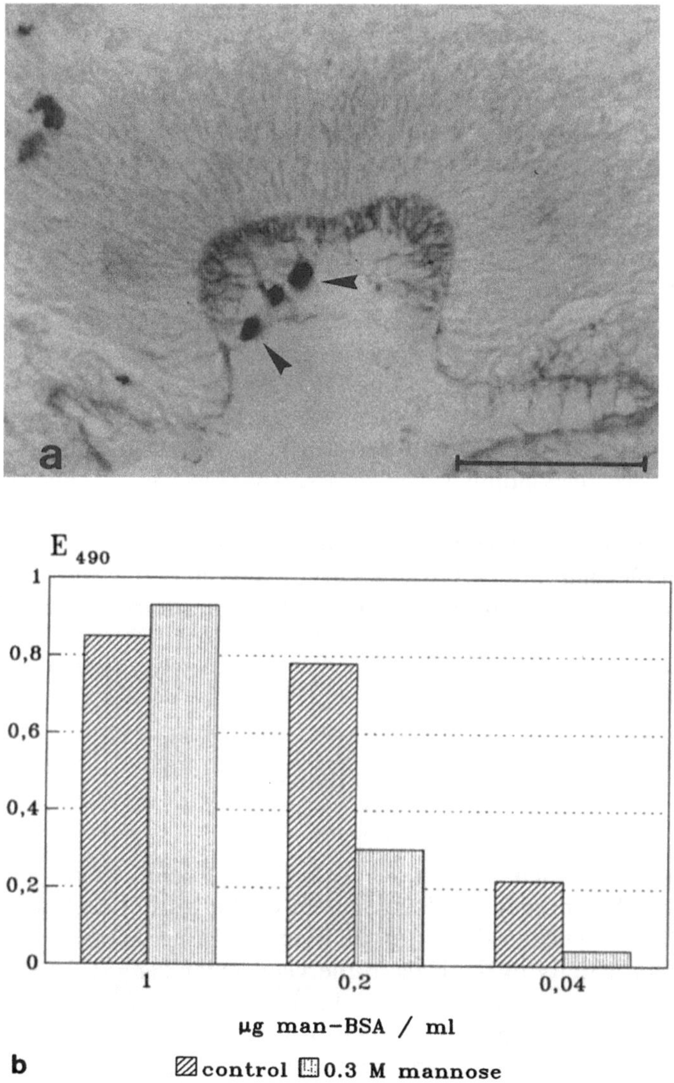

Fig. 5. Structural and functional properties of SCO secretory products (**a**) stained immunocytochemically in the mouse using an anti-*Phaseolus* agglutinin-L serum and (**b**) as revealed in a neoglycoprotein binding assay based on indirect ELISA with solid phase immobilized RF proteins. Immunostaining marks the apical ventricular lining of the SCO and the RF (*arrowheads*) but fails to label the remainder of the ependymocyte cytoplasm; the hypendymocytes are entirely unstained (**a** *Bar*: 100 µm). Binding of mannose bovine serum albumin (*BSA*) to RF proteins was inhibited by the corresponding sugar (**b**). *E490*, extinction 490 nm

histochemical preparations of the mouse SCO, after staining with a polyclonal antiserum directed against the *Phaseolus vulgaris* agglutinin-L (PHA-L) (P. Debbage et al., manuscript in preparation). In contrast to normal rabbit serum, anti-PHA-L labeled immobilized bovine RF proteins at several dilutions as well as material at the ventricular border of the mouse SCO (Fig. 5a). This anti-PHA-L immunoreactivity and the binding to the mouse SCO of anti-*Ricinus* agglutinin-2 in a similar pattern and of anti-wheat-germ agglutinin in a modified pattern indicates that the SCO secretes proteins bearing antigenic determinants characteristic of specific phythaem-agglutinins. Recently, proteins homologous to lectins were included among the protein superfamily of the affinitins (Franz 1988); it is a fascinating idea that these proteins of a highly conservative structure might be responsible for mediating ontogenetic processes and for forming RF.

The functional properties of a lectin are based on carbohydrate binding. Indeed, we observed that bovine RF proteins are capable of binding to neoglycoproteins (NGP), i.e., glycosylated bovine serum albumin. We could show that mannose- (Fig. 5b), galactose-, and fucose-containing NGP bound to immobilized RF proteins and that this system was susceptible for inhibition by the corresponding sugar in ELISA (Fig. 5b). When *N*-acetyl-D-galactosamine-NGP was tested in this assay, its binding was enhanced rather than inhibited by the corresponding sugar, suggesting a certain complexity of these molecular interactions. Lactose-NGP and deglycosylated bovine serum albumin did not bind to RF proteins, underlining the specifity of the test.

Conclusions

In conclusion, we believe that RF bears constituents which may have an influence on the ontogenetic development of the vertebrate brain: (1) There is a correlation between the differentiation of distinct areas of the brain and the secretory activity of the cells in the ventricular zone. (2) SCO secretory products, including compounds of RF, may be a ubiquitous constituent of the brain extracellular matrix. (3) RF protein solutions have functional and structural lectin properties, and are capable of aggregation of embryonic brain cells.

References

Franz H (1988) The ricin story. In: Franz H (ed) Advances in lectin research, vol 1. Springer, Berlin Heidelberg New York, pp 1–25

Fujita S (1969) Synthesis and secretion of biomacromolecules in the matrix cell and its progeny. In: Sterba G (ed) Zirkumventrikuläre Organe und Liquor. Fischer, Jena, pp 197–200

Hauser R (1969) Abhängigkeit der normalen Schwanzregeneration bei *Xenopus*-Larven von einem diencephalen Faktor im Zentralkanal. Acta Anat (Basel) 99:234–230

Hauser R, Murbach V (1979) Achsenverkrümmungen bei Amphibien und Fischen als Folge der Abwesenheit des Reissnerschen Fadens im Zentralkanal. Acta Anat (Basel) 99:234

Leonhardt H, Krisch B, Erhardt H (1987) Organization of the neuroglia in the midsagittal plane of the central nervous system: a speculative report. In: Scharrer B, Korf HW, Hartwig HG (eds) Functional morphology of neuroendocrine systems. Springer, Berlin Heidelberg New York, pp 175–187

Lösecke W, Naumann W, Sterba G (1986) Immuno-electron-microscopic analysis of the basal route of secretion in the subcommissural organ of the rabbit. Cell Tissue Res 244:449–456

Meiniel A, Molat JL, Meiniel R (1988) Complex-type glycoproteins synthesized in the subcommissural organ of mammals. Light- and electron-microscopic investigation by use of lectins. Cell Tissue Res 53:383–395

Naumann W (1986) Immunhistochemische Untersuchungen zur Ontogenese des Subcommissuralorgans. Acta Histochem Suppl (Jena) 33:265–272

Naumann W, Müller G, Kloss P (1987) Immunoreactive glycoprotein of the subcommissural organ in the embryonic stage of the vertebrate brain. Wiss Z Karl-Marx-Univ Leipzig Math-Naturwiss R 36:17–20

Olsson R (1956) The development of Reissner's fibre in the brain of the salmon. Acta Zool (Stockh) 37:235–250

Puusepp L, Voss HEV (1924) Studien über das Subcommissuralorgan beim Menschen. Folia Neuropathol Eston 2:13–21

Rodríguez EM, Herrera H, Peruzzo B, Rodríguez S, Hein S, Oksche A (1986) Light- and electron-microscopic immunocytochemistry and lectin histochemistry of the subcommissural organ: evidence for processing of the secretory material. Cell Tissue Res 243:545–559

Rühle HJ (1971) Anomalien im Wachstum der Achsenorgane nach experimenteller Ausschaltung des Komplexes Subcommissuralorgan-Reissnerscher Faden. Untersuchungen am Rippenmolch (*Pleurodeles waltli* MICHAH. 1830). Acta Zool (Stockh) 52:23–68

Schoebitz K, Garrido O, Heinrichs M, Speer L, Rodríguez EM (1986) Ontogenetical development of the chick and duck subcommissural organ. An immunocytochemical study. Histochemistry 84:31–40

Sterba G, Kiessig C, Naumann W, Petter H, Kleim I (1982) The secretion of the subcommissural organ. A comparative immunocytochemical investigation. Cell Tissue Res 226:427–439

Takeuchi IK, Takeuchi YK (1986) Congenital hydrocephalus following X-irradiation of pregnant rats on an early gestational day. Neurobehav Toxicol Teratol 8:143–150

III. Biochemistry

Protein Glycosylation in Mammalian Cells

C.B. Hirschberg

Introduction

The principal organelles in mammalian cells where protein glycosylation occurs are the endoplasmic reticulum and Golgi apparatus (Kornfeld and Kornfeld 1985). Among the glycosylation reactions occurring in the endoplasmic reticulum are the initiation of N-glycosylation of proteins and the incorporation of mannose and glucosamine into the glycosylphosphatidylinositol anchor of proteins (Kornfeld and Kornfeld 1985; Doering et al. 1990). Following these reactions, proteins move via vesicles to the Golgi apparatus where N-linked glycoproteins undergo processing and further glycosylation and where also O-glycosylation reactions occur (Kornfeld and Kornfeld 1985; Pfeffer and Rothman 1987; Hirschberg and Snider 1987; Abeijon and Hirschberg 1987). We will briefly summarize in this review the current important features regarding the above glycosylation reactions, among them topographical issues as well as the cloning of some of the enzymes catalyzing the above reactions.

Glycosylation Reactions of the Endoplasmic Reticulum

The sequence of reactions which lead to N-glycosylation of nascent polypeptide chains in the lumen of the rough endoplasmic reticulum can be seen in Fig. 1. Briefly, dolichol becomes the acceptor molecule for a growing oligosaccharide chain, which upon reaching 14 sugars is transferred en bloc to the nascent polypeptide chain in the lumen of the endoplasmic reticulum. The first seven sugars added to dolichol are directly derived from the corresponding nucleotide sugars, UDPGlcNAc and GDPMan. The last seven sugars are first transferred from the corresponding nucleotide sugar to dolichol monophosphate, which then serves as the direct precursor for glycosylation to the growing dolichol oligosaccharide chain.

The reactions leading to the assembly of the dolichol oligosaccharide chain containing 14 sugars have proven to be a complex topographical feature as can be growing dolichol oligosaccharide chain on the *cytoplasmic*

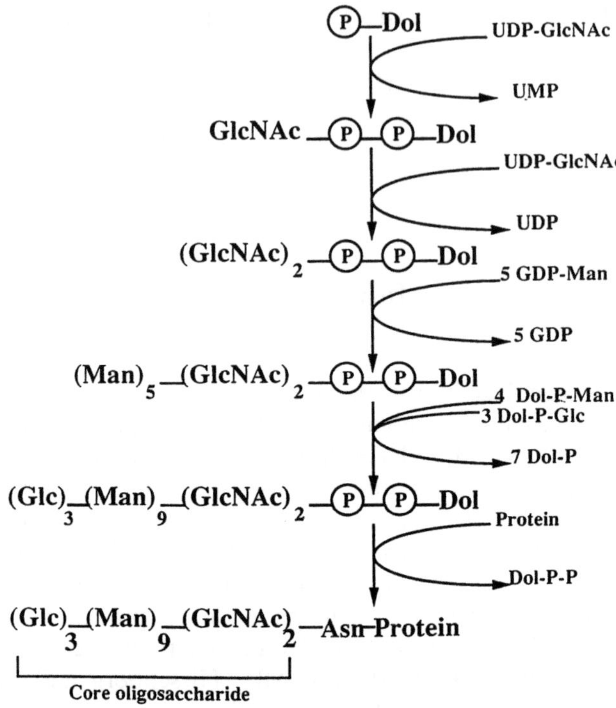

Fig. 1. Reaction pathway leading to the biosynthesis of dolichol $GlcNAc_2Man_9Glc_3$

side of the endoplasmic reticulum membrane. Among the principal experiments supporting this is the inability for GDPMan to cross the endoplasmic reticulum membrane (Hirschberg and Snider 1987), the demonstration that the UDPGlcNAc which crosses the membrane does *not* serve as a precursor for the dolichol chain in the lumen of the endoplasmic reticulum (Abeijon and Hirschberg 1990), and the demonstration that dolichol-P-P-$GlcNAc_2$ can be detected facing the cytosolic side of the endoplasmic reticulum (Abeijon and Hirschberg 1990). The subsequent addition of the last seven sugars (four mannoses and three glucoses) occurs on the lumenal side of the endoplasmic reticulum membrane. This requires that dolichol-P-P-$GlcNAc_2Man_5$ be translocated from the cytosolic side towards the lumen. While direct demonstration of this event has not been forthcoming, studies with the lectin Concanavalin A, which binds terminal mannoses, have shown that derivatives with five mannoses face the cytosol while those with nine face the lumen (Snider and Robbins 1982; Snider and Rogers 1984). $GlcNAc_2Man_9Glc_3$ are then transferred in the lumen of the rough endoplasmic reticulum to nascent polypeptide chains containing the consensus motif for N-linked glycosylation [Asn-X-Ser(threo)]. Thereafter these proteins migrate via vesicles to the Golgi apparatus (Pfeffer and Rothman 1987; Fig. 2).

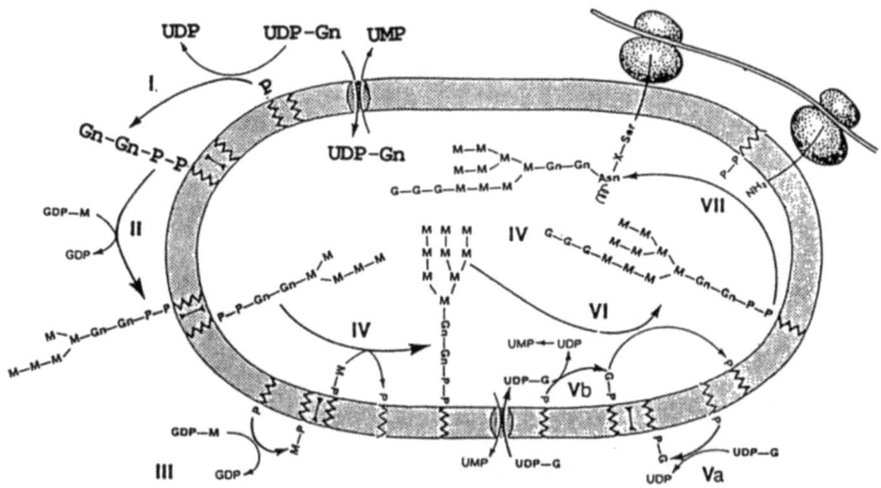

Fig. 2. Topography of glycosylation in the rough endoplasmic reticulum. ∿∿, Dolichol; *Gn*, N-acetylglucosamine; *M*, mannose; *G*, glucose; Roman numerals *I–VII* indicate sequence

Future important studies will require the elucidation of the mechanism by which dolichol oligosaccharides can be translocated across the membrane of the endoplasmic reticulum and the cloning of the different enzymes involved in these reactions. These latter studies have just begun with the recent cloning of the enzyme leading to the formation of dolichol-P-P-$GlcNAc_2$ (Zhu and Lehrman 1990; Scocca and Krag 1990).

Glycosylation Reactions of the Golgi Apparatus

N-Glycosylation

Following the arrival at the Golgi apparatus of the N-linked glycoproteins from the endoplasmic reticulum a series of mannoses are removed in those glycoproteins destined to become complex glycoproteins, while so-called high mannose glycoproteins remain intact. It is not known what regulatory features cause oligosaccharide chains to become high mannose or complex although it is most likely the protein backbone conformation. Subsequently, additional sugars such as N-acetylglucosamine, galactose, fucose, and sialic acid are added to the oligosaccharide chains to form the complex oligosaccharides (Kornfeld and Kornfeld 1985).

The precursors for these glycosylation reactions are the corresponding nucleotide sugars which are synthesized in the cytosol, with the exception of CMP-sialic acid which is synthesized in the nucleus (Hirschberg and Snider

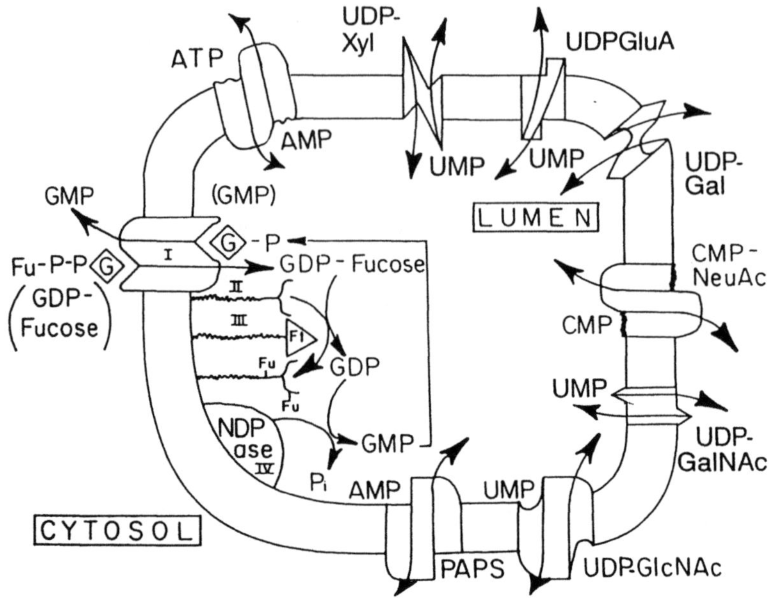

Fig. 3. Topography of glycosylalion in the Golgi apparatus

1987). The nucleotide sugars enter the lumen of the Golgi apparatus via specific carrier proteins, which are antiporters. Following entry into the lumen of the Golgi apparatus and glycosylation, the nucleoside diphosphates are converted to the corresponding nucleoside monophosphate in a reaction catalyzed by a lumenal nucleoside diphosphatase. The nucleoside monophosphates then exit the lumen of the Golgi apparatus in a reaction coupled with entry of additional corresponding nucleotide sugars (Fig. 3). These antiporters appear to be of physiologic significance as mutant Chinese hamster ovary cells have been isolated with a specific defect in Golgi transport of CMP-sialic acid and UDP-Gal; this results in a corresponding decrease in the specific sugar in both proteins and lipids. Because the mutant defective in UDP-Gal is not defective in Golgi transport of other uridine-containing nucleotide sugars, it appears that there are individual uridine nucleotide sugar transporters in the Golgi membrane as opposed to a common one for all the uridine nucleotide sugars (Hirschberg and Snider 1987).

It is likely that a potential site for regulation of glycosylation in the Golgi apparatus is at the level of these specific membrane nucleotide sugar transporters. In addition, in recent studies, several laboratories have began to clone several glycosyltransferases. Some of the future questions in this field that are beginning to be addressed include: the sub-Golgi localization of these glycosyltransferases, the sub-Golgi localization of the different nucleotide sugar transporters, as well as the relationship within the membrane between these two types of proteins and their regulation.

O-Glycosylation

There is considerable evidence suggesting that in many cells O-glycosylation of proteins occurs in the Golgi apparatus. Among the approaches used to reach this conclusion are electron microscopy and subcellular fractionation studies (Roth 1984; Abeijon and Hirschberg 1987). The sugar precursors of O-glycosylation are the corresponding nucleotide sugars which enter the Golgi lumen via the same antiporter system as described above for N-linked glycoproteins. There appears to be no simple concensus motif in the protein for O-glycosylation reactions of proteins or O-xylosylation of proteoglycans. Future studies of glycosylation of proteins in the subcommissural organ should concentrate in determining to what extent these reactions occur in this organ.

In addition to the above-described glycosylation reactions, sulfation and phosphorylation of proteins and sugars are also known to occur in the Golgi apparatus (Capasso et al. 1989; Hirschberg and Snider 1987). In these instances, the corresponding precursors, PAPS and ATP are transported into the Golgi lumen via similar antiporters as those described in the above glycosylation reactions.

An area that also could be explored in relation to the glycoproteins within the subcommissural organ is the question of whether glycoproteins from this organ are phosphorylated and/or sulfated and the possible occurrence of proteoglycans.

References

Abeijon C, Hirschberg CB (1987) Subcellular site of synthesis of the N-acetylgalactosamine (α1-0) serine (or threonine) linkage in rat liver. J Biol Chem 262: 4153–4159

Abeijon C, Hirschberg CB (1990) Topography of initiation of N-glycosylation reactions. J Biol Chem 265:14691–14695

Capasso JM, Keenan TW, Abeijon C, Hirschberg CB (1989) Mechanism of phosphorylation in the lumen of the Golgi apparatus. J Biol Chem 264: 5233–5240

Doering TL, Masterson WJ, Hart GW, Englund PT (1990) Biosynthesis of glycosylphosphatidylinositol membrane anchors. J Biol Chem 265:611–614

Hirschberg CB, Snider MD (1987) Topography of glycosylation in the rough endoplasmic reticulum and Golgi apparatus. Annu Rev Biochem 56:63–88

Kornfeld R, Kornfeld S (1985) Assembly of asparagine-linked O-oligosaccharides. Annu Rev Biochem 54:631–664

Pfeffer SR, Rothman JE (1987) Biosynthetic protein transport and sorting by the endoplasmic reticulum and Golgi. Annu Rev Biochem 56:829–852

Roth J (1984) Cytochemical localization of terminal N-acetyl-D-galactosamine residues in cellular compartments of intestinal goblet cells: implications for the topology of O-glycosylation. J Cell Biol 98:399–406

Scocca JR, Krag SS (1990) Sequence of a cDNA that specifies the uridine diphosphate N-acetyl-D-glucosamine: dolichol phosphate N-acetylglucosamine-1-phosphate transferase from Chinese hamster ovary cells. J Biol Chem 265:20621–20626

Snider MD, Robbins PW (1982) Transmembrane organization of protein glycosylation. J Biol Chem 257:6796–6801

Snider MD, Rogers OC (1984) Transmembrane movement of oligosaccharide-lipids during glycoprotein synthesis. Cell 36:753–761

Zhu X, Lehrman MA (1990) Cloning, sequence, and expression of a cDNA encoding hamster UDP GlcNAc: dolichol phosphate N-acetylglucosamine-1-phosphate transferase. J Biol Chem 265:14250–14255

Partial Characterization of the Secretory Products of the Subcommissural Organ

S. HEIN, F. NUALART, E.M. RODRÍGUEZ, and A. OKSCHE

Introduction

Histochemically, the secretory material of the subcommissural organ (SCO) has been characterized as a polysaccharide–protein complex (Oksche 1962; Naumann 1968; Diederen 1970); most probably it represents N-linked complex-type glycoproteins (Meiniel and Meiniel 1985; Meiniel et al. 1986, 1988; Rodríguez et al. 1986, 1987, 1990; Herrera and Rodríguez 1990). The immunocytochemical study of the bovine SCO using antibodies raised against bovine Reissner's fiber (RF) showed that the secretory material present in the rough endoplasmic reticulum (RER), secretory granules, and RF is strongly immunoreactive (Hein 1988; Nualart et al. 1991). Concanavalin A (ConA, affinity = mannose, glucose)-binding sites have been only found in the RER, whereas *Limax flavus* agglutinin (LFA, affinity = sialic acid) bound to the secretory material is present in the secretory granules and RF (Nualart et al. 1991).

These findings led us to study the secretory compounds present in SCO extracts and the secretory compounds released into the ventricle (RF extracts) by means of blots processed for immunostaining and lectin binding. Our preliminary results (Rodríguez et al. 1987; Hein 1988) suggest that: (a) The SCO releases into the ventricle a nonglycosylated protein and several glycoproteins. (b) Some glycoproteins are synthesized as precursors, stored in the RER, and then processed during transport to the apical pole of the secretory cells, along the secretory pathway. The glycoproteins present in the secretory granules and those that have been released into the ventricle and aggregate to form pre-RF and RF correspond to "mature" forms. (c) Other glycoproteins are of the high mannose-type oligosaccharide; they might be released into the ventricle from cisternae of the RER.

Other authors have also attempted to identify and characterize the glycoproteins secreted by the SCO (Wolf and Sterba 1972; Meiniel et al. 1986; Rodríguez et al. 1987; Karoumi et al. 1990). However, there is disagreement with respect to the number and molecular weights of the secretory glycoproteins. Furthermore, our own new evidence confirms only part of our preliminary findings. The present investigation was designed for further identification and characterization of the probable precursor and processed forms of the SCO secretion.

Using sodium dodecyl sulfate-polyacrylamide gel electrophoresis (SDS-PAGE) and immunostaining and lectin binding (with or without previous treatment with endoglycosidase F) in blots, we have identified and partially characterized the secretory proteins of the SCO in their intracellular location (SCO extracts) and after they have been released (RF extracts). Evidence was obtained indicating the probable existence, in the bovine SCO, of two different precursors whose processed forms are released into the cerebrospinal fluid (CSF) to form RF.

Materials and Methods

RF Extract. Bovine spinal cords were perfused through the central canal with artificial CSF. RF were collected in the perfusate and extracted in 50 m*M* ammonium bicarbonate, pH 8.0, containing 0.5 m*M* PMSF (for details, see Nualart et al. 1991).

SCO Extracts. Pools of about 150 bovine SCOs were homogenized with a glass–Teflon homogenizer containing 5 ml 50 m*M* ammonium bicarbonate, pH 8.0, 0.5 m*M* PMSF, and 1 m*M* EDTA. Then they were sonicated (15 s, eight times). The extracts were centrifuged at 9200x *g* for 30 min and the supernatants were regarded as crude extracts. All procedures were performed at 4°C. For further details, see Nualart et al. (1991).

Polyacrylamide Gel Electrophoresis and Blotting. Samples of the SCO and RF extracts (80 µg and 20 µg protein, respectively) were prepared and submitted to electrophoresis according to the method of Laemmli (1970). The gels were stained with Coomassie blue or transferred onto nitrocellulose sheets (Towbin et al. 1979). Some blots were processed for immunoperoxidase (PAP) staining using one of the following polyclonal antibodies: (1) Anti-bovine RF, developed in rabbits (AFRU, Rodríguez et al. 1984); (2) Ab-540, Ab(1)-450, Ab-320. SCO extract (30 mg protein) was submitted to preparative electrophoresis using the Laemmli system (Laemmli 1970). The RF-immunoreactive bands (540, 450, and 320 kDa; see below) were cut out according to the procedure described by Cozzani and Hartmann (1980) and used, separately, to raise antibodies in rats. They were labeled Ab-540, Ab(1)-450 and Ab-320. Parallel blots of SCO and RF extracts were processed for ConA and LFA binding using the lectin-antilectin procedure (Nualart et al. 1991). RF extracts were also submitted to electrophoresis under nondenaturing conditions (Davis 1964).

Incubation with Endoglycosidase F. Samples of RF extract (25 µg protein) were incubated with endoglycosidase F (0.2 enzyme units, SIGMA) in 100 m*M* sodium phosphate buffer, pH 6.1, containing 50 m*M* EDTA, 1% β-mercaptoethanol, and 0.1% SDS at 37°C overnight. Then, these extracts and nontreated extracts of RF were run in parallel for SDS-PAGE. Gels were stained with Coomassie blue or processed for blotting analysis using anti-RF serum (AFRU) and LFA.

Immunocytochemistry. Paraplast sections of the bovine SCO + RF and of dog, rat, snake (*Natrix maura*), and turtle (*Mauremys caspica*) SCO were processed for the immunoperoxidase method of Sternberger et al. (1970), using one of the following primary antibodies: AFRU; Ab-540, Ab(1)-450, and Ab-320 (see above); Ab-700; Ab(2)-450. The material corresponding to the 700-kDa band present in nonreduced RF extract (see below) was obtained from preparative gels after SDS-PAGE and used to raise antibodies in rats (Ab-700). The last antibody [Ab(2)-450] was raised in rats against the material corresponding to the 450-kDa band obtained from preparative gels after SDS-PAGE of RF extracts.

Results

Bovine SCO and RF Extracts

Immunoblotting Using AFRU and Lectin Binding. SDS-PAGE of SCO extracts, followed by immunoblotting using AFRU showed three immunoreactive bands with apparent mass of 540, 450 and 320 kDa (Fig. 1B). Occasionally, a 50-kDa band was observed. The 540- and 320-kDa polypeptides showed affinity for ConA but not for LFA. The 450-kDa polypeptide displayed affinity for ConA *and* for LFA. These three bands were weakly stained with Coomassie blue (Fig. 1A).

Most of the immunoblots of RF extracts revealed six bands with apparent mass of 450, 300, 230, 190, 140, and 89 kDa (Fig. 1C). All these bands bound LFA (Fig. 1D) but none of them bound ConA. The only band consistently found in all 40 blots performed was that of 450 kDa.

Immunoblotting Using Ab-540, Ab(1)-450, and Ab-320. The immunoblot analysis of SCO extracts using these three antisera showed that: (a) Ab-540 reacted with the 540-, 450-, and 320-kDa polypeptides; (b) Ab(1)-450 reacted with the 450- and 320-kDa bands; and (c) Ab-320 reacted with the 320- and 190-kDa bands. This 190-kDa band was Con A- and LFA-positive. A band with a similar molecular weight was seen in blots of RF immunostained with AFRU (see above).

Effect of Endoglycosidase F. Treatment of RF extracts with endoglycosidase F produced a decrease of the molecular weights of the sialylated peptides, as was shown in blots of RF processed for LFA binding and immunostaining

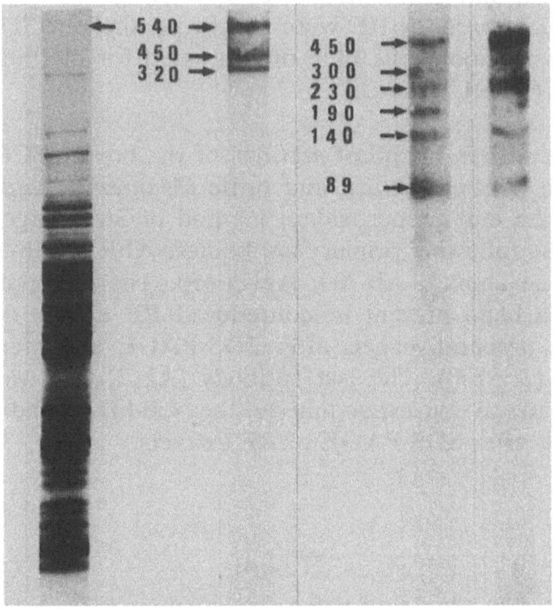

Fig. 1A–D. Immunoreactive secretory proteins and their LFA-binding capacity, as present in bovine subcommissural organ (SCO) and Reissner's fiber (RF) extracts. SCO (**A**, **B**) and RF (**C**, **D**) materials were processed for electrophoresis on a SDS-polyacrylamide gel and transferred to nitrocellulose sheets. **A** Gel stained with Coomassie blue; **B**, **C** immunoblotting using AFRU; **D** LFA binding; *numbers* represent mass in kilodaltons

with AFRU. The 450-kDa compound shifted to 380-kDa polypeptide, that of 300-kDa to 240-kDa, that of 230-kDa to 200-kDa, and that of 190-kDa to 170-kDa. The four "new bands" have no affinity for LFA, but they continue to be immunoreactive with AFRU.

Nonreduced RF Extract

When RF extracted in ammonium bicarbonate was processed for SDS-PAGE as described by Laemmli (1970), but excluding β-mercaptoethanol from the sample buffer of Laemmli, one band with an apparent mass of 700 kDa was observed (Fig. 2,D). When the reducing reagent was added to the sample buffer, a 450-kDa band mainly occurred (Fig. 2, D-R). Antibodies against the material corresponding to these two bands were obtained [Ab-700 and Ab(2)-450] as described in "Materials and Methods."

Fig. 2. Electrophoresis of Reissner's fiber (RF) extracts under native and denaturing conditions. RF extracts were processed for electrophoresis under native conditions (*N*); denaturing conditions (*D*, *D-R*), excluding β-mercaptoethanol from the sample buffer (*D*) or in the presence of this reducing reagent (*D-R*). Gels were stained with Coomassie blue. *Number* indicates mass in kilodaltons; *arrowhead* indicates the 700-kDa protein; −, cathode; +, anode

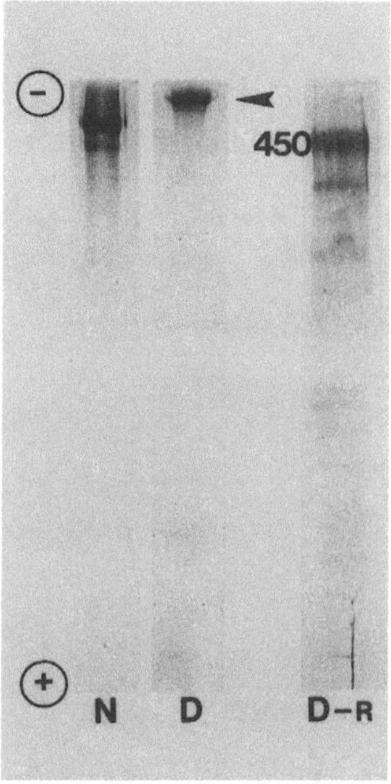

PAGE Under Nondenaturing Conditions

When RF extracts were submitted to electrophoresis according to the procedure of Davis (1964), two bands stained with Coomassie blue were observed (Fig. 2N).

Immunocytochemistry

Ab-540, Ab(1)-450, and Ab-320. Ab-540 and Ab(1)-450 showed immunoreactive material in the SCO of the bovine and dog, but they did not immunoreact with the SCO secretion in the rat and snake. Ab-320 immunoreacted with the bovine, dog and rat SCO, but not with the snake SCO.

In the bovine SCO, Ab-540 and Ab(1)-450 strongly stained secretory material stored in RER and in the apical granules of all secretory cells. The hypendymal cells were also stained by these two antisera. The immunostaining of the material stored in the RER was stronger with Ab-540 than with Ab-450. The apical secretory granules, however, reacted with the same intensity with both antisera. The Ab-320 stained strongly the apical secretory granules of the ependymal and hypendymal cells, but reacted only weakly with the material stored in the RER. The three antisera reacted with the bovine RF.

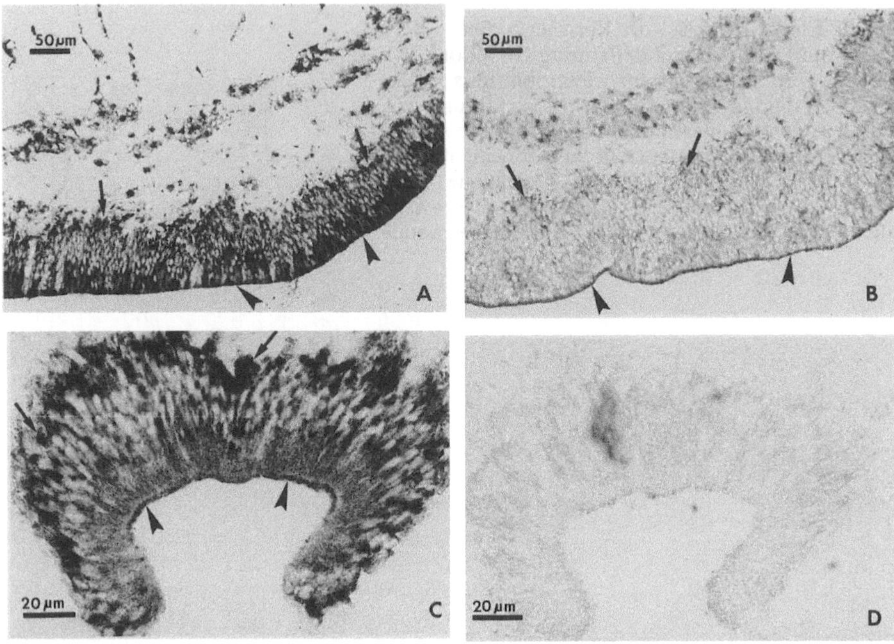

Fig. 3A–D. Bovine and turtle (*Mauremys caspica*) subcommissural organ (SCO). Immunoperoxidase staining of the bovine (**A**, **B**) and turtle (**C**, **D**) SCO using Ab-700 (**A**, **C**) and Ab(2)-450 (**B**, **D**). *Arrows*, RER; *arrowheads*, apical secretory granules

Table 1. Immunostaining of the subcommissural organ

	Bovine		Dog		Rabbit		Rat		Turtle		Snake	
	RER	SG	RER	SG	RER	SG	RER	SG	RER	SG	RER	SG
Ab-540	+++	+++	++	−			−	−			−	−
Ab(1)-450	++	+++	+	+			−	−			−	−
Ab-320	+	++	++	+++			++	+++			−	−
Ab-700	+++	+++	+++	+++	+++	+++	+++	+++	+++	+++	++	++
Ab(2)-450	++	+++	+	+	++	++	−	−	−	−	−	−

RER, rough endoplasmic reticulum; SG, apical secretory granules; Ab-540, Ab(1)-450, Ab-320, antibodies against the 540-, 450- and 320-kDa bands of SDS-PAGE of a bovine SCO extract, respectively; Ab-700, antibody against the 700-kDa band of SDS-PAGE of a nonreduced extract of bovine RF; Ab(2)-450, antibody against the 450-kDa band of SDS-PAGE of bovine RF under reducing conditions. +++, Strong reaction; ++, moderate reaction; +, weak reaction; −, no reaction

In the dog SCO, Ab-540 stained weakly the perinuclear region of a population of secretory cells, but did not react with the apical secretory granules. Ab-540 stained only some fibrous structures within RF. Ab(1)-450 produced essentially the same pattern of immunoreaction as Ab-540, but the former also reacted with a small population of apical secretory granules. Ab-

320 reacted with material present in the RER of a population of secretory cells and in numerous apical secretory granules. Ab-320 immunoreacted very strongly with RF.

Ab-700 and Ab(2)-450. Ab-700 strongly stained material present in the RER and in apical secretory granules of the SCO of all species studied (bovine, rat, dog, rabbit, snake, *Natrix maura*, and turtle, *Mauremys caspica*) (Fig. 3A,C). Ab(2)-450 reacted strongly with material stored in the RER and secretory granules of bovine (Fig. 3B) and rat SCO; it reacted weakly with the dog SCO. It did not react with the snake and the turtle SCO (Fig. 3D).

The immunocytochemical findings obtained with the different polyclonal antibodies used are summarized in Table 1.

Discussion

The use of polyclonal antibodies against secretory products of the bovine SCO has allowed to identify four secretory glycoproteins in bovine SCO extracts with an apparent mass of 540, 450, 320, and 190 kDa. A 50-kDa polypeptide has been detected only in some extracts. Six immunoreactive polypeptides have been found in bovine RF extracts with an apparent mass of 450, 300, 230, 190, 140, and 89 kDa. Meiniel et al. (1986) reported the occurrence of a 54-kDa glycoprotein in the sheep SCO; this compound might be homologous to the 50-kDa polypeptide detected in the bovine SCO.

According to Karoumi et al. (1990), in the SCO of chick embryos 10 glycoproteins have been observed, their molecular weights ranging between 240 000 and 42 000. The discrepancy to our findings could reflect species differences in the secretory products (see below) or be due to the different SDS-PAGE conditions used by these authors.

Immunocytochemical and lectin-histochemical studies (Hein 1988) as well as radioautographic studies (Diederen et al. 1987) have provided evidence to postulate that the protein and carbohydrate moieties of the SCO-secreted glycoproteins are processed along the secretory pathway. The following findings may be considered strong evidence indicating that the 540-kDa glycoprotein is a precursor form: (a) it is present in the bovine SCO and missing from RF; (b) it is a "pre-Golgi" glycoprotein (ConA positive, LFA negative); (c) Ab-540 immunoreacts with the secretion stored in the RER of the dog SCO, but not with the apical secretory granules. It may also be postulated that the 450-kDa glycoprotein is a processed form: (a) this compound is present in SCO and RF; (b) it is a "post-Golgi" glycoprotein (ConA and LFA positive); (c) Ab(1)-450 reacts with material stored in the RER and apical granules of the dog SCO. Following the same line of reasoning, it is concluded that the 320-kDa glycoprotein is a precursor form: (a) it is present in the SCO and absent in RF; (b) it is a "pre-Golgi" glycoprotein (ConA positive, LFA negative); (c) Ab-320 reacts with a 190-kDA polypeptide present in the SCO and RF. The 190-kDa glycoprotein may be a processed form of the putative 320-kDa precursor, since it is present in SCO and RF and can be regarded as a "post-Golgi" glycoprotein.

The results obtained after treatment of RF extracts with endoglycosidase F confirm that the secretory glycoproteins released by the SCO correspond to N-linked glycoproteins and suggest that the number of N-glycosylation sites ranges between nine (for the smallest polypeptide, 190 kDa) and 27 (for the largest polypeptide, 450 kDa).

All the immunocytochemical findings using Ab-540, Ab(1)-450 and Ab-320 could be best explained if: (a) the bovine SCO harbored two precursor forms; (b) in the rat, only one of them were present; and (d) neither of them would occur in the snake SCO.

The existence of glycoproteins that are present in the RF extracts (300- and 230-kDa compounds) but missing from the SCO may suggest a post-release processing. The possibility that they are RF components originating from a source different from the SCO has to be considered (see Rodríguez et al. 1992; Olsson, this volume).

The possibility that the 540-, 450-, and 320-kDa glycoproteins result from polymerization or aggregation of polypeptides of low molecular weight appears highly unlikely. These glycoproteins are visualized in gels after electrophoresis in denaturing conditions. Furthermore, the presence in RF extracts processed for SDS-electrophoresis, under nonreducing conditions, of a 700-kDa band, and the fact that under reducing conditions the 700-kDa band is not present but those of lower molecular weights are now visualized, suggest that (a) the 700-kDa glycoprotein is the main component of RF and (b) the other six compounds detected in blots of RF extracts may correspond to subunits or artificial fragments of this 700-kDa compound. Furthermore, the different immunocytochemical properties of Ab-700 and Ab(2)-450 may indicate that both compounds are different, or that the 700-kDa compound has a component missing from the 450-kDa polypeptide. This, again, does not support the assumption that the compounds of high molecular weight might result from polymerization of a single compound of low molecular weight.

The fact that the bovine RF contains at least two different compounds is supported by the presence in the bovine SCO of two precursor forms. Our preliminary results obtained after electrophoresis under nondenaturing conditions speak in favor of this possibility.

Acknowledgements. This work was supported by Grant I/63 476 from the Volkswagen-Stiftung, FRG; Grant 91/0956 from FONDECYT, Chile, and Grant S-89-01 from the Dirección de Investigaciones, Universidad Austral de Chile.

References

Cozzani C, Hartmann B (1980) Preparation of antibodies specific to choline acetyltransferase from bovine caudate nucleus and immunohistochemical localization of the enzyme. Proc Natl Acad Sci USA 77:7453–7457

Davis B (1964) Disc electrophoresis-II method and application to human serum proteins. Ann NY Acad Sci 121:404–427

Diederen JH (1970) The subcommissural organ of *Rana temporaria*. A cytological, cytochemical, cytoenzymological and electronmicroscopical study. Z Zellforsch 111:379–403

Diederen JH, Vullings HG, Legerstee-Oostveen (1987) Autoradiographic study of the production of secretory material by the subcommissural organ of frogs (*Rana temporaria*) after injection of several radioactive precursors, with special reference to the glycosylation and turnover rate of the secretory material. Cell Tissue Res 248:215–222

Hein S (1988) Analysis of the secretion of the subcommissural organ. Thesis, Doctor in Biological Sciences. University of Chile, Chile

Herrera H, Rodríguez EM (1990) Secretory glycoproteins of the rat subcommissural organ are N-linked complex type glycoproteins. Demonstration by combined use of lectin and specific glycosydases, and by the administration of Tunicamycin. Histochemistry 93:607–615

Karoumi A, Croisille Y, Croisille F, Meiniel R, Belin MF, Meiniel A (1990) Glycoprotein synthesis in the subcommissural organ of the chick embryo. II. An immunochemical study. J Neural Transm (Gen Sect) 80:203–212

Laemmli UK (1970) Cleavage of structural proteins during the assembly of the head of bacteriophage T4. Nature 227:680–685

Meiniel A, Molet JL, Meiniel R (1988) Complex-type glycoproteins synthesized in the subcommissural organ of mammalians. Light- and electron-microscopic investigations by use of lectins. Cell Tissue Res 253:383–395

Meiniel R, Meiniel A (1985) Analysis of the secretions of the subcommissural organs of several vertebrate species by use of fluorescent lectins. Cell Tissue Res 239:359–364

Meiniel R, Molat JL, Meiniel A (1986) Concanavalin A-binding glycoproteins in the subcommissural and the pineal organ of the sheep (*Ovis aries*). Cell Tissue Res 245:605–613

Naumann W (1968) Histochemische Untersuchungen am Subkommissuralorgan und Reissnerschen Faden von *Lampetra planeri* (Bloch). Z Zellforsch 87:571–591

Nualart F, Hein S, Rodríguez EM, Oksche A (1991) Identification and partial characterization of the secretory glycoproteins of the bovine subcommissural organ-Reissner's fiber complex. Evidence for the existence of two precursor forms. Mol Brain Res 111:227–238

Oksche A (1962) Histologische, histochemische und experimentelle Studien am Subkommissuralorgan von Anuren (mit Hinweisen auf den Epiphysenkomplex). Z Zellforsch 57:230–326

Rodríguez EM, Oksche A, Hein S, Rodríguez S, Yulis CR (1984) Comparative immunocytochemical study of the subcommissural organ. Cell Tissue Res 237:427–441

Rodríguez EM, Herrera H, Peruzzo B, Rodríguez S, Hein S, Oksche A (1986) Light- and electron-microscopic immunocytochemistry and lectin histochemistry of the subcommissural organ: evidence for processing of secretory material. Cell Tissue Res 243:545–559

Rodríguez EM, Hein S, Rodríguez S, Herrera H, Peruzzo B, Nualart F, Oksche A (1987) Analysis of the secretory products of the subcommissural organ: In: Scharrer B, Korf H, Hartwig H (eds) Functional morphology of neuroendocrine systems. Springer, Berlin Heidelberg New York, pp 189–202

Rodríguez EM, Garrido O, Oksche A (1990) Lectin histochemistry of the fetal subcommissural organ. Cell Tissue Res 262:105–113

Rodríguez EM, Oksche A, Hein S, Yulis CR (1992) Cell biology of the subcommissural organ. Int Rev Cytol 135:39–121

Sternberger LA, Hardy PH, Cuculis JJ, Meyer HG (1970) The unlabeled antibody enzyme method of immunohistochemistry. Preparation and properties of soluble antigen-antibody complex (horseradish peroxidase-anti peroxidase) and its use in identification of spirochetes. J Histochem Cytochem 18:315–333

Towbin H, Staehelin T, Gordon J (1979) Electrophoretic transfer of proteins from polyacrylamide gels to nitrocellulose sheets: procedure and some applications. Proc Natl Acad Sci USA 76:4350–4354

Wolf G, Sterba G (1972) Zur stofflichen Charakteristik des Reissnerschen Fadens. Acta Zool 53:147–154

Biochemical and Immunochemical Analysis of Specific Compounds in the Subcommissural Organ of the Bovine and Chick

A. MEINIEL, R. MEINIEL, I. CREVEAUX, and J.L. MOLAT

Introduction

The best known activity of the subcommissural organ (SCO) is the synthesis of proteins and their release into the ventricular cavity, where they condense to form Reissner's fiber (RF), a threadlike structure that extends throughout the fourth ventricle and the central canal of the spinal cord (see Oksche 1969; Leonhardt 1980; Hofer et al. 1980; Olsson 1986). Production of specific antibodies against RF (Sterba et al. 1982; Rodríguez et al. 1984) or against SCO extracts (Karoumi et al. 1990a, 1991; Meiniel et al. 1988b) has proven that RF is produced, at least mainly, by the secretory ependymal cells of the SCO. In addition, application of lectins to this secretion (Meiniel and Meiniel 1985) has allowed to demonstrate the synthesis of the complex type glycoproteins (Rodríguez et al. 1986; Meiniel et al. 1988a) which contribute to the formation of RF.

During the last few years, the combined use of these specific tools and the application of biochemical techniques has promoted the knowledge of the complex biochemical evolution of the SCO (see Meiniel et al. 1992), and the idea that RF is merely the condensed secretory product of the SCO appears more and more to be an oversimplication.

The aim of this paper is to analyze some biochemical results obtained in two species of vertebrates: the bovine and the chick embryo. To find out specific compounds of the SCO, we compared the SCO samples with other brain structures serving as reference tissues.

Results

Lectin Reactivity to SCO Compounds

Concanavalin A (ConA), which strongly reacts with mannosyl and glucosyl residues, reveals in the entire SCO epithelium of several vertebrate species a high content in glycoproteins (Meiniel and Meiniel 1985). One of these compounds probably corresponds to high mannose-type glycoproteins, the precursor form of the secreted complex-type glycoprotein. The latter can be visualized in the borderline (surface) area of the SCO epithelium and in RF with the use of various lectins, e.g., wheat-germ agglutinin (WGA), *Lens culinaris* agglutinin (LCA) and *Phaseolus vulgaris* agglutinin (PHA) (Meiniel et al. 1988b).

With regard to the characteristic affinities of the SCO secretory products toward lectins, we attempted to identify after electrophoresis of either crude extracts or eluted fractions (lectin-affinity or immunoaffinity chromatography) specific glycosylated compounds to the SCO profile. Thus, after Western blotting, we revealed glycoproteins rich in mannosyl residues using ConA and sialylated glycoproteins using WGA.

Antibodies Specific to SCO Compounds

These antibodies were generated after immunization with crude SCO extracts. Polyclonal antibodies (A99 anti-bovine SCO, A74 anti-chick SCO) were adsorbed with brain tissue homogenates to precipitate antibodies raised against compounds common to both tissues. The specificity of these antibodies to SCO compounds was tested by various techniques (histofluorescence, cross-dot technique; Karoumi 1990). The monoclonal antibody $C_1B_8A_8$ was selected from an immunohistofluorescence screening test. This IgG recognizes the different glycosylated forms of the complex-type glycoprotein (Meiniel et al. 1988b). After coupling to an activated gel (AcA22-LKB) these antibodies served to purify the corresponding antigens.

Electrophoretic Analysis in the Bovine

The electrophoretic patterns of the SCO crude extracts were compared with those of typical ependymal formations. After staining with Coomassie blue, only few differences could be revealed between the two tissues (Fig. 1A; Table 1A). On Western blots stained with ConA, the SCO profiles revealed a high content in glycoproteins, and a number of glycosylated compounds appeared to be more extensively represented in SCO profiles (Fig. 1B; Table 1B). After purification on immobilized ConA, similar observations could be made (Fig. 2; Table 2).

Using two-dimensional (2D) electrophoresis, we also compared the pattern of SCO- and ependymal soluble extracts. Several polypeptides were found to be specific of the SCO tissue (Fig. 3).

From these electrophoretic analyses of crude soluble extracts and purified fractions after ConA affinity chromatography, it appears that the SCO possesses a high content in glycosylated compounds, not all of which appear to be related to the secreted complex-type glycoproteins. Using sodium dodecyl sulfate-polyacrylamide gel electrophoresis (SDS-PAGE), the occurrence of glycoproteins specific to the SCO profile was difficult to demon-

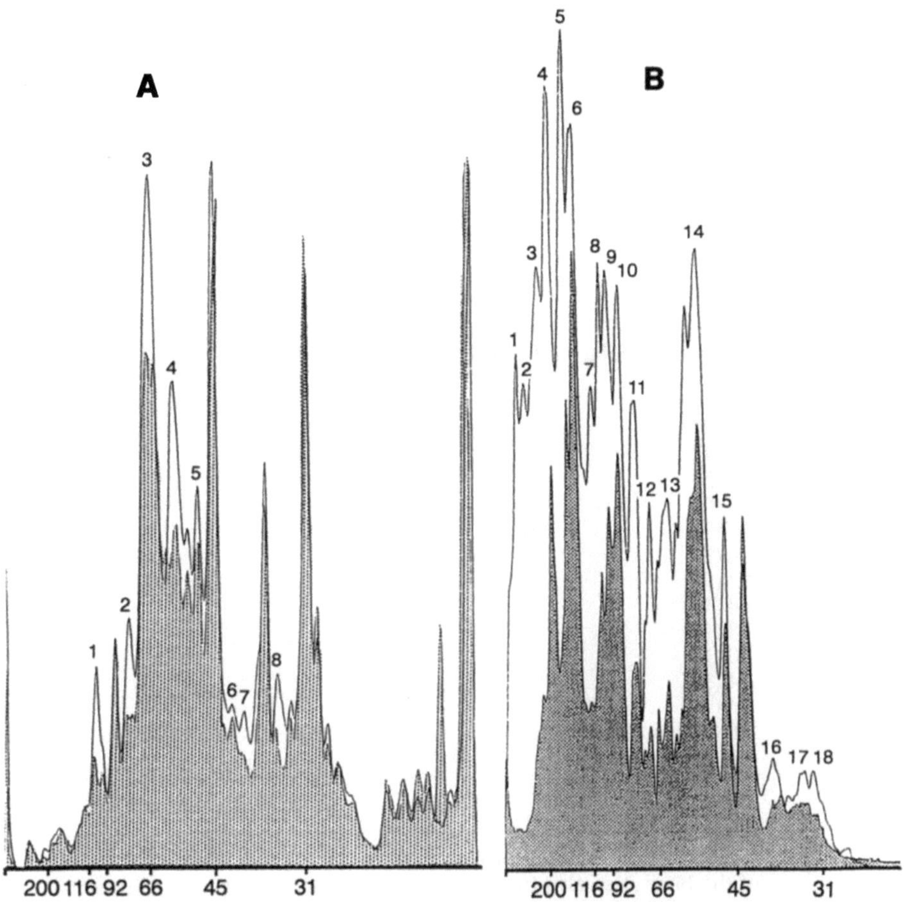

Fig. 1A,B. SDS-PAGE of SCO soluble extracts □ and ependymal soluble extracts ■. The electrophoretic patterns were analyzed by transmission or reflexion densitometry. **A** Coomassie-blue stain; **B** Western blot of the same electrophoresis stained with ConA. See Tables 1A,B for the molecular weight and intensity of the bands

strate, probably due to comigration of specific and unspecific compounds. Nevertheless, by 2D electrophoresis, a high-resolution technique, several polypeptides were found to be specific to the SCO tissue but, to date, it is difficult to establish a clear-cut correlation of this material with compounds related to the secretory process.

It can be concluded that the SCO tissue, due to its specific development, displays marked biochemical differences in comparison to the typical, common ependyma. For future work, a set of highly specific antibodies will be necessary for a clear-cut biochemical characterization of these specific components.

Table 1. SCO and ependymal (EPE) soluble extracts using SDS-PAGE and Coomassie blue, and Western blotting and ConA

Molecular weight (kDa)	SCO	EPE
A *SDS-PAGE and Coomassie blue*		
100 (1)	++	+
78 (2)	++	+/−
66 (3)	+++	+
54 (4)	++	+
48 (5)	+	+
42 (6)	+/−	+/−
38 (7)	+	+/−
34 (8)	+	−
B *Western blotting and ConA*		
350 (1)	+++	+/−
280 (2)	+++	+/−
250 (3)	+++	+/−
200 (4)	+++	+
160 (5)	+++	+
150 (6)	+++	++
110 (7)	+++	+/−
100 (8)	+++	+
92 (9)	+++	+
88 (10)	+++	+
78 (11)	+++	+
70 (12)	++	+/−
64 (13)	++	+
54 (14)	+++	++
48 (15)	++	+
38 (16)	+	+/−
34 (17)	+	+/−
32 (18)	+	+/−

Symbols + to +++ are proportional to the staining intensity of the band. See also Fig. 1.

Table 2. Eluted fractions of the SCO and ependyma (EPE) after ConA affinity chromatography using Western blots and ConA

Molecular weight (kDa)	SCO	EPE
480 (1)	++	+/−
200 (2)	+++	+/−
160 (3)	++	+/−
150 (4)	++	+
115 (5)	+	+/−
100 (6)	+++	+
92 (7)	++	+
88 (8)	++	+/−
64 (9)	+++	++
54 (10)	+++	++

Symbols + to +++ are proportional to the staining intensity of the band.

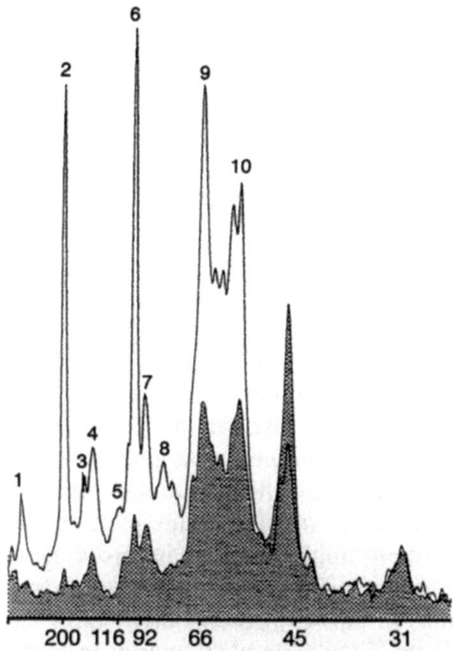

Fig. 2. ConA affinity chromatography. Western blot (stained with ConA) of eluted fractions of SCO □ and ependyma ■. The electrophoretic patterns were analyzed by reflexion densitometry. See Table 2 for the molecular weight and intensity of the bands

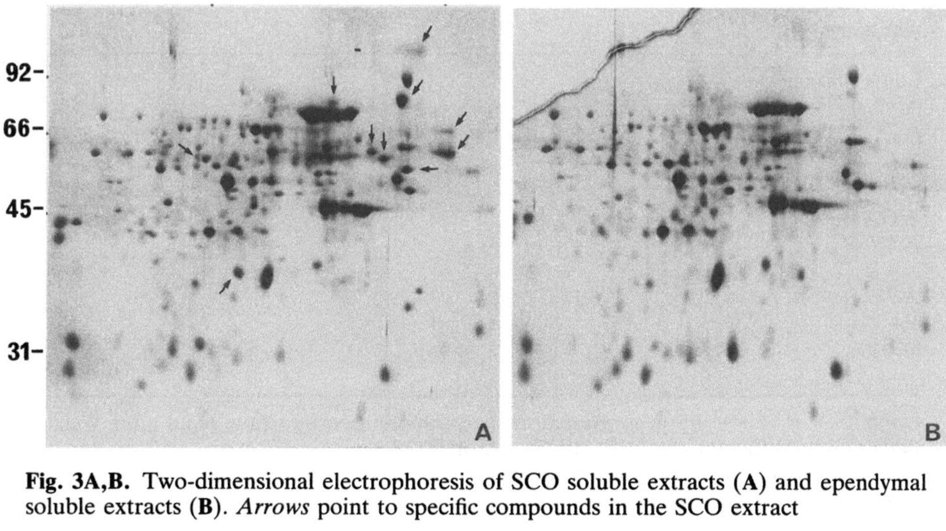

Fig. 3A,B. Two-dimensional electrophoresis of SCO soluble extracts (**A**) and ependymal soluble extracts (**B**). *Arrows* point to specific compounds in the SCO extract

Immunoaffinity Chromatography

Several glycopeptides were purified after performing A99 and A74 immunoaffinity chromatography (Karoumi et al. 1990b, 1991). The electrophoretic characteristics of these compounds and their affinity to the lectins ConA and WGA are reported in Table 3. After chromatography on $C_1B_8A_8$ immunoadsorbent, only three glycopeptides could be identified (Table 3) (Meiniel et al. 1991).

In the eluted fractions of ependymal extracts that served as control, no glycopeptide could be visualized while in the eluted fractions of cerebral extracts, a ConA- and WGA-positive glycopeptide located at 52/54 kDa could be detected in the bovine and in the chick embryo.

From this comparative immunochemical study it can be assumed that the 88-, 52/54- and 32/34-kDa glycopeptides purified on $C_1B_8A_8$ imunoadsorbent correspond to the complex-type glycoproteins at various steps of glycosylation. Thus, only few specific compounds appear to be related to the particular secretory process, and the other glycopeptides isolated on A99 and A74 imunoadsorbent might mirror, in more general terms, the expression of the phenotype of the SCO ependymocytes. These compounds could include enzymes or carrier proteins involved in the secretory process.

The presence of the 52/54 kDa glycopeptide in eluted fractions of brain extracts after A99 and A74 immunoaffinity chromatography remains open to discussion. To date, a cross-reactivity of brain proteins having similar properties cannot be excluded. Further analyses are required after purification on $C_1B_8A_8$ immunoadsorbent of brain extracts.

Table 3. Immunoaffinity chromatography using A 99, A 74 and $C_1B_8A_8$

Antibody \ Band in kDa	240	220	150	98	88	84	77	66	54/52	50	46	42	34/32
ConA *staining*													
A 99	++	+		++	++			+	+++	+			
A 74	+		±	++	++	++	±	±	+++		±	±	
$C_1B_8A_8$					++				++				++
WGA *staining*													
A 99				+++	±			±	++	+			
A 74				+	±				++				
$C_1B_8A_8$									±				+++

Symbols + to +++ are proportional to the staining intensity of the band after Western blots of eluted fractions stained with ConA or WGA.
A 74 anti-chick SCO; *A 99* anti-bovine SCO; $C_1B_8A_8$ monoclonal antibody specific to the complex type glycoproteins.

From the comparison of the results obtained in the bovine and chick embryo, several immunopurified glycopeptides appear to be common to both species. This suggests a certain permanence in the biochemical features of this structure in the course of phylogeny as indicated previously by the use of histochemical and immunohistochemical techniques (see Rodríguez et al. 1984).

Reissner's Fiber

In the bovine, RF was isolated from the spinal cord and analyzed after staining of Western blots with WGA. In this connection, the denaturation efficacy of various detergents or combined detergents was tested. After treatment with SDS or SDS+urea or SDS+CHAPS, a WGA-positive smear was found in the upper part of the blots (Fig. 4), indicating the presence of polymers. In addition, in a few cases, a band located at 150 kDa could also be detected.

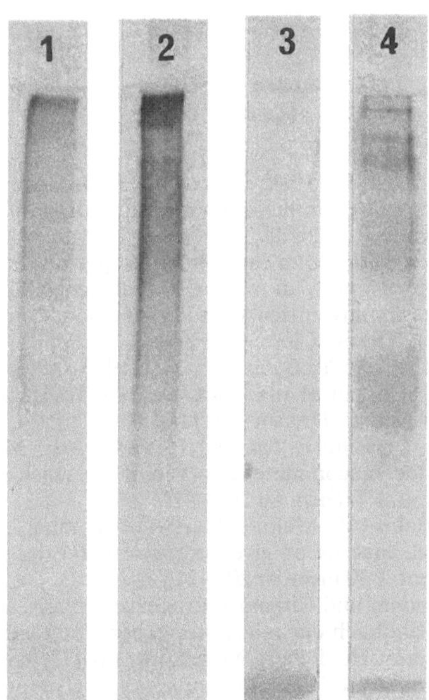

Fig. 4. Western blots of RF extracts stained with WGA: *1*, after preincubation in SDS 4%; *2*, after preincubation in SDS 4% + CHAPS 2%; *3*, after preincubation in trypsin 0.5%; *4*, after preincubation in trypsin 0.01%

The WGA-positive smear disappeared when RF material was preincubated in various proteases (pronase, pepsin, papain or trypsin; Fig. 4). A partial cleavage of the RF polymer was obtained using a low dose of trypsin (Fig. 4).

Thus, in the adult bovine, RF can be regarded as a stable polymer, only partly denatured in the presence of detergents. Analysis of the composition of RF must now be accomplished using other methods.

Conclusions

In conclusion, the combined use of specific reagents and biochemical techniques indicates a complex and specific character and development of the SCO-ependymocytes in comparison to the common ventricular ependyma. The application of the monoclonal antibody $C_1B_8A_8$ led to the identification of three glycopeptides linked to the secreted glycoprotein(s). Other specific probes will be needed to characterize the other components present in the SCO tissue.

References

Hofer H, Meinel W, Erhard H (1980) Electron microscopic study of the origin and formation of Reissner's fiber in the subcommissural organ of *Cebus apella* (Primates, Platyrrhina). Cell Tissue Res 205:292–301

Karoumi A (1990) Mise en évidence et caractérisation par des méthodes immunologiques de sécrétions glycoprotéiques au niveau du diencéphale dorsal (Etude particulière d'une structure circumventriculaire: l'organe sous-commissural du boeuf et de l'embryon de poulet). Thesis, Clermont-Ferrand

Karoumi A, Meiniel R, Croisille Y, Belin MF, Meiniel A (1990a) Glycoprotein synthesis in the subcommissural organ of the chick embryo. I: An ontogenetical study using specific antibodies. J Neural Transm 79:141–153

Karoumi A, Croisille Y, Croisille F, Meiniel R, Belin MF, Meiniel A (1990b) Glycoprotein synthesis in the subcommissural organ of the chick embryo. II: An immunochemical study. J Neural Transm 80:203–212

Karoumi A, Meiniel R, Belin MF, Meiniel A (1991) A comparative immunocytochemical and immunochemical analysis of glycoproteins synthesized in the bovine subcommissural organ. J Neural Transm 86:205–216

Leonhardt H (1980) Ependym und Circumventriculäre Organe. In: Oksche A, Vollrath L (eds) Neuroglia I. Handbuch der mikroskopischen Anatomie des Menschen, vol IV, part 10. Springer, Berlin Heidelberg New York, pp 117–665

Meiniel A, Molat JL, Meiniel R (1988a) Complex-type glycoproteins synthesized in the subcommissural organ of mammals. Light- and electron-microscopic investigations by use of lectins. Cell Tissue Res 253:383–395

Meiniel A, Meiniel R, Karoumi A, Duchier-Liris N, Molat JL (1992) The subcommissural organ of the bovine: cytochemical and immunochemical characterization of the secretory process. Prog Brain Res 91:331–342

Meiniel R, Meiniel A (1985) Analysis of the secretions of the subcommissural organs of several vertebrate species by use of fluorescent lectins. Cell Tissue Res 239:359–364

Meiniel R, Duchier N, Meiniel A (1988b) Monoclonal antibody $C_1B_8A_8$ recognizes a ventricular secretory product elaborated in the bovine subcommissural organ. Cell Tissue Res 254:611–615

Meiniel R, Duchier-Liris N, Molat JL, Meiniel A (1991) The complex type glycoprotein secreted by the bovine subcommissural organ: an immunological study using $C_1B_8A_8$ monoclonal antibody. Cell Tissue Res 266:483–490

Oksche A (1969) The subcommissural organ. J Neuro Visc Relat [Suppl] 9:111–139

Olsson R (1986) Basic design of the chordate brain. In: Uyeno T, Arai R, Taniuchi T, Matsuura K (eds) Proceedings of the 2nd international conference on Indo-Pacific fishes. Ichthyological Society of Japan, pp 86–93

Rodríguez EM, Oksche A, Hein S, Rodríguez S, Yulis R (1984) Comparative immunocytochemical study of the subcommissural organ. Cell Tissue Res 237:427–441

Rodríguez EM, Herrera H, Peruzzo B, Rodríguez S, Hein S, Oksche A (1986) Light- and electron-microscopic immunocytochemistry and lectin histochemistry of the subcommissural organ: evidence for processing of the secretory material. Cell Tissue Res 243:545–559

Sterba G, Kiessig C, Naumann W, Petter H (1982) The secretion of the subcommissural organ. A comparative immunocytochemical investigation. Cell Tissue Res 226:427–439

Immunochemical Analysis of the Dogfish Subcommissural Organ

J. Pérez, J.M. Grondona, M. Cifuentes, F. Nualart,
P. Fernández-Llebrez, and E.M. Rodríguez

Introduction

Several antisera against the subcommissural organ (SCO) and its ventricular secretion, Reissner's fiber (RF), have been obtained over the last decade. These include polyclonal antisera against the bovine RF (Sterba et al. 1981, 1982; Rodríguez et al. 1984) and the SCO of bovine (Rodríguez et al. 1988) and chick embryo (Karoumi et al. 1990). Likewise, a number of monoclonal antibodies against the bovine SCO and RF have been developed (Meiniel et al. 1988; Rodríguez et al. 1991). However, to date, no antibodies against the SCO of lower vertebrates were available. The aim of the present work was: (a) to produce a polyclonal antiserum against the SCO of the dogfish, *Scyliorhinus canicula*, and (b) to characterize, at least partially, the molecular components of its secretory material. This species has been chosen because of its rather primitive position in the phylogenetic scale of the vertebrates, the conspicuous morphologic development of its SCO, and the fact that dogfish are commercially available, making it possible to obtain a large number of specimens for immunological and biochemical studies.

The availability of antisera against the secretory material of the bovine SCO and their combined use with lectins in blots allowed Nualart et al. (1991) and Hein et al. (this volume) to identify the secretory compounds in a crude extract of SCO, and to distinguish between precursor and processed forms.

A similar methodology is used in our laboratories to analyze the secretory material of the dogfish SCO by means of an antiserum against the bovine RF (AFRU), which cross-reacts with the dogfish SCO and an antiserum against the dogfish SCO (ADSO).

Materials and Methods

SCO Dissection and Extraction

The study was carried out according to the procedure described by Nualart et al. (1991) with minor modifications. Approximately 3500 SCO of dogfish were used for the present study. They were collected over the last 2 years.

The heads of dogfish were obtained at the fish market (port of Málaga) approximately 8 h after death and kept on ice until they were brought to the laboratory. The brains were removed and immersed in cold acetone containing 1 m*M* EDTA and 50 m*M* PMSF, for 48 h at 4°C. Then, the SCO region (SCO plus posterior commissure) was dissected out under a stereoscopic microscope.

The SCO was immersed in the extraction medium. Three different solutions were tested for extraction of the secretory material of the SCO: (1) 50 m*M* ammonium bicarbonate, pH 8 – this type of SCO extract was used to obtain a polyclonal antiserum and for the electrophoretic study; (2) 50 m*M* ammonium bicarbonate, 0.1% SDS, 0.5% Triton X-100 – this extract was employed exclusively for the electrophoretic analysis; and (3) TRIS buffer, HCl 0.1 *M*, pH 8.6, 0.01 *M* DTT, 8 *M* urea (for further details, see Rodríguez et al. 1984; Nualart et al. 1991) – this latter SCO extract was also used to raise an antiserum and then to compare its immunocytochemical properties with those of an anti-bovine RF serum (AFRU; see Rodríguez et al. 1984). In all cases, the extraction medium contained 1 m*M* EDTA, 0.5 m*M* PMSF, 1 μ*M* pepstatin A, and 1 μ*M* leupeptin as protease inhibitors. The pool of SCOs was homogenized in the extraction medium with the use of a homogenizer at 8000 rpm for 2 min. Then, it was sonicated for 15 s and cooled down in ice. This sonication-cooling cycle was repeated four times. The extract was kept overnight at 4°C and then centrifuged at 10 000 rpm for 20 min at 4°C. The supernatant was centrifuged again at 15 000 rpm for 3 min at 4°C. The latter supernatant was regarded as a crude extract of the dogfish SCO.

The protein content of the extract was determined according to Bradford (1976). Aliquots were lyophilized and stored at −20°C.

In addition, extracts of other regions of the dogfish brain (optic tectum, cerebellum, telencephalon, and choroid plexus) were obtained to be used for the absorption of the anti-SCO serum (see below) and for comparison in the electrophoresis and blotting studies.

The extracts of dogfish SCO were used: (a) to raise a polyclonal antiserum against the dogfish SCO, and (b) to analyze the secretory material of the SCO by sodium dodecyl sulfate-polyacrylamide gel electrophoresis (SDS-PAGE) and Western blot.

Antisera

Antisera against two types of extracts of the dogfish SCO (see above) were developed in rat and rabbit. An extract containing 1 mg protein (equivalent to approximately 50 SCOs) was emulsified with complete Freund's adjuvant and injected intradermally into multiple sites in the back of each rat. Thirty days later, each rat was injected in the same way with an extract containing 0.5 mg SCO protein emulsified with incomplete Freund's adjuvant. Finally,

10 days later, the rats received an intraperitoneal injection of an extract of SCO containing 0.5 mg protein dissolved in phosphate-buffered saline (PBS). A few days after the last injection a sample of blood was collected from the retroocular sinus. The titer of the antiserum was estimated by immunostaining sections of dogfish SCO at different dilutions. If the titer was high enough, the rat was anesthetized and the bulk of blood was obtained by heart puncture; otherwise, the animal received an additional intraperitoneal injection of extract before the final bleeding.

Rabbits were immunized following the same protocol as described for rats, but they received 5 mg protein in the first injection and 2.5 mg in the other two sessions of immunization.

Since the extract used as immunogen was a crude extract, the antisera obtained contained unwanted antibodies. In order to remove the latter, the antiserum was sequentially absorbed with crude extracts of telencephalon, cerebellum, and optic tectum, in the presence of carrageenan and Triton X-100.

Characterization of antisera was performed by: (a) immunoblot analysis; (b) immunocytochemical staining of sections of dogfish brain containing the SCO, sections of dogfish spinal cord containing RF, and sections of SCO of the following species belonging to different groups of vertebrates:

1. Cyclostomata: *Myxine glutinosa*
2. Elasmobranchii: *Scyliorhinus canicula*, *Squalus blainvillei*, *Centrophorus granulosus*, *Galeorhinus galeus*, *Heptranchius griseus*, *Raja clavata*
3. Teleostei: *Sardina pilchardus*, *Carassius auratus*, *Sparus auratus*, *Liza aurata*
4. Amphibia: *Rana perezi*
5. Reptilia: *Mauremys caspica*, *Natrix maura*
6. Aves: *Alectoris rufa*
7. Mammalia: Rat, pig, bovine

Histology

Blocks of tissue containing the SCO were fixed by immersion in Bouin's fluid for 2 days. Embedding was in paraffin and methacrylate. Thin and semithin sections were processed for lectin histochemistry and immunocytochemistry.

Lectin Histochemistry. Concanavalin A (ConA) and wheat-germ agglutinin (WGA) (Sigma) labeled with peroxidase were used at a concentration of 10 and 20 µg/ml. Unlabeled *Limax flavus* agglutinin (LFA) (Calbiochem) was used at a concentration of 3 µg/ml. After incubation in the lectin for 45 min, sections were incubated in an anti-LFA serum (Rodríguez et al. 1990) at a dilution 1:5000 and then processed as for immunostaining.

Immunocytochemistry. Sections were processed according to the immunoperoxidase method of Sternberger et al. (1970). Two different primary antisera were used: anti-dogfish SCO (ADSO) at a dilution of 1:2000, and anti-bovine RF (AFRU) at two different dilutions (1:1000 and 1:200000). Secondary antiserum (anti-rat immunoglobulin G, Ig-G, anti-rabbit IgG) was used at 1:20 dilution and the PAP complex at a 1:75 dilution, for 30 min. The reaction was visualized with diaminobenzidine.

Immunocytochemical controls included: (a) immunostaining using a preimmune serum as primary antiserum; (b) immunostaining with ADSO immunoabsorbed with a dogfish SCO extract. In both cases, the immunoreactivity in the SCO was completely abolished.

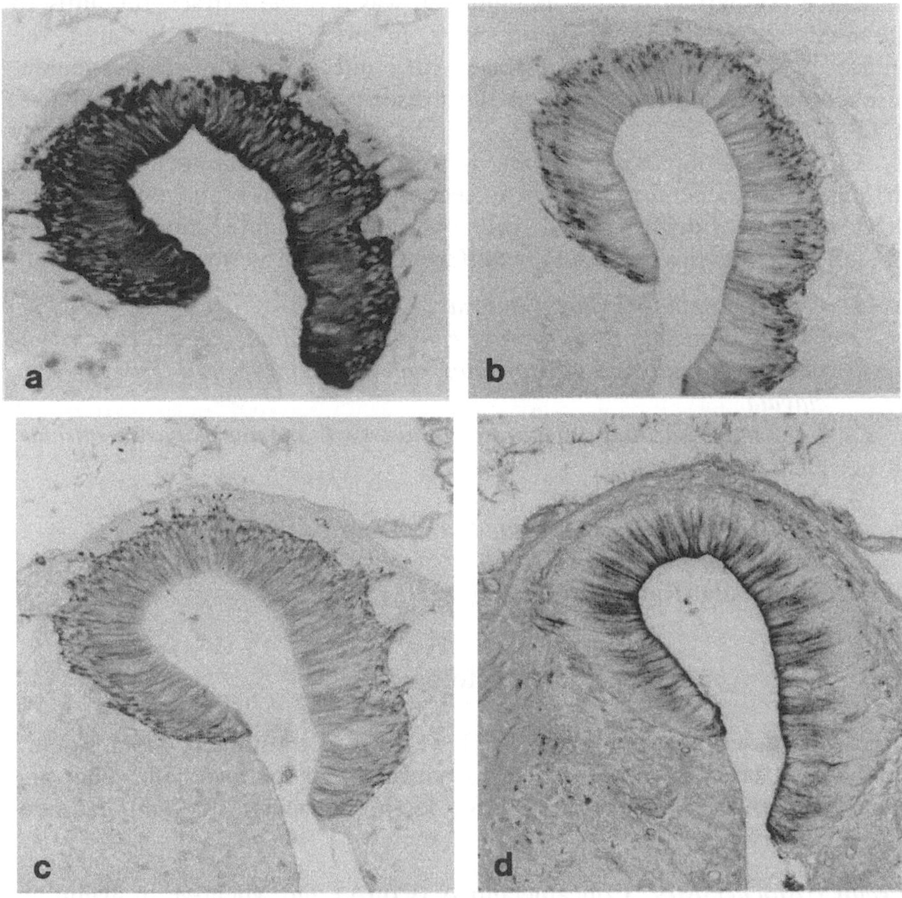

Fig. 1a–d. Frontal serial sections through the rostral region of the SCO of the dogfish, *Scyliorhinus canicula*, processed for immunoperoxidase staining using ADSO (1:2000; **a**) and AFRU (1:200000; **b**), as well as for Concanavalin A (**c**) and wheat-germ agglutinin (**d**) binding. The ADSO-IR material is distributed throughout the SCO cells. While the AFRU- and the ConA-positive material is mainly located in the perinuclear region, WGA-binding sites occupy the supranuclear cytoplasm

Electrophoresis and Blotting

SDS-PAGE according to Laemmli (1970) was performed using a 5%–15% polyacrylamide gradient. Samples of dogfish SCO extract containing 100 µg protein (equivalent to 25–30 SCOs) were applied to each well. Extract of dogfish optic tectum and choroid plexuses were run in parallel lines to that of the SCO extract.

Gels were electrotransferred onto nitrocellulose or Inmobilon-P membranes following the method of Towbin et al. (1979). The transfer was accomplished at 80 mA for 16 h in 25 m*M* TRIS-HCl, pH 8.3, 192 m*M* glycine, 0.2% SDS, and 20% methanol.

Immunostaining of the blots was carried out according to the following sequence: (1) saturation of the protein-binding sites with 5% nonfat milk in PBS; (2) immunoperoxidase staining using ADSO and AFRU as primary antisera, at a dilution of 1:2000 for 18 h, and 4-chloro-1-naphtol as electron donor. The primary and secondary antisera and PAP were diluted in PBS containing 0.7% carrageenan (Sigma).

Results and Discussion

As in most submammalian species, the ependymal cells of the dogfish SCO display prominent basal processes, the majority of which terminate on the external limiting membrane of the brain. Only some of them seem to establish contact with blood vessels. Hypendymal cells are only occasionally found.

Lectin Histochemistry

The properties of the carbohydrate moieties of the secretory glycoproteins of the dogfish SCO were studied by the use of ConA (affinity = mannose, glucose), WGA (affinity = glucosamine, sialic acid), and LFA (affinity = sialic acid).

ConA-binding material occupied most of the cellular regions of the SCO cells. The material located in perinuclear droplets (RER) was strongly stained by this lectin. In the apical cytoplasm the reaction was weak and the apex of the cells was completely devoid of ConA-positive material (Fig. 1c).

WGA and LFA reacted, both in a very similar binding pattern, with materials located in the supranuclear cytoplasm and in apical granules (Fig. 1d).

Considering (1) that LFA selectively binds to sialic acid and that this sugar residue is exclusively added in the Golgi apparatus and (2) that the

LFA binding sites occur throughout the supranuclear cytoplasm, it may be postulated that in the dogfish SCO, a large proportion of the secretory glycoproteins is stored in post-Golgi compartments (secretory granules). This is a relevant difference with respect to the situation in the mammalian SCO, where the bulk of secretion is stored in the RER (see Rodríguez et al. 1991).

Immunocytochemisty and Immunoblotting

The immunocytochemical results are summarized in Table 1.

After absorption with extracts of telencephalon, cerebellum and optic tectum of dogfish, ADSO reacted strongly and selectively with the SCO of *Scyliorhinus canicula*. The immunoreaction was very intense both in the basal and apical portion of the cells (Fig. 1a). RF was strongly stained in the brain ventricles and in the central canal of the spinal cord. Thus, ADSO recognizes the material throughout the entire secretory pathway of the dogfish SCO.

Table 1. Immunocytochemical affinity of AFRU and ADSO for the secretory material in the SCO of different vertebrate species (cf. p. 101)

Species	AFRU	ADSO
Myxine glutinosa	+++	–
Heptranchius griseus	+++	–
Scyliorhinus canicula	+++	+++
Centrophorus granulosus	+++	+++
Squalus blainvillei	+++	++
Galeorhinus galeus	+++	++
Raja clavata	+++	+
Sardina pilchardus	+++	–
Carassius auratus	+++	–
Sparus auratus	+++	–
Liza aurata	+++	–
Rana perezi	+++	–
Natrix maura	+++	–
Mauremys caspica	+++	–
Alectoris rufa	+++	–
Rat	+++	–
Pig	+++	–
Bovine	+++	–

AFRU, antiserum against bovine RF extracted in DTT, EDTA, and urea (Rodríguez et al. 1984); ADSO, antiserum against an ammonium bicarbonate extract of the SCO of the dogfish, *Scyliorhinus canicula*, purified by sequential absorption with extracts of telencephalon, cerebellum and optic tectum.

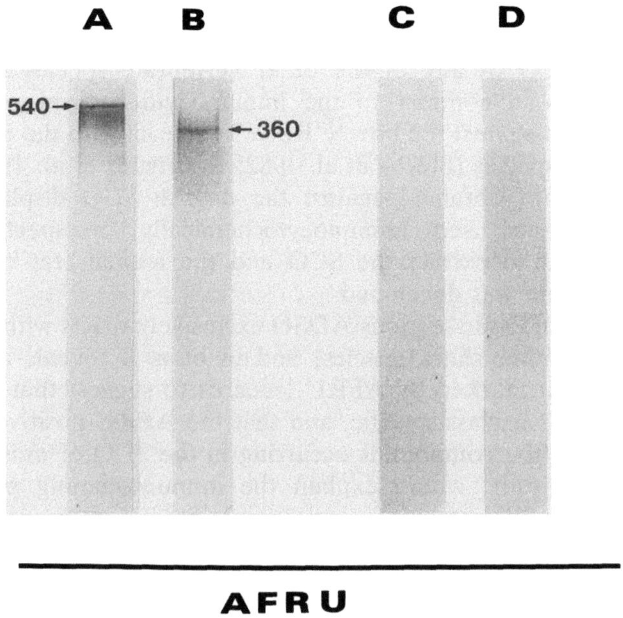

Fig. 2A–D. Immunoblotting, using AFRU, of bovine SCO (**A**), dogfish SCO (**B**), optic tectum (**C**), and choroid plexus (**D**). Molecular weights in kilodaltons

AFRU reacted distinctly with the secretory material of the dogfish SCO. When used at a 1:1000 dilution, this antiserum revealed the secretory material present in most of the cellular regions. The basal processes displayed immunoreactivity exclusively in their proximal portion. AFRU at 1:200000 dilution strongly stained perinuclear secretory material (most likely stored in the cisternae of the RER), but the apical cytoplasm reacted weakly; apical secretory granules did not react at all (Fig. 1b).

When used on blots of dogfish SCO, AFRU revealed a band with an apparent molecular weight of 360 kDa (Fig. 2). No comparable bands were seen in blots of the optic tectum and choroid plexus. The AFRU-positive band was not revealed by ADSO. Only recently we succeeded in immunostaining blots of dogfish SCO using ADSO. Three bands were distinctly stained; two of them were also present in the optic-tectum blot and one of them in the choroid-plexus blot. The band found exclusively in the SCO blot had an apparent molecular weight of 225 kDa. These facts suggest that AFRU and ADSO react with different secretory compounds: the former with a high molecular weight polypeptide (precursor form ?), the latter with a compound of lower molecular weight (processed form ?). This agrees well with the immunocytochemical findings described above.

ADSO also immunostained the SCO of other shark species (*Galeorhinus*, *Squalus*, *Centrophorus*). In these species, the antiserum mostly reacted with the perinuclear secretory material. The SCO of a primitive shark, *Heptranchius*, was completely negative to this antiserum. ADSO did not react with the SCO of any of the other vertebrate species studied (see Table 1). This is in contrast to the immunocytochemical properties of antisera developed against the bovine RF, which react with the SCO of most of the vertebrate species (Sterba et al. 1982; Rodríguez et al. 1984).

All the antisera obtained against the dogfish SCO displayed similar characteristics when used immunocytochemically, irrespective of the medium employed to extract the SCO and the animal (rat or rabbit) in which the antiserum was developed.

The fact that in tissue sections ADSO exclusively reacts with the SCO of the dogfish and other shark species, and in blots it reveals a compound different from that marked by AFRU leads us to suggest that the ADSO-positive compound is class-specific, and that the AFRU-positive compound may be similar to the compounds occurring in the SCO of most vertebrate species. This, in turn, would explain the immunostaining properties of AFRU.

Acknowledgements. This work was supported by grants from: DGICYT, PB90/0804 and CICYT, MAR 88-0751, Madrid, Spain, to P.F.LL.; Instituto de Cooperación Iberoamericana (1989–90) to J.M.P.F., Madrid; Volkswagen-Stiftung I/63 476, FRG, FONDECYT 089188, Chile, and Dirección de Investigaciones de la Universidad Austral de Chile, S-89-01, to E.M.R.

References

Bradford MM (1976) A rapid and sensitive method for the quantitation of microgram quantities of protein utilizing the principle of protein-dye binding. Anal Biochem 72:248–254

Karoumi A, Meiniel R, Croisille Y, Belin MF, Meiniel A (1990) Glycoprotein synthesis in the subcommissural organ of the chick embryo. J Neural Transm 79:141–153

Laemmli UK (1970) Cleavage of structural proteins during the assembly of the head of bacteriophage T4. Nature 227:680–685

Meiniel A, Molat JL, Meiniel R (1988) Complex-type glycoproteins synthesized in the subcommissural organ of mammals: light and electron-microscopic investigations by use of lectins. Cell Tissue Res 253:383–395

Nualart F, Hein S, Rodríguez EM, Oksche A (1991) Identification and partial characterization of the secretory glycoproteins of the bovine subcommissural organ-Reissner's fiber complex. Evidence for the existence of two precursor forms. Mol Brain Res 11:227–238

Rodríguez EM, Oksche A, Hein S, Rodríguez S, Yulis R (1984) Comparative immunocytochemical study of the subcommissural organ. Cell Tissue Res 237:427–441

Rodríguez EM, Korf HW, Oksche A, Yulis R, Hein S (1988) Pinealocytes immunoreactive with antisera against secretory glycoproteins of the subcommissural organ: a comparative study. Cell Tissue Res 254:469–480

Rodríguez EM, Garrido O, Oksche A (1990) Lectin histochemistry of the human fetal subcommissural organ. Cell Tissue Res 262:105–113

Rodríguez EM, Oksche A, Hein S, Yulis R (1991) Cell biology of the subcommissural organ. Int Rev Cytol 135:39–121

Sterba G, Kleim I, Naumann W, Petter H (1981) Immunocytochemical investigation of the subcommissural organ in the rat. Cell Tissue Res 218:659–662

Sterba G, Kiessing C, Naumann W, Petter H, Kleim I (1982) The secretion of the subcommissural organ. A comparative immunocytochemical investigation. Cell Tissue Res 226:427–439

Sternberger LA, Hardy PH Jr, Cuculis JJ, Meyer HG (1970) The unlabeled antibody enzyme method of immunohistochemistry. Preparation and properties of soluble antigen-antibody complex (horseradish-peroxidase-antiperoxidase) and its use in identification of spirochetes. J Histochem Cytochem 18:315–333

Towbin H, Staehelin T, Gordon J (1979) Electrophoretic transfer of proteins from polyacrylamide gels to nitrocellulose sheets: procedure and some applications. Proc Natl Acad Sci USA 76:4350–4354

IV. Secretory Process, Including Model Systems

Dynamic Aspects of the Secretory Process in the Amphibian Subcommissural Organ*

J.H.B. Diederen and H.G.B. Vullings

Introduction

The amphibian species which will be dealt with, namely *Rana temporaria* and *Rana esculenta*, belong to the anuran family of the Ranidae. The morphology of the subcommissural organ (SCO) in these species is essentially alike, except for the position of the secretory material within the ependymal SCO cells. In the SCO of *Rana esculenta*, most secretory material is situated apically and perinuclearly, whereas in the SCO of *Rana temporaria*, most secretory material lies subnuclearly, in the basal parts of the cells (Oksche 1962; Diederen 1970, 1973). In both species, a well-developed Reissner's fiber (RF) originates from the SCO. The secretory substance produced by the frog SCO is a glycoprotein (Diederen 1970; Diederen et al. 1987), immunologically related to that of mammals; it is immunoreactive with an antiserum against bovine RF (Sterba et al. 1982; Vullings and Diederen 1983).

Spatiotemporal Aspects of the Intracellular Secretory Process

Electron-microscopic investigation of the SCO cells of *Rana temporaria* revealed the existence of two distinct compartments of (rough) endoplasmic reticulum: one consisting of flattened cisternae regularly studded with ribosomes and situated mainly peri- and supranuclearly, and another consisting of spherical cisternae very irregularly studded with ribosomes and situated mainly subnuclearly, with much secretory material stored in it (Diederen 1970). Incorporation of radioactive cysteine, a precursor of the secretory substance, initially takes place in the apical regions of the SCO cells of *Rana temporaria*; later on, the radioactivity of the subnuclear secretory material gradually increases to a degree equal to that of the apical material, as was established autoradiographically (Diederen 1972). Apparently, the apical and perinuclear regions of the SCO cells are primarily concerned with synthesis and the basal regions with storage of secretory material. The Golgi complexes are situated in between the region

* Dedicated to the memory of Professor Dr. J.C. Van de Kamer (1916–1991)

with well-developed rough endoplasmic reticulum and the apical border of the SCO cells, far away from the storage compartment.

The synthetic processes were studied in more detail by following autoradiographically the incorporation of several radioactive precursors of the protein moiety (amino acids) and of the oligosaccharide moiety (monosaccharides) of the secretory material in the SCO cells of *Rana temporaria* (Diederen et al. 1987). The amino acids are incorporated already in the rough endoplasmic reticulum. The incorporation site of the monosaccharides depends on whether the precursor monosaccharide used is part of the core region of the oligosaccharides or part of their terminal regions, which means that the site of their incorporation is the rough endoplasmic reticulum and the Golgi apparatus, respectively. The frogs were kept alive for 72 h after injection of the radioactive precursors. ^{3}H-glucosamine produced an excellent labeling of the RF, as did ^{3}H-fucose (after a longer autoradiographic exposure time). In comparison, the radioactive amino acids (^{3}H-leucine, ^{3}H-cystine, and ^{35}S-cysteine) produced only a moderate labeling of the RF. In the frogs injected with the ^{3}H-monosaccharides, the apical region of the SCO was more strongly labeled than any other region. The subnuclear region, with its large content of stored secretory material, was distinctly labeled only in the ^{3}H-glucosamine-injected frogs, whereas in the ^{3}H-fucose-injected frogs it was practically unlabeled. In the frogs injected with the radioactive amino acids, however, the subnuclear region was always distinctly labeled, as was the apical region. In these frogs, the degree of labeling of the apical secretory product, the RF, did not differ clearly from that of the secretory material stored subnuclearly, whereas in the frogs injected with the radioactive monosaccharides, the degree of labeling of the RF was much higher than that of the subnuclear secretory material, especially in the ^{3}H-fucose-injected frogs. This indicates that glucosamine molecules are incorporated into the oligosaccharides already within the rough endoplasmic reticulum, whereas additional glucosamine molecules and the fucose molecules are being added later on in the supranuclearly situated Golgi apparatus prior to apical release. As a consequence, the subnuclear material and the material that is being released into the brain ventricle to form the RF are chemically different, at least in so far as the oligosaccharide part is concerned. This is in agreement with findings of Meiniel et al. (1988) and Herrera and Rodríguez (1990) for the secretory substances in the SCO of higher vertebrates.

Intracellular Turnover Rate of the Secretory Material

None of the above-mentioned radioactive precursors labeled the subnuclear secretory material over its entire sectional area, 72 h after injection of these precursors. In particular, the parts situated most basally were practically unlabeled. This indicates a slow intracellular turnover of this material. To investigate the turnover rate more specifically, frogs (*Rana temporaria*) were subjected to an increasing number of daily injections (maximally 10) of ^{3}H-leucine, eventually followed by an increasing number of injections (maximally 9) of an excess (×1000) of unlabeled leucine every 2 days. The latter injections were meant to suppress further radioactive labeling of the secretory substance. The injection schedule thus was different for each individual frog. This schedule was such that all frogs were killed on the same day, 24 h after the last injection. The SCO was studied autoradiographically (Diederen et al. 1987). The number of silver grains over the subnuclear region of the SCO cells increased continuously during the injections of ^{3}H-leucine. It took about 5.5 days after the start of the injections of excess unlabeled leucine for the number of silver grains over the subnuclear region to decrease to half the maximal value reached after 10 daily injections of ^{3}H-leucine. This indicates that the half-life of the radioactively labeled macromolecular subnuclear material was approximately 5.5 days. Since, however, considerable portions of the subnuclear secretory material were still unlabeled after 10 daily injections of ^{3}H-leucine, it may be concluded that the intracellular half-life of this material in its entirety was even longer than 10 days.

Silver grain counts over the apical regions of the SCO cells indicated that the half-life of the apical radioactively labeled macromolecular material was also 5.5 days. Earlier studies on the SCO of frogs have indicated that secretory material synthesized in the apical regions is released into the cerebral ventricle soon (within a few hours) after synthesis. Some of it, however, may be transported towards subnuclear regions of the SCO cells (Diederen 1970, 1972, 1973; Ermisch et al. 1971; Diederen and Vullings 1980a). Consequently, an intracellular half-life considerably shorter than 5.5 days has to be expected for the apical secretory material. That the established half-life still was 5.5 days may be considered to be a reflection of the half-life of radioactive secretory material that was on its way from the original subnuclear site of storage to the site of release at the apical cell border.

Histological Parameters of the Secretory Process

The influences of experimental treatments on the dynamics of the secretory process in the SCO were investigated quantitatively in histological sections by measuring three parameters which are closely related to the processes of synthesis and release of secretory material by the SCO. These parameters are: (a) the amount of selectively stained secretory material in the SCO; (b) the amount of radioactivity in the SCO in the form of autoradiographically produced silver grains after the secretory material has been labeled by a radioactive precursor; and (c) the growth rate of the RF (Diederen and Vullings 1980a). The amounts of stained secretory material and of silver grains were measured densitometrically by using a computer-controlled scanning cytophotometer or an automatic image analysis system (Diederen et al. 1981). The growth rate of the RF was measured according to the method developed by Sterba and coworkers (Sterba et al. 1967; Ermisch et al. 1971). Autoradiographs were made of sagittal sections of brain and spinal cord of frogs that had been killed at various points of time (hours to days) after injection of a radioactive precursor of the substance of the RF. The rate of passage through the brain ventricles and central canal of this precursor built into the RF was measured. This was done by visual observation of the autoradiographs and determining at what distance from the SCO the RF was no longer covered with silver grains.

The growth rate of the RF may be considered to be the most direct and unequivocal measure of the secretory activity of the SCO, since the RF is the main secretory product of the SCO. The two other parameters are intracellular parameters. They are more difficult to interpret in terms of degree of increase or decrease of secretory activity of the SCO, depending on the experimental design, which may be either a long- or short-term precursor incorporation study, as will be explained below. Both intracellular parameters, however, are more sensitive than the growth rate of the RF. This appeared from the measurements of all three parameters in frogs belonging to three very closely related species within the so-called *Rana esculenta* complex, namely, *Rana esculenta*, *Rana lessonae*, and *Rana ridibunda* (Diederen and Vullings 1980a,b). The differences in growth rate of the RF were clearly smaller than the differences in the intracellular parameters. These measurements also revealed the existence of an inverse relationship between the amount of stained secretory material and the growth rate of the RF. This means that in the SCO of frogs with a higher growth rate of the RF, secretory material is relatively rapidly released after synthesis, keeping storage of secretory material within the SCO cells low. In the SCO of frogs with a lower growth rate of the RF, on the other hand, it takes more time before the newly synthesized secretory material is released by the SCO cells; it is maintained in storage for some time, causing greater accumulation of secretory material within the cells.

The amount of stained secretory material may thus be considered to be a measure of the degree of storage of secretory material; the larger this amount, the larger the storage, and the lower the growth rate of the RF. The differences in the amounts of radioactive material in this experiment were similar to those of the stained material. This is due to the fact that first the SCO cells were allowed to incorporate the injected radioactive precursor (^{3}H-cystine) over many (35) hours. Next the frogs were injected with a high dose of unlabeled precursor, which suppressed strongly the incorporation of radioactivity by the SCO cells from that moment on. The frogs were kept alive for several hours (6) after the injection of excess unlabeled precursor.

It is to be expected that during these hours the amount of radioactive material within the SCO decreased more rapidly in frogs with a higher secretory activity than in frogs with a lower secretory activity. This experimental design is similar to that used to establish the turnover rate of secretory material discussed earlier. It can be characterized as a long-term precursor incorporation study, which is focussed on the release of radioactive secretory material rather than on its synthesis. Within this experimental design, the amount of radioactive material within the SCO cells, just like that of stained material, may be considered to be an inverse measure of the release of secretory material; the larger this amount, the smaller the release, and the lower the growth rate of the RF.

Influence of Temperature on the Secretory Process

The growth rate of the RF was measured in several experiments on frogs of both species (*Rana esculenta* and *Rana temporaria*) adapted for several weeks to an environmental temperature of 18°C. The average time needed for complete renewal of the RF varied between the individual experiments, approximately from 2.5 to 4 days (Diederen 1973, 1975a,b; Hess et al. 1977; Vullings et al. 1983; Vullings and Diederen 1985). This variation in passage time between successive experiments, which were carried out in different months and different years, may point to influences of season or age. The growth rate of the RF strongly depends on the environmental temperature; it was considerably faster (1.5 times) in frogs (*Rana esculenta*) adapted for 2 weeks to 24°C than in frogs adapted to 18°C (Diederen 1975a). In poikilothermic animals like frogs, increase in environmental temperature may easily stimulate the intracellular metabolism of the SCO cells, thereby stimulating the production and release of secretory substances, which then results in enhancement of the growth rate of the RF.

Influence of Light on the Secretory Process

In frogs, a close topographic relationship exists between the pineal tract and the SCO. The pineal tract is a prominent bundle of nerve fibers originating from the light-perceptive pineal complex (=pineal organ and frontal organ; Diederen 1970, 1973). This raises the question of whether there might be a functional relationship between the SCO and pineal complex. To answer this, the effects on the growth rate of the RF of elimination of the frontal organ, of the entire pineal complex, and also of the eyes were investigated. This was done in combination with long-term adaptation (6 weeks) to light or darkness (*Rana esculenta*; Diederen 1975b). When all light-perceptive organs were intact, no significant differences in growth rate of the RF were detected between frogs that were adapted to long daily light periods and frogs that were adapted to continuous darkness. Elimination of the frontal organ, or the entire pineal complex, or the eyes, or the eyes in combination with the frontal organ did not induce differences in growth rate of the RF. When all these light-perceptive organs were eliminated together, however, then the growth rate of the RF was clearly higher in light-adapted than in dark-adapted frogs. Apparently, light exerted an inhibitory influence on the growth rate of the RF in intact frogs via these organs. Light that acted on the SCO directly, via skin and skull, or perhaps influenced the SCO via other, still unknown light-sensitive structures enhanced the growth rate of the RF in the operated frogs. The presence of only one of the known light-perceptive organs was apparently sufficient to neutralize this stimulatory effect of light on the growth rate of the RF. These investigations have led to the hypothesis that the pineal complex and the eyes exert a spontaneous stimulatory influence on the SCO in darkness. Inhibition of this stimulatory influence by light may compensate for the assumed direct or indirect stimulatory effect of light on the SCO. As a result of this complex regulatory mechanism, the secretory process in the SCO, and thereby the growth rate of the RF, appear to be independent of the actual conditions of light and darkness.

This hypothesis is supported by the results of investigations on the influence of light and darkness on the amount of radioactive material within the SCO (Diederen 1975b). Frogs (*Rana esculenta*) were injected with a radioactive precursor (^{35}S-cysteine) and killed after relatively short survival times (30 min, 2 and 8 h). These experiments, therefore, can be characterized as short-term precursor incorporation studies, which are focussed on the synthesis rather than on the release of radioactive secretory material. Within this experimental design, a larger amount of radioactive material in the SCO indicates a higher synthetic activity. This experimental design is not suitable to measure the growth rate of the RF, which demands longer survival times. The amounts of radioactivity in the SCO were about equal in light-adapted and dark-adapted intact frogs at all survival times. Only after elimination of the entire pineal complex, was a stimulatory influence of light on the synthetic activity within the SCO cells observed. This was reflected by the amount of radioactive material in the SCO being higher in the light-adapted than in the dark-adapted operated frogs at 2 and 8 h after injection of the radioactive precursor. Elimination of either the frontal organ alone or the eyes had no such consequences. The stimulatory influence of light on the intracellular secretory process that became obvious after elimination of the inhibitory light-perceptive pineal complex did not yet result in a measurable increase in the growth rate of the RF, as was already mentioned. Apparently, this influence needed to be strengthened by the additional elimination of the eyes before differences in the growth rate of the RF became manifest. Simultaneous elimination of the pineal complex and the eyes indeed permitted light to stimulate the growth rate of the RF, as was evidenced before.

The consequences of simultaneous elimination of both the pineal complex and the eyes for the influence of light on the secretory process within the SCO were also investigated with the intracellular amount of radioactive material as the parameter. In this experiment, however, operated frogs (*Rana esculenta*) were exposed to light of short (blue) or long (red) wavelengths for long daily periods over 9 days and compared with operated frogs kept in darkness (Vullings et al. 1983). Moreover, it was a long-term precursor incorporation study, focussed on the release of radioactive secretory material by the SCO. Light indeed stimulated this release in the operated frogs, as was to be expected. The amount of radioactive material had decreased in the red-illuminated frogs in particular, which indicates a clear enhancement of the secretory process in these frogs. The influence of light was much weaker in blue-illuminated frogs, which may be explained by the finding that blue light penetrates considerably less effectively into the brain than red light (Hartwig and Van Veen 1979). Another explanation may be found in the spectral sensitivity of an as yet unknown encephalic photoreceptor that might be involved.

Influence of Changes in the Chemical Composition of the Internal Environment on the Secretory Process

Olsson (1958) put forward the hypothesis that the RF detoxicates the cerebrospinal fluid. Sterba and coworkers focussed this hypothesis on an involvement of the RF in the regulation of the composition of the cerebrospinal fluid, especially in so far as the catecholamine concentration is concerned. This was based on the observation that the RFs of bovine and cat selectively bind (nor)epinephrine (Hess and Sterba 1973). This was confirmed for the RF of *Rana esculenta*, as illustrated by the binding of ^{3}H-norepinephrine to the fiber visualized autoradiographically (Diederen 1975a; Diederen et al. 1983). Raising the norepinephrine concentration in the cerebrospinal fluid (injection of 0.1 μg norepinephrine dissolved in 0.5 μl amphibian saline, followed by 8-h survival time), however, did not clearly influence the secretory activity of the SCO (Hess et al. 1977).

Gilbert (1956) put forward the hypothesis that the SCO is actively involved in osmoregulation, which was based on results of experiments with rats. Dehydration of frogs (*Rana temporaria*) over 1 week stimulated the secretory activity of the SCO significantly, as was reflected by an enhanced release of radioactive material from the SCO in a long-term precursor incorporation study, accompanied by a small but significant increase in growth rate of the RF (Vullings and Diederen 1985). Since dehydration is known to stimulate the secretion of aldosterone by the frog interrenal tissue, the influence of this osmoregulatory hormone on the secretory process in the SCO was investigated. Pieces of the brain of *Rana temporaria* containing the SCO were incubated for 30, 60, and 120 min in a medium containing aldosterone (0.8 μg/ml) and ^{3}H-leucine (10 μCi/ml). Some SCOs were from frogs that had been daily injected with aldosterone (50 μg) over 4 days. After 60 and 120 min of incubation, the amount of autoradiographically established radioactivity was higher in both aldosterone-treated groups of SCOs than in the non-treated group, this difference being much larger at 120 than at 60 min. These data do not provide an unambiguous answer to the question of whether the influence of aldosterone on the secretory activity of the SCO was either stimulatory or inhibitory. The growing difference in radioactivity may have been caused by aldosterone inhibiting the release of newly synthesized radioactive secretory substances from the SCO. Such an inhibition might indicate that aldosterone exerts a negative feedback regulatory influence on the production of an aldosteronotropic factor by the SCO. A similar relationship between the SCO and aldosterone has been proposed to exist in mammals (Palkovits 1968; Dundore et al. 1987).

References

Diederen JHB (1970) The subcommissural organ of *Rana temporaria* L. A cytological, cytochemical, cyto-enzymological and electronmicroscopical study. Z Zellforsch 111:379–403

Diederen JHB (1972) Influence of light and darkness on the subcommissural organ of *Rana temporaria* L. A cytological and autoradiographical study. Z Zellforsch 129:237–255

Diederen JHB (1973) Influence of light and darkness on secretory activity of the subcommissural organ and on growth rate of Reissner's fibre in *Rana esculenta* L. A cytological and autoradiographical study. Z Zellforsch 139:83–94

Diederen JHB (1975a) Influence of ambient temperature on growth rate of Reissner's fibre in *Rana esculenta*. Cell Tissue Res 156:267–271

Diederen JHB (1975b) A possible functional relationship between the subcommissural organ and the pineal complex and lateral eyes in *Rana esculenta* and *Rana temporaria*. Cell Tissue Res 158:37–60

Diederen JHB, Vullings HGB (1980a) Comparison of several parameters related to the secretory activity of the subcommissural organ in European green frogs. Cell Tissue Res 212:383–394

Diederen JHB, Vullings HGB (1980b) A comparative study on the secretory activity of the subcommissural organ in the European green frogs: *Rana esculenta*, *Rana lessonae*, and *Rana ridibunda*. Comp Biochem Physiol 66A:593–597

Diederen JHB, Vullings HGB, Van der Vlist EJ, Legerstee-Oostveen GG (1981) Densitometric measurement of two cellular parameters related to the secretory activity of the subcommissural organ. Microsc Acta 85:25–44

Diederen JHB, Vullings HGB, Rombout JHWM, De Gunst-Schoonderwoerd ATM (1983) The subcommissural organ-liquor fibre complex: the binding of catecholamines to the liquor fibre in frogs of the *Rana esculenta* complex. Acta Zool (Stockh) 64:47–53

Diederen JHB, Vullings HGB, Legerstee-Oostveen GG (1987) Autoradiographic study of the production of secretory material by the subcommissural organ of frogs (*Rana temporaria*) after injection of several radioactive precursors, with special reference to the glycosylation and turnover rate of the secretory material. Cell Tissue Res 248:215–222

Dundore RL, Wurpel JND, Balaban CD, Harrison TS, Keil LC, Seaton JF, Severs WB (1987) Site-dependent central effects of aldosterone in rats. Brain Res 401:122–131

Ermisch A, Sterba G, Mueller A, Hess J (1971) Autoradiographische Untersuchungen am Subcommissuralorgan und dem Reissnerschen Faden. I. Organsekretion und Parameter der Organleistung als Grundlagen zur Beurteilung der Organfunktion. Acta Zool (Stockh) 52:1–21

Gilbert GJ (1956) The subcommissural organ. Anat Rec 126:253–265

Hartwig HG, Van Veen T (1979) Spectral characteristics of visible radiation penetrating into the brain and stimulating extraretinal photoreceptors. J Comp Physiol 130:277–282

Herrera H, Rodríguez EM (1990) Secretory glycoproteins of the rat subcommissural organ are N-linked complex-type glycoproteins. Demonstration by combined use of lectins and specific glycosidases, and by the administration of Tunicamycin. Histochemistry 93:607–615

Hess J, Sterba G (1973) Studies concerning the function of the complex subcommissural organ-liquor fibre: the binding ability of the liquor fibre to pyrocatechin derivatives and its functional aspects. Brain Res 58:303–312

Hess J, Diederen JHB, Vullings HGB (1977) Influence of changes in composition of the cerebrospinal fluid on the secretory activity of the subcommissural organ in *Rana esculenta*. Cell Tissue Res 185:505–514

Meiniel A, Molat J-L, Meiniel R (1988) Complex-type glycoproteins synthesized in the subcommissural organ of mammals. Light- and electron-microscopic investigations by use of lectins. Cell Tissue Res 253:383–395

Oksche A (1962) Histologische, histochemische und experimentelle Studien am Subkommissuralorgan von Anuren (mit Hinweisen auf den Epiphysenkomplex). Z Zellforsch 57:240–326

Olsson R (1958) Studies on the subcommissural organ. Acta Zool (Stockh) 39:71–102

Palkovits M (1968) Karyometrische Untersuchungen zur Klärung der osmo- bzw. volumenregulatorischen Rolle des Subcommissuralorganes und seiner funktionellen Verbindung mit der Nebennierenrinde. Z Zellforsch 84:59–71

Sterba G, Ermisch A, Freyer K, Hartmann G (1967) Incorporation of sulphur-35 into the subcommissural organ and Reissner's fibre. Nature 216:504

Sterba G, Kiessig C, Naumann W, Petter H, Kleim I (1982) The secretion of the subcommissural organ. A comparative immunocytochemical investigation. Cell Tissue Res 226:427–439

Vullings HGB, Diederen JHB (1983) A comparative histochemical and immunocytochemical study on the secretory material in the subcommissural organ of *Rana temporaria* L. Histochemistry 77:405–414

Vullings HGB, Diederen JHB (1985) Secretory activity of the subcommissural organ in *Rana temporaria* under osmotic stimulation. Cell Tissue Res 241:663–670

Vullings HGB, Diederen JHB, Smeets AJM (1983) Influences of light on the subcommissural organ in European green frogs. Comp Biochem Physiol 74A:455–458

Evidence for the Release of CSF-Soluble Secretory Material from the Subcommissural Organ, with Particular Reference to the Situation in the Human

E.M. Rodríguez, P. Jara, H. Richter, H. Montecinos, B. Flández, R. Wiegand, and A. Oksche

General Considerations

According to substantial evidence the secretion of the subcommissural organ (SCO) is primarily released into the cerebrospinal fluid (CSF) (see Sterba 1969; Oksche 1969; Leonhardt 1980; Rodríguez et al. 1987). It is also well established that the bulk of this secretion, upon release into the CSF, becomes densely packed in the form of Reissner's fiber (RF) (Fig. 1). In this respect, the most convincing evidence has been gained from immunocytochemical investigations. Antibodies raised against RF extracts react specifically with RF and the secretory material stored in the SCO proper (Sterba et al. 1982; Rodríguez et al. 1984). On the other hand, antibodies against secretory products extracted (Rodríguez et al. 1985; Meiniel et al. 1988; Karoumi et al. 1990) or purified (Nualart et al. 1991) from the SCO immunoreact with RF.

The question of whether the entire secretion of the SCO released into the ventricular CSF polymerizes to form an insoluble structure such as RF or whether part of this material remains soluble in the CSF appears to be justified. The answer is relevant with respect to functional considerations. The demonstration of a component of SCO secretion soluble in CSF would provide the research of this brain gland with a novel, stimulating perspective. The evidence that will be presented and discussed in this contribution strongly points toward the existence of such CSF-soluble secretion (Fig. 1).

Does the SCO Release Secretion Without Forming a RF?

Immunocytochemical and lectin-histochemical studies of the developing SCO of chick embryos have provided evidence indicating that the SCO starts to secrete into the ventricle at day 7 of incubation (Schoebitz et al. 1986; Karoumi et al. 1990). Despite this early ventricular release of SCO secretion, a RF proper is first found in the sylvian aqueduct only at day 11

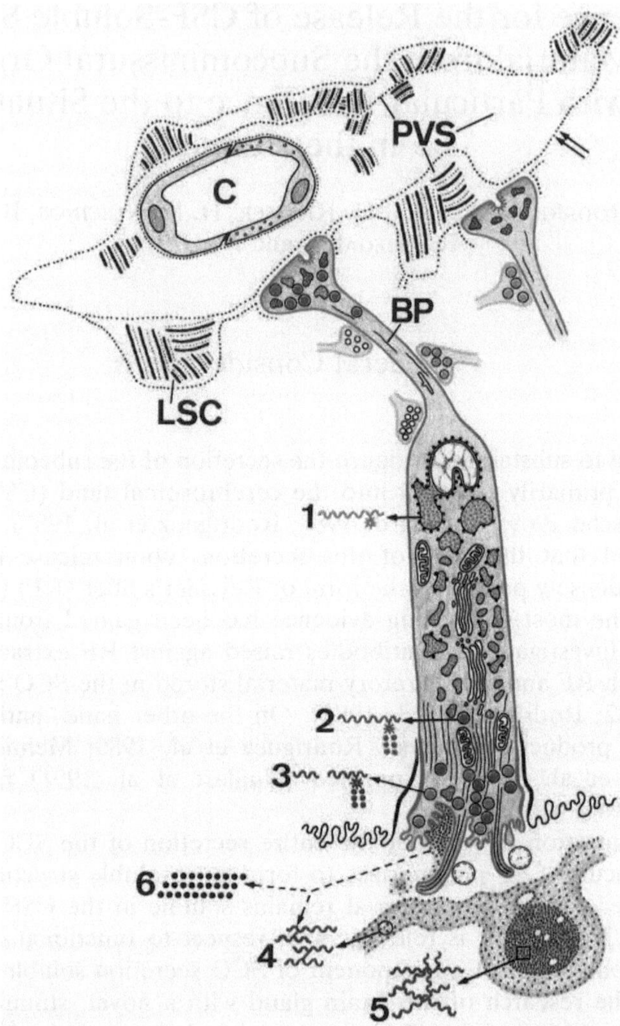

Fig. 1. Schematic representation of a secretory ependymal cell in the rat SCO. *1* N-linked, high mannose-type glycoproteins in RER; *2* precursor forms of complex-type secretory glycoproteins in secretory granules; *3* processed secretory glycoproteins; *4* partially packed secretory glycoproteins forming pre-RF; *5* densely packed glycoproteins forming RF; *6* secretory material remaining soluble in the cerebrospinal fluid. *BP*, Basal process ending on an expanded area of the perivascular space (*PVS*) partially filled with long-spacing collagen (*LSC*); *double arrows*, extensions of the perivascular basal lamina; *C*, capillary. Nerve endings contacting the secretory cell (*yellow structures*)

of incubation (Schoebitz et al. 1986). Thus, during 4 days of the chick embryonic life, the SCO appears to discharge into the ventricular system a material that is not packed into a distinct threadlike structure visible at the light-microscopic level. It appears, therefore, possible that during this limited period of embryonic development the secretion of the SCO dissolves in the CSF.

A similar, although more clear-cut and convincing situation is observed in the rat. As shown by Schoebitz et al. (this volume), polymerized, immunocytochemically detectable SCO secretory material appears in the aqueduct and fourth ventricle no earlier than during the first postnatal day, and in the central canal of the spinal cord only during the second postnatal day. A RF proper is not even visible until the sixth day after birth. However, the SCO appears as a fully developed and highly secretory active structure already during the last six days of embryonic life. This leads to two crucial suppositions: (1) in the rat, during the last third of the embryonic life the entire material secreted by the SCO into the ventricle might be CSF soluble; (2) if the SCO of the rat does play a significant role during embryonic life, as has been repeatedly suggested, its biological function would be maintained by a CSF-soluble secretion and not by the RF.

Two other experimental conditions are known in which a RF is not formed despite the existence of an active SCO.

(1) Rats suffering from postnatal hydrocephalus induced either by injection of kaolin into the cisterna magna or by intracerebral inoculation with Borna disease virus possess an SCO displaying ultrastructural and immunocytochemical features indicative of an enhanced secretory activity (Irigoin et al. 1990; Irigoin 1991). Although fibrous, loosely packed secretory material may occur attached to the walls of the sylvian aqueduct, a RF proper was never observed in these animals. Thus, it appears most likely that, in hydrocephalic rats, the bulk of the SCO secretion released into the CSF remains soluble. Under these circumstances, the CSF-soluble material may correspond to RF material which for unknown reasons (changes in hydrodynamics of the CSF?) has been unable to form a RF.

(2) The SCO of the rat grafted under the kidney capsule becomes organized in such a way that its ependymal secretory cells and the accompanying ciliated ependymal elements line one to three cavities. In our material, these cavities were filled with a flocculent secretory material that during the 3-month period after transplantation was never seen to form a RF-like structure. The small size of the newly formed cavities and the apparent hypersecretory performance of the grafted SCO point toward the possibility that the material released into these cavities may have finally reached the local blood vessels (Rodríguez et al. 1989).

Thus, there are several physiological and experimental conditions under which the release of secretory products from the SCO does not result in the formation of a RF. This conclusion has led us to the following question.

Do RF and CSF-Soluble Material Secreted by the SCO Coexist Under Physiological Conditions?

In order to answer this question we developed a sensitive, noncompetitive solid-phase enzyme immunoassay. A sandwich assay method was used (Tijssen 1987). For this purpose anti-bovine RF sera were obtained in rats and rabbits. Bovine RF extracted in ammonium bicarbonate was applied as a standard preparation. An extract of bovine SCO, also produced by use of ammonium bicarbonate, was applied as an interassay reference preparation. The sensitivity of the assay was in the range of 1–10 ng of RF-immunoreactive material, and the interassay variations averaged 10%.

CSF samples were obtained from the cisterna magna and the lateral ventricle of six rabbits. In five of these rabbits the concentration of RF-immunoreactive material in the cisternal CSF ranged between 9 ng/ml and 23 ng/ml, with an average of 17 ng/ml. The values for the ventricular CSF were higher: range 27–67 ng/ml, average 40 ng/ml. The sixth rabbit presented much higher values of RF-immunoreactive material: 943 ng/ml in the CSF of the lateral ventricle and 292 ng/ml in the cisternal CSF. This latter animal (rabbit no. 1) and one of the other rabbits (rabbit no. 2) displaying lower concentrations of RF-immunoreactive material received an injection into a lateral ventricle of 80 μl of an anti-RF serum (immunoglobulin G, IgG, fraction). Cisternal CSF was collected 1, 3, 5, 7, and 9 days after the antibody injection. In both rabbits the concentration of immunoreactive material in the CSF increased considerably at all postinjection intervals in comparison to the basal values: rabbit no. 1, from 292 ng/ml (basal value) to 5250 ng/ml at the ninth postinjection day; rabbit no. 2, from 23 ng/ml (basal value) to 1555 ng/ml 9 days after the injection of the antibody. A sham-operated rabbit did not show major variations in the concentration of RF material in the cisternal CSF (11 ng/ml, basal value; 19 ng/ml 9 days after the sham operation).

Cisternal CSF (8 ml) was collected from eight different rabbits weighing 2–3 kg (about 1 ml of CSF/rabbit). The total protein concentration, determined for each CSF sample, was 300–400 μg/ml. Aliquots of CSF containing 250 μg/ml of proteins were lyophilized and kept at −5°C. They were prepared and submitted to electrophoresis according to Laemli (1970) (sodium dodecyl sulfate-polyacrylamide gel electrophoresis, SDS-PAGE, 5%–15% gradient) and then to immunoblotting. Blots were immunostained using a serum raised in rabbits against a bovine RF extract. Control tests included (a) immunostaining with an anti-keratin serum developed in rabbits; (b) immunostaining using anti-rabbit IgG; (c) parallel SDS-PAGE of rabbit serum samples and immunoblotting with the three antisera.

In the CSF runs, three bands with an apparent molecular weight of 350, 300, and 250 kDa were stained by the use of the anti-RF serum; they did not react with the anti-keratin and anti-rabbit IgG sera. All three bands were missing from the rabbit blood serum. Until new and more direct evidence is obtained, we suggest that the three above-mentioned compounds correspond to CSF-soluble polypeptides secreted by the SCO. Unfortunately, to date we are unable to present immunoblotting data concerning the secretory compounds existing in the SCO proper and RF of the rabbit. However, the above molecular weights are close to the compounds identified in the bovine RF (Nualart et al. 1991; Hein et al., this volume). Furthermore, since all three compounds present in the CSF immunoreact with an antiserum against RF, it seems likely that they correspond to soluble forms of RF material. Still, the possibility that the three CSF-soluble compounds are chemically related but not identical to those forming RF must be considered.

The compounds detected in immunoblots of cisternal CSF may correspond to the CSF-immunoreactive material detected by use of enzyme-immunoassay (EIA) in similar CSF samples. There is, however, a discrepancy between the immunoblot and EIA findings. The assayable material in the cisternal CSF represents 0.005%–0.01% of the total protein. This implies that in each well of the gel only 12–25 ng of immunoreactive proteins were run. This amount should, in turn, be distributed between the three compounds visualized in the immunoblot. Such small amounts cannot be detected by immunoblotting. The best explanation we have at present is that several epitopes of the secretory polypeptides are reactive to the two anti-RF sera (one developed in rats, the other in rabbits) used with the present sandwich EIA method. This would result in artificial low concentrations of RF-immunoreactive material in the CSF when detected by such an EIA.

Does the Human SCO Secrete a CSF-Soluble Material?

The human SCO reaches its maximum morphological development during the fetal life (Bargmann 1943; Oksche 1956, 1961, 1969; Palkovits 1965). After birth, the SCO undergoes regressive changes, and in the adult human only remnants of the specialized SCO cells can be found (Pesonen 1940; Oksche 1961). According to Palkovits (1965) the human SCO remains highly differentiated during the first postnatal year.

Whether or not the human SCO discharges a secretory material into the CSF has been a matter of controversy. Although the ultrastructural characteristics of the human fetal SCO strongly point toward a distinct secretory activity (Oksche 1969), "stainable" secretory material has never been demonstrated in these SCO cells at a site other than the apical borderline. Furthermore, a representative series of antibodies available in our laboratory, which immunoreact with the SCO of virtually all vertebrate species investigated, do not immunostain the SCO of man and anthropoid apes (Rodríguez et al. 1984, 1990). The absence of RF in the human (see Oksche 1969; Leonhardt 1980) has also contributed toward the uncertainty with respect to the secretory capacity of the human SCO.

In a recent lectin-histochemical study of the human fetal SCO we obtained evidence that this organ secretes glycoproteins with a carbohydrate chain similar to that of the secretions elaborated by the SCO of other species (Rodríguez et al. 1990). We also postulated that the material released by the human SCO into the ventricle does not form a condensate (pre-RF or RF) but becomes soluble in the CSF (Rodríguez et al. 1990). This assumption gained strong support by the detection in the rabbit of RF-immunoreactive, CSF-soluble material (see above) and the fact that in rat embryos a RF is missing although a highly differentiated SCO is present (see above, and Schoebitz et al., this volume).

The attractive possibility that in the human, especially in fetuses and newborn infants, the bulk of secretory material released by the SCO becomes soluble in the CSF led us to the following investigation. CSF was collected weekly (for therapeutic reasons) from 10 children (four of them 1–4 months old) suffering from congenital hydrocephalus. All these CSF samples and serum samples were run in SDS-PAGE and then transferred to nitrocellulose sheets. The blots were processed for concanavalin A (ConA) and wheat-germ agglutinin (WGA) binding, using the lectin-anti-lectin method (see Nualart et al. 1991). The aim of this experiment was to identify glycoproteins which might be present in the CSF and missing (or not detectable) in the serum. Indeed, four WGA-positive bands present in the CSF and missing in the serum were identified; their apparent molecular weights were 235, 150, 73, and 45 kDa. We have regarded them as "CSF-specific glycoproteins." The 235-, 150-, and 45-kDa bands were missing from a pool of CSF samples of adult humans (Fig. 2).

The CSF from a hydrocephalic child that showed the four WGA-positive bands more distinctly was used to run a preparative SDS-PAGE. Material from each of these four bands was utilized to immunize rats. Subsequently, the antisera were used for immunostaining of serial sagittal sections through the brain of 11- to 14-week-old human fetuses (from legal abortions), and of adult rats.

The antisera against the 235- and 73-kDa bands did not immunostain any structure either in the human or rat brain. The antiserum against the 45-kDa band immunoreacted with certain structures of the fetal human brain, but did not react with the rat CNS. In the fetal human brain the anti-45-kDa serum immunostained the following structures: (1) The SCO: About half of the population of the ependymal cells of the SCO displayed a very distinct immunoreaction (Fig. 3a,c). The immunoreactive material appeared in the form of granules embedded in the ground cytoplasm and filling the main portion of the cell body from the nucleus to the apical pole that generally protruded into the ventricle (Fig. 3c). Furthermore, this granular material also marked the slender basal processes projecting to and

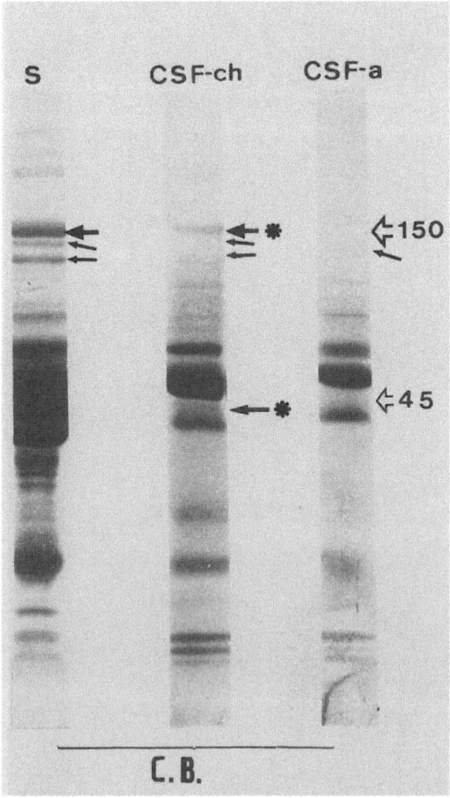

Fig. 2. SDS-PAGE of human blood serum (*S*), cerebrospinal fluid from a 1-month-old child suffering from congenital hydrocephalus (*CSF-ch*), and a pool of CSF from adult humans (*CSF-a*). Coomassie Blue (*C.B.*) staining. Bands with apparent molecular weight of 150 and 45 kDa, also binding wheat-germ agglutinin (WGA), are present in the CSF from the hydrocephalic child (*large black arrows*), and missing in the CSF of adults (*open arrows*). The 150-kDa band is present in the serum, but does not bind WGA. *Small arrows*, Coomassie Blue-positive, WGA-positive bands are present in the three samples, however, at different concentrations. *Asterisks*, CSF bands from the hydrocephalic child used to raise antibodies

penetrating the posterior commissure (Fig. 3b). (2) The pineal organ: The immunoreaction was evenly scattered throughout the pineal cells, giving the entire hollow pineal primordium a brown colour (Fig. 3a). (3) Choroid epithelium: About one third of the population of epithelial cells in the primordia of the telencephalic and myelencephalic choroid plexuses reacted strongly with the antiserum against the 45-kDa band. The glycocalyx of the choroid epithelium displayed a distinct immunoreaction.

The antiserum against the 150-kDa band immunostained, in the fetal human brain, the same structures as revealed by the antiserum against the 45-kDa compound. In the rat, however, the antiserum against the 150-kDa

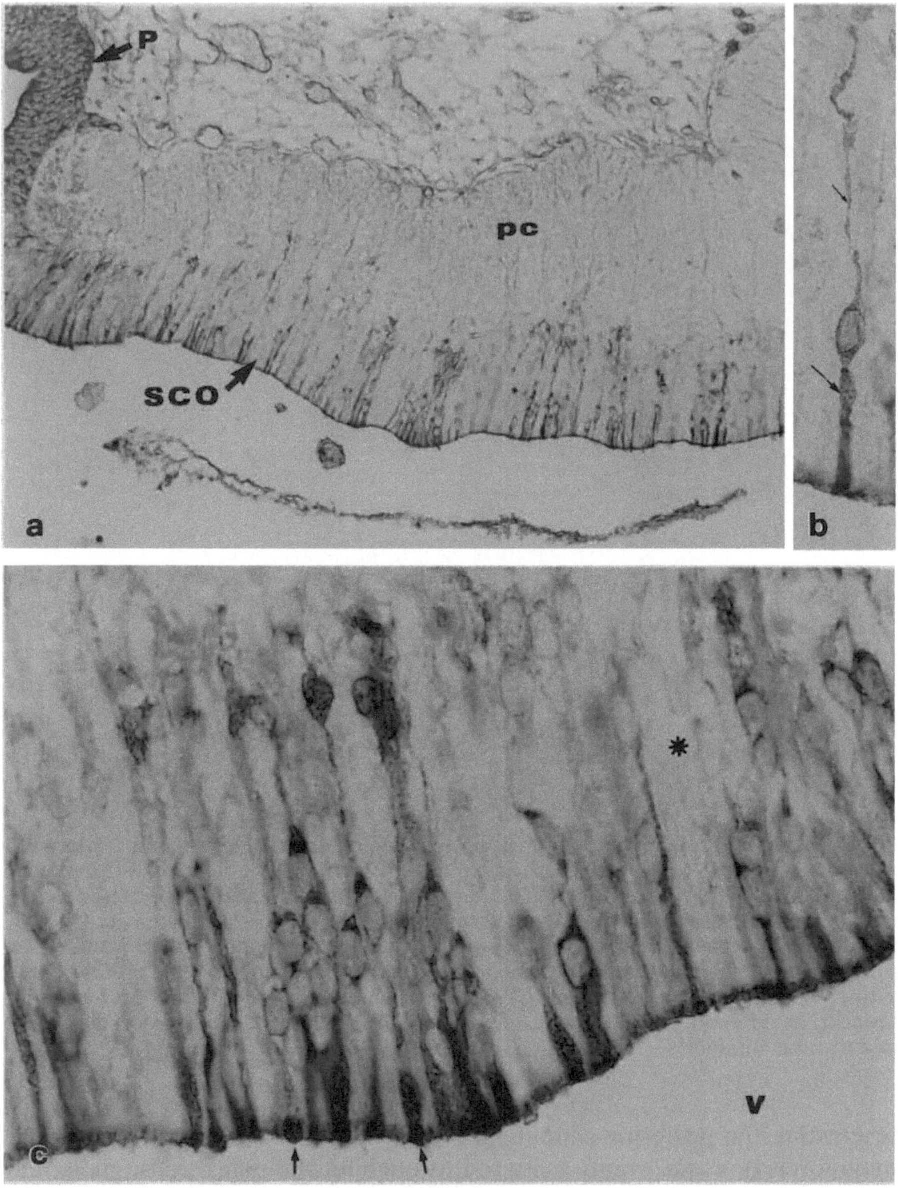

Fig. 3a–c. Sagittal section through the epithalamus of a 13-week-old human fetus immunostained with an antiserum against the 45-kDa band in SDS-PAGE of the CSF from a hydrocephalic child (see Fig. 2). **a** Note immunoreaction in the pineal gland (*P*) and in a population of the ependymal cells of the subcommissural organ (*SCO*); *PC*, posterior commissure. ×130. **b** Higher, detailed magnification of **a** showing a bipolar immunoreactive cell of the SCO, with a distinct ventricular process (*large arrow*) and a fine basal extension (*small arrow*). ×500. **c** Most immunoreactive cells of the subcommissural organ display an apical protuberance (*arrows*) extending into the ventricle (*V*). *Asterisk*, Cells lacking immunoreactive material. ×550

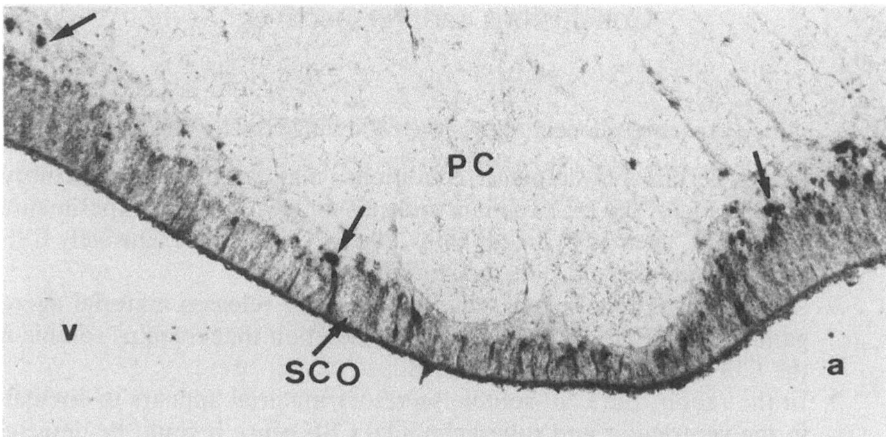

Fig. 4. Sagittal section through the subcommissural organ (*SCO*) of a 3-month-old rat. Immunoperoxidase staining using an antiserum against the 150-kDa band from the CSF of a hydrocephalic child (see Fig. 2). Immunoreactive material in ependymal cells; note large droplets located in the perinuclear region (*arrows*). *V*, Ventricle; *a* sylvian aqueduct; *PC*, posterior commissure. ×250

band stained two structures, namely, the glycocalyx of the choroid epithelium of all choroid plexuses, and the SCO. In the latter, the reaction was especially strong in the area of the large cisternae of the rough endoplasmic reticulum located in the vicinity of the cell nucleus (Fig. 4).

CSF blots from hydrocephalic children treated with the antiserum against the 45-kDa compound showed a positive immunoreaction of the 45-kDa band. In addition, the fraction of IgG with an electrophoretic mobility slightly faster than the 45-kDa compound also immunoreacted with the anti-45-kDa serum. This indicates that this latter antiserum contains antibodies against a component of the IgG molecule. However, an anti-human IgG serum did not stain any structure of the fetal brain. Thus, it seems highly probable that the immunostaining of certain structures of the human fetal brain with the antiserum against the 45-kDa band is a reaction produced by antibodies directed against a 45-kDa polypeptide present in the CSF and probably secreted by the SCO.

Thus, for the first time, a successful immunoreaction against a proteinaceous compound of the fetal human SCO has been performed. The fact that the antibodies exhibiting this reaction were raised against CSF-soluble glycoproteins may be taken as new and strong evidence that the human SCO does possess a secretory activity, resulting in a discharge of a CSF-soluble material. Whether this property is actively shared by some choroid epithelial cells or the immunoreaction in the latter mirrors an uptake of SCO-secretion from the CSF remains open to discussion.

Conclusions and Perspectives

The evidence presented herein allows the following conclusions to be drawn:

1. Under certain physiological conditions, manifest during the embryonic period of the rat and man, and also under particular experimental conditions, such as hydrocephalus, the SCO secretes exclusively CSF-soluble material and does not form an RF.
2. In mature mammals, e.g., rabbits, the SCO releases material aggregating into RF and a fraction of the secretion that remains soluble in the CSF (Fig. 1).
3. In the rabbit, the CSF-soluble secretory material appears to circulate in the ventricular and subarachnoidal CSF, since it could be detected in the CSF from the lateral ventricles and cisterna magna. This raises the question concerning the target site(s) for the secretory products of the SCO circulating in the CSF.
4. Identification in the CSF of glycoproteins that might correspond to SCO secretions appears to provide a useful analytic tool. This approach might enable the detection of SCO secretory products not related to the clearly outlined structure of the RF in the CSF of RF-bearing animals such as the bovine.
5. The human fetal SCO, and probably also that of young infants, appears to secrete exclusively CSF-soluble compounds.
6. The availability of antisera reactive with the human SCO opens new avenues in the investigation of the SCO and CSF of fetuses and infants under normal and defined pathological conditions.

Acknowledgements. This work was supported by Grant I/63 476 from the Volkswagen-Stiftung, Federal Republic of Germany; Grant 91/0956 from FONDECYT, Chile, and Grant S-89-01 from the Dirección de Investigaciones, Universidad Austral de Chile.

References

Bargmann W (1943) Die Epiphysis cerebri. In: Von Möllendorff W (ed) Handbuch der mikroskopischen Anatomie des Menschen, vol IV, part 4. Springer, Berlin, pp 307–502

Irigoin C (1991) Analisis inmunocitoquímico y ultraestructural del órgano subcomisural de ratas hidrocefálicas. Thesis, University Austral of Chile

Irigoin C, Rodríguez EM, Heinrichs M, Frese K, Herzog S, Oksche A, Rott R (1990) Immunocytochemical study of the subcommissural organ of rats with induced postnatal hydrocephalus. Exp Brain Res 82:384–392

Karoumi A, Meiniel R, Croisille Y, Belin MF, Meiniel A (1990) Glycoprotein synthesis in the subcommissural organ of the chick embryo. I. An ontogenetical study using specific antibodies. J Neural Transm (Gen Sect) 79:141–153
Laemmli UK (1970) Cleavage of structural proteins during the assembly of the head of bacteriophage T4. Nature 227:680–685
Leonhardt H (1980) Ependym und circumventriculäre Organe. In: Oksche A, Vollrath L (eds) Neuroglia. I. Handbuch der mikroskopischen Anatomie des Menschen, vol IV, part 10. Springer, Berlin Heidelberg New York, pp 117–665
Meiniel R, Duchier N, Meiniel A (1988) Monoclonal antibody $C_1B_8A_8$ recognizes a ventricular secretory product elaborated in the bovine subcommissural organ. Cell Tissue Res 254:611–615
Nualart F, Hein S, Rodríguez EM, Oksche A (1991) Identification and partial characterization of the secretory glycoproteins of the bovine subcommissural organ-Reissner's fiber complex. Evidence for the existence of two precursor forms. Mol Brain Res 11:227–238
Oksche A (1956) Funktionelle histologische Untersuchungen über die Organe des Zwischenhirndaches der Chordaten. Anat Anz 102:404–419
Oksche A (1961) Vergleichende Untersuchungen über die sekretorische Aktivität des Subkommissuralorgans und den Gliacharakter seiner Zellen. Z Zellforsch 54:549–612
Oksche A (1969) The subcommissural organ. J Neuro Visc Relat [Suppl] 9:11–139
Palkovits M (1965) Morphology and function of the subcommissural organ. Stud Biol Hung 4:1–105
Pesonen A (1940) Über das Subcommissuralorgan beim Menschen. Acta Soc Med Fenn Duodecim, Ser A 22:79–114
Rodríguez EM, Oksche A, Hein S, Rodríguez S, Yulis R (1984) Comparative immunocytochemical study of the subcommissural organ. Cell Tissue Res 237:427–441
Rodríguez EM, Oksche A, Rodríguez S, Hein S, Peruzzo B, Schoebitz K, Herrera H (1987) The subcommissural organ and Reissner's fiber. In: Gross PM (ed) Circumventricular organs and body fluids, vol II. CRC Press, Boca Raton, pp 1–41
Rodríguez EM, Rodríguez S, Schoebitz K, Yulis CR, Hoffmann P, Manns V, Oksche A (1989) Light- and electron-microscopic investigation of the rat subcommissural organ grafted under the kidney capsule, with particular reference to immunocytochemistry and lectin histochemistry. Cell Tissue Res 258:449–514
Rodríguez EM, Garrido O, Oksche A (1990) Lectin histochemistry of the human fetal subcommissural organ. Cell Tissue Res 262:105–113
Rodríguez S, Yulis CR, Siegmund I, Rodríguez EM (1985) Reissner's fiber and the wall of the central canal in the lumbo-sacral region of the bovine spinal cord. Cell Tissue Res 240:649–662
Schoebitz K, Garrido O, Heinrichs M, Speer L, Rodríguez EM (1986) Ontogenetical development of the chick and duck subcommissural organ. An immunocytochemical study. Histochemistry 84:31–41
Sterba G (1969) Morphologie und Funktion des Subcommissuralorgans. In: Sterba G (ed) Zirkumventriculäre Organe und Liquor. Internationales Symposium, Schloss Reinhardsbrunn 1968. Fischer, Jena, pp 17–32
Sterba G, Kiessig C, Naumann W, Petter H, Kleim I (1982) The secretion of the subcommissural organ. A comparative immunocytochemical investigation. Cell Tissue Res 226:427–439
Tijssen P (1987) Practice and theory of enzyme immunoassay. Elsevier, Amsterdam

The Subcommissural Organ In Vitro

W. Lehmann and G. Sterba

Introduction

Morphological studies have revealed that the mammalian SCO is composed of an ependymal and a hypendymal division (see Oksche 1961). The secretory capacity of the SCO cells has been a central subject of most work in this field, in an attempt to gain more insight into its functional role. Möller (1978) had already noted that tissue culture may be a useful tool for studying the secretory activity of the SCO in vitro. With this aim in mind we have introduced tissue culture techniques for the SCO, in order to find conditions under which SCO tissue is capable of maintaining stable morphological and secretory properties for long periods in vitro.

In Vitro Procedures

In vitro studies were performed mainly using mouse SCO cultured on a feeder layer of embryonic mouse fibroblasts (Lehmann et al. 1989) or rat and bovine SCO culture without any feeder layer (Lehmann and Sterba 1989; Lehmann et al. 1989). Further studies included culture of rat SCO on feeder layer, or of rat choroid plexus and mouse SCO maintained without a feeder layer. With the exception of one experimental design, when postnatal rat SCO was cultured without feeder layer, only tissues of adult animals were examined. Tissue explants were cultured in Dulbecco's modified Eagle medium (SERVA) containing 10% fetal calf serum (Staatliches Institut für Immunpräparate und Nährmedien, Berlin) (DMEM) for mouse and rat, or in Wissler BM86 (BM86), a serum-free defined culture medium (Boehringer). The conditions applied in vitro and the preparation of the explants, with a maximum edge length of 1 mm, were described by Lehmann et al. (1989; Lehmann et al. 1993, unpublished results) and Lehmann and Sterba (1989).

Examination of Living Explants

Living explants were examined weekly using an inverted microscope. Immediately after explantation, SCO tissue of all three species showed a strong ciliary activity persisting over longer periods in vitro.

Freshly explanted SCO tissue adhered quickly to the bottom of the culture vessel only if a fibroblast feeder layer was used. This could be attributed to fibronectin deposition at the fibroblast surface (McDonald 1988).

Although the absolute size of the original explants did not change extensively, all explants developed a more or less characteristic globular shape during the first 30 days in vitro (div) except rat choroid plexus tissue. In some cultures of rat and bovine SCO single cells lost contact with the explant and migrated to the bottom of the culture vessel. Some of those bore long processes and may be regarded as fibrous astrocytes (Raff et al. 1983). In feeder layer cultures of adult mouse SCO and in rat choroid plexus cultures without feeder layer, disaggregation of the explants was more common, but these cells never exhibited long processes. The possibility has to be kept in mind that migrating cells may form island-like subcolonies (Lehmann et al. 1989) at some distance from the original explants; these colonies have diameters of 200–300 μm in feeder layer cultures of mouse SCO (see below).

All explants appeared transparent, and only after longer periods in vitro did some dark spots develop, indicating partial tissue degeneration. These dark spots occurred preferentially in rat SCOs after approximately 40 div and in choroid plexus tissue after about 21 div. Without feeder layers, however, SCO tissue of all species examined survived for many weeks in vitro. Only the feeder-layer cultures (10 div) and adult rat plexus cultures without feeder layer (30 div) displayed shorter periods of survival under the conditions applied.

Histology

Regardless of the alterations in explant shape (see above), the typical histological properties of the SCO remained unchanged. It was possible to distinguish SCO ependyma and, to a certain extent, the subependymal tissue at any given stage of culture. In particular, the structural stability of SCO ependyma consisting of tall epithelial cells was checked after 96 div (not shown). Furthermore, SCO ependymocytes exhibit apical cilia and microvilli, with their nuclei occupying a more basal portion of the cell (Figs. 1, 2).

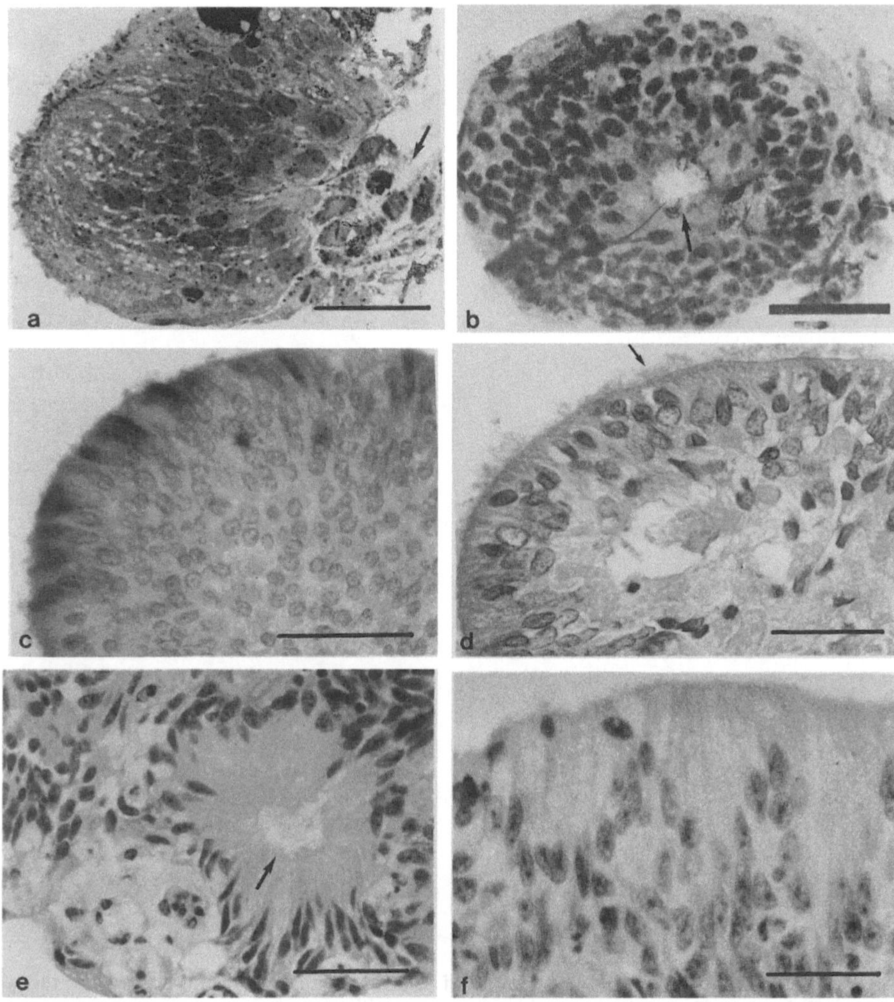

Fig. 1a–f. Histological examination of SCO explants in vitro. **a** Toluidine blue-stained median tissue section of a bulblike eversion of a mouse SCO explant cultured on embryonic mouse fibroblast feeder layer, after 10 div. The basal aspect of the ependymocytes is bordered by loosely arranged subependymal tissue (*arrow*). *Bar*, 50 µm. **b** Haemalum-eosin-stained, whole-mount preparation of a subcolony after 10 div. *Arrow* indicates a small central cavity. *Bar*, 30 µm. **c** Tissue section of a rat SCO explant after 21 div in DMEM, immunostained (peroxidase-anti-peroxidase technique) with an antiserum raised against Reissner's fiber and counterstained with haemalum-eosin. Only few ependymal cells in the outer cell layer are strongly immunostained. *Bar*, 35 µm. **d** Tissue section of a rat SCO explant after 96 div in BM86 culture medium stained with haemalum-eosin. Only a few nuclei are stained in the inner cell mass. *Arrow* indicates cilia and microvilli. *Bar*, 25 µm. **e** Tissue section of bovine SCO explant cultured for 14 div in BM86. Ependymal tissue borders the surface of the lumen of a crypt (*arrow*). *Bar*, 50 µm. **f** Ependymal tissue bordering the external surface of a bovine SCO explant after 56 div in BM86. *Bar*, 20 µm

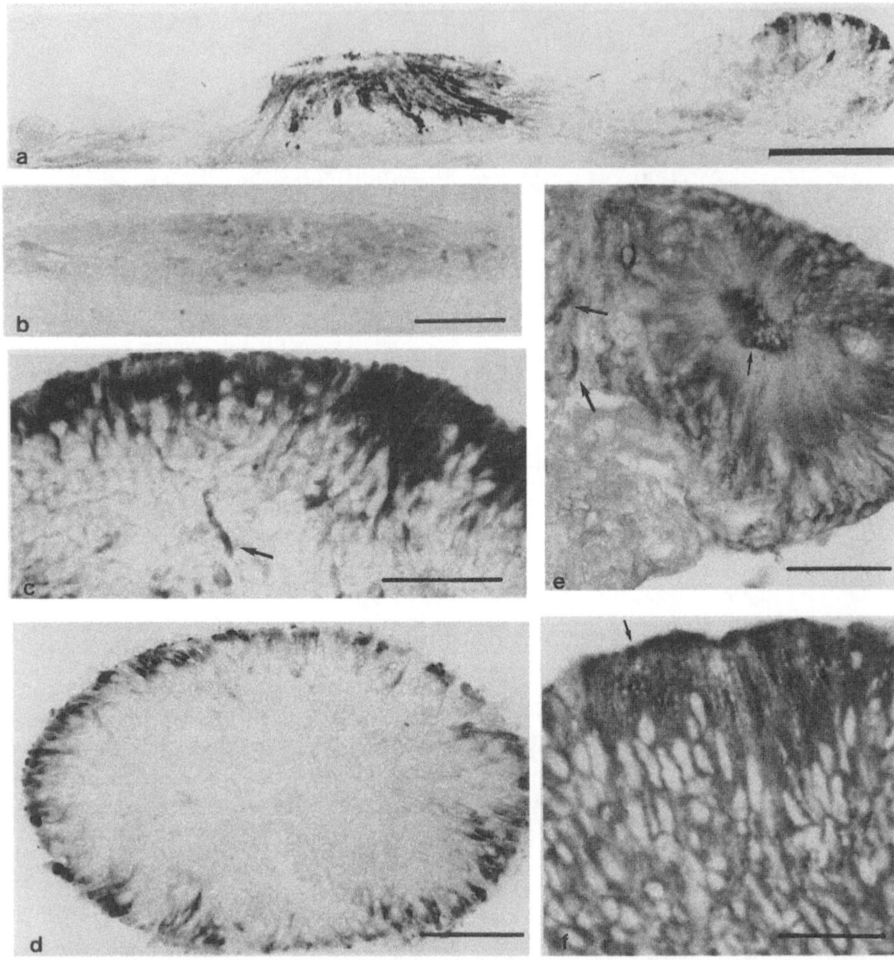

Fig. 2a–f. Immunocytochemical staining of SCO explants with an antiserum raised against Reissner's fiber using peroxidase-anti-peroxidase technique. **a** Explanted mouse SCO cultured on a fibroblast feeder layer in DMEM for 10 div. Immunoreactive material covers the apical cytoplasm of cells of an eversion (*right*) and a solid explant (*center*). Round profiles of cells migrating on the feeder layer remained unstained (*left*). *Bar*, 100 µm. **b** Paramedian cross section of a subcolony (mouse SCO). Immunoreactive material is localized mainly in the central portion. *Bar*, 50 µm. **c**, **d** Tissue section of a rat SCO after 21 (**c**) or 96 (**d**) div in BM86. Strong immunostaining is restricted to the ependymocytes after 21 and 96 div and to hypendymocytes (*arrow*) after 21 div. *Bars*, 50 µm. **e**, **f** Tissue section of a bovine SCO explant after 14 (**e**) and 56 (**f**) div in BM86. Immunoreactive material covers ependymal tissue bordering the external surface of the explant (*arrow* in **f**) and the lumen of a crypt (*small arrow* in **e**). *Large arrows* in **e** indicate individual hypendymocytes. *Bars*, 50 µm (**e**); 20 µm (**f**)

The arrangement of the ependyma in the explants depends on the culture technique (feeder layer or not) and on the species studied. In mouse SCO culture, the SCO ependyma could be demonstrated in the bulblike everted parts (Fig. 1a) but also in the remainder of the original explant, i.e., in the so-called eversions and solid explants (Lehmann et al. 1989). Subependymal tissue underlying the ependyma adhered basally to the feeder layer and appeared in some portions to be loosely arranged (Fig. 1a). The histoarchitecture of the subcolonies differed from the SCO ependyma proper in that the arrangement of their tall epithelial cells was altered considerably (see whole-mount preparation in Fig. 1b). The apices of the subcolony cells were directed centrally to form a small cavity (arrow in Fig. 1b) whereas the cell bases, and thus also the nuclei, occupied a more peripheral position. The height of the subcolonies measured about 30–50 μm (Fig. 2b).

Explanted rat SCOs which are already globular show an outer layer of ependymal cells often completely covering an inner cell mass; the latter corresponds to the subependymal tissue and often contains hypendymocytes. After 40 div, degeneration clearly affected the cells of the inner cell mass as indicated by the loss of haemalum stainability of their nuclei (compare Fig. 1c,d); this change increased during the course of cultivation.

In bovine SCO explants, the external surface was bordered by subependymal and ependymal tissue (Fig. 1e,f). Invaginations of the explant surface, the so-called crypts (Sterba et al. 1982), were also covered by ependymal cells (Fig. 1e, arrow). The subependymal tissue was intermingled with residues of capillaries (for details, see Lehmann et al. 1992, unpublished results) and sometimes with tissue of the posterior commissure. Hypendymocytes could only be identified in haemalum-eosin-stained tissue sections if they were arranged in hypendymal rosettes, characteristic structures of the normal noncultured bovine SCO (Talanti 1958).

Immunocytochemistry

Explants of all three species were studied by light-microscopic immunocytochemistry using the PAP-procedure according to Sternberger et al. (1970) and an antiserum raised against Reissner's fiber according to Sterba et al. (1981).

Mouse SCO explants cultured exclusively in DMEM for about 10 div displayed only a weak reaction in immunocytochemical preparations. Finely dispersed immunoreactive material was located mostly in the apical cytoplasm of the ependymocytes of eversions, solid explants, and subcolonies (Fig. 2a,b). Immunoreactive hypendymocytes were rarely found, and then exclusively in solid explants. Single cells migrating from the explants remained unstained (Fig. 2a).

Rat SCOs were studied immunocytochemically after culture in DMEM or BM86 for 21 div or in BM86 for 84 div. In any given preparation at least some cells of the outer layer were immunostained. Then DMEM was used as culture medium; a faint immunoreaction product covered only a few cells (Fig. 1c). Cells of the inner cell mass were never stained. In contrast, BM86 seems to provide better conditions, since besides nearly all cells of the outer layer and some cells of the inner cell mass, the hypendymocytes were also immunostained (Fig. 2). After 84 div only cells of the outer layer were stained immunocytochemically, probably because the inner cell mass had degenerated (see above) (Fig. 2d). Dense immunoreactive material was restricted mainly to the supranuclear and apical cytoplasm (Fig. 2c,d). In the perinuclear and basal cytoplasm faint, finely granulated staining was found. No extracellular representation of immunoreactive material could be observed.

Explants of bovine SCO were studied by immunocytochemistry after 14 and 70 div in BM86 culture medium. After 14 and 70 div, ependymal cells bordering the external explant surface or the lumen of a crypt were immunolabeled almost throughout the cytoplasm (Fig. 2e,f). In addition to the ependyma, hypendymal cells, which were sometimes arranged in rosettes, were also immunostained. Extracellular immunoreactive material was clearly demonstrated in the lumen of the crypts (Fig. 2e).

Conclusions

SCO tissue survives well for about 100 div, a period much longer than that observed for choroid plexus tissue of adult rats. Culture of adult choroid plexus tissue has been reported for shorter periods of about 20 div (Meller and Wagner 1968); however, further studies are needed before general conclusions can be drawn.

Feeder layer cultures are not suitable for long-term studies; they are limited because X-ray-inactivated fibroblasts lose their contact to the floor of the culture vessel. In addition, feeder cells may influence the in vitro differentiation in an unknown way, and for this reason we also tended to avoid serum-supplemented culture medium. Since subcolonies develop as a completely new ependymalike tissue in mouse-SCO feeder-layer cultures, this model may be generally appropriate for studying ependymal or epithelial differentiation in vitro. In this context it may be an effective tool in addition to studying the alteration of SCO tissue after grafting beneath the kidney capsule (Rodríguez et al. 1989) or within the brain ventricles (Didier-Bazes et al., this volume).

Tissue culture of the SCO of different adult mammals, with the use of the serum-free defined medium BM86, proved to be a simple procedure;

a stable condition of the explanted tissue, isolated from physiological regulation, could be observed in vitro over long periods. Even species-specific features in the structural arrangement of the ependyma in situ, such as the crypts of the bovine SCO, were preserved. The most drastic rearrangement of the SCO ependyma was found in the formation of a closed outer cell layer in rat SCO explants after long periods in vitro. The formation of this ependymal layer probably leads to an insufficient nutrient supply to the inner cell mass, which degenerates in the course of culture. Insufficient nutrient supply has been considered to cause the death of central tissue portions in the avian pineal organ (*Passer domesticus*); this phenomenon was, however, observed already after only a few days in vitro (Möller 1978).

Especially the ependymal cells of all three species were clearly immunoreactive even after long periods in vitro, indicating continued synthesis of SCO secretion.

References

Chouaf L, Didier-Bazes M, Hardin H, Aguera M, Fevre-Montagne, Voutsinos B, Belin MF (1991) Developmental expression of glial markers in ependymocytes of the rat subcommissural organ: role of the environment. Cell Tissue Res 266:553–561

Isomäki AM, Kivalo E, Talanti S (1965) Electron-microscopic structure of the subcommissural organ in the calf (*Bos taurus*) with special reference to the secretory phenomena. Ann Acad Sci Fenn (Med) 111:1–64

Lehmann W (1990) Untersuchungen zur Gewebekultur des Subcommissuralorgans verschiedener Säugerspecies in vitro. Dissertation, University of Leipzig

Lehmann W, Sterba G (1989) Tissue culture of rat subcommissural organ in vitro. Biomed Res [Suppl 3] 10:11–18

Lehmann W, Sterba G, Wobus AM (1989) Primary culture of the mouse subcommissural organ in vitro. Acta Zool (Stockh) 70:199–203

Lösecke W, Naumann W, Sterba G (1984) Preparation and discharge of secretion in the subcommissural organ in the rat. Electron-microscopic immunocytochemical study. Cell Tissue Res 235:201–206

Lösecke W, Naumann W, Sterba G (1986) Immuno-electron-microscopic analysis of the basal route of secretion in the subcommissural organ of the rabbit. Cell Tissue Res 244:449–456

McDonald JA (1988) Extracellular matrix assembly. Annu Rev Cell Biol 4:183–207

Meller K, Wagner HH (1968) Vergleichende elektronenmikroskopische Untersuchungen des Plexus choroideus der Maus in vivo und in vitro. Z Zellforsch 91:507–518

Möller W (1978) Circumventricular organs in cell culture. Adv Anat Embryol Cell Biol 54:3–95

Oksche A (1961) Vergleichende Untersuchungen über die sekretorische Aktivität des Subkommissuralorgans und den Gliacharakter seiner Zellen. Z Zellforsch. 54:549–612

Raff MC, Abney ER, Cohen J, Lindsay R, Noble M (1983) Two types of astrocytes in cultures of developing rat white matter: differences in morphology, surface gangliosides and growth characteristics. J Neurosci 3:1289–1300

Rodríguez EM, Rodríguez S, Schoebitz K, Yulis CR, Hoffman P, Manns V, Oksche A (1989) Light and electron-microscopic investigation of the rat subcommissural organ graft under the kidney capsule with particular reference to immunocytochemistry and lectin histochemistry. Cell Tissue Res 258:499–514

Sterba G, Kleim I, Naumann W, Petter H (1981) Immunocytochemical investigation of the subcommissural organ in the rat. Cell Tissue Res 218:659–662

Sterba G, Kiessig C, Kleim I, Naumann W, Petter H (1982) Immunzytochemische Untersuchungen an den Subkommissuralorganen von Rind, Ziege und Schaf. Biol Zbl 101:241–248

Sternberger LA, Hardy PH, Cuculis JJ, Meyer HG (1970) The unlabeled antibody enzyme method of immunohistochemistry; preparation of properties of soluble antigen-antibody complex (horseradish peroxidase-antiperoxidase) and its use in identification of spirochetes. J Histochem Cytochem 18:315–333

Talanti S (1958) Studies on the subcommissural organ in some domestic animals. Ann Med Exp Fenn [Suppl 9] 36:1–97

V. Relation to Other Circumventricular Organs: Conceptual Aspects

Circumventricular Organs and Modulation in the Midsagittal Plane of the Brain*

H. Leonhardt and B. Krisch

General Aspects

The phylogenetically persistent circumventricular organs (CVOs) develop in the midsagittal plane at those sites of the prosencephalic vesicle where the prospective brain will remain thin-walled. The *anlagen* of the CVOs develop in this plane free from influences from the telencephalic vesicles and their developmental transformations. Only the choroid plexus of the lateral ventricles follow the developing telencephalon, extending in the form of a ram's horn.

Frequently, the CVOs are characterized as parts of the brain that lack a blood–brain barrier (BBB). Such a generalized statement is incorrect. Only the neurohemal regions of the CVOs are devoid of a BBB. The term "neurohemal region" implicates the existence of two compartments or milieus. One of them belongs to the brain and is dominated by the cerebrospinal fluid (CSF). The other one is a hemal milieu as it is found inside the capillaries and, in the case of a lacking BBB, at least in the pericapillary spaces and their vicinity. Consequently, the idea of a borderline between both compartments arises. Such borderline structures separate the hemal milieu within the CVO from its remaining areas that may belong to the CSF compartment. Whenever a CVO is the source of neuronal projections, a CSF-dominated compartment containing the perikarya of the projecting nerve fibers must be expected.

Hence, a CVO is a meeting point for neuronal and humoral in- and outputs. The neuronal afferents and efferents may belong to the central or peripheral nervous system; the humoral inputs or outputs may be blood borne or, at least theoretically, CSF borne. Different pathways are used for information transfer. Attention should be paid to the differing extension of the neurohemal regions in some CVOs and the corresponding functional differences. In relation to the extension of neurohemal regions it can be stated that humoral inputs and outputs require a more excessive neurohemal region than neuronal afferents and efferents. This does, however, not contradict the fact that a high number of neuronal afferents and efferents are connected with an extended CSF-dominated part of the CVO.

---

* Dedicated to Prof. Dr.med Dr.rer.nat. Dr.h.c. J.-H. Scharf on the occasion of his 70th birthday.

Special Features

In the following, these statements will be examined for some of the CVOs considering their major afferent and efferent connections to other central nervous structures. It will be shown that the offspring of the afferents and the target structures of the efferents are constituents of the midsagittal plane. The choroid plexus will be excluded from the discussion because even in the adult choroid plexus its pallial component is reduced to a single epithelial layer dominated by the hemal milieu. The choroid plexus contains only a very few axonal endings probably innervating the secretory choroid epithelia.

A word about terminology in advance: the terms "afferents" or "efferents" will always be related to the particular CVO discussed. Neuronal afferents invade the organ and innervate it; neuronal efferents leave the organ and their respective perikarya may be located within or outside the CVO. Humoral afferents or inputs penetrate the organ from the blood via a neurohemal region or possibly also via the CSF. Humoral efferents or outputs leave the organ via the blood from the neurohemal region or perhaps also via the CSF. The terms "afferents" or "efferents" will be used exclusively for neuronal projections. If humoral afferents and humoral afferents are involved, they will be particularly emphasized and indicated as "inputs" and "outputs".

1. The *organon vasculosum laminae terminalis* (OVLT) is, like the subfornical organ (SFO), a receptor site for blood-borne angiotensin II (A II) (see Lind and Ganten 1990). The organ is involved in the regulation of water–electrolyte balance and arterial blood pressure as well as in the control of luliberin secretion (Piva et al. 1982). Additionally, the OVLT acts as a mediator for the fever response (Blatteis et al. 1983). The OVLT is composed of two functionally different neurohemal regions, separated from one another by tanycytic processes and leptomeningeal cells. Even though both regions possess fenestrated capillaries surrounded by wide perivascular spaces, they differ in their permeability for horseradish peroxidase. In

Table 1. Afferent (A) and efferent (E) connections of the OVLT (McKinley and Oldfield 1990)

Major connection	*Hypothalamic nuclei*
A + E, subfornical organ	A, dorsomedial nucleus
Preoptic nuclei	A, ventromedial nucleus
A + E, lateral preoptic area	E, anterior hypothalamic nuclei
A + E, median preoptic area	E, paraventricular nucleus
Distant connections	E, supraoptic nucleus
A, central gray	
A, locus coeruleus	

a subependymal position a few neuronal perikarya are embedded in a CSF-dominated network of tanycytic and glial processes. The neuronal connections of the OVLT are part of the network within the lamina terminalis. The major *neuronal afferents* originate from the SFO (see Lind and Ganten 1990), the lateral preoptic and lateral hypothalamic areas, the anterior, dorsomedial and ventromedial hypothalamic nuclei, and from the locus coeruleus and central gray. *Humoral input* is given by AII via the circulation. *Neuronal efferents* project to the dorsomedial, supraoptic, and paraventricular nuclei (Luiten et al. 1987). *Humoral output* involves several peptides, e.g., vasopressin, luliberin, and, in particular, somatostatin. However, their target areas are unknown (Table 1).

2. The SFO (Fig. 1) is involved in balancing body fluid. Its neurohemal region consists of the pericapillary spaces of the vascular network. They are particularly wide near the ventricular surface of the organ and decrease towards its center. Among the capillaries tanycytes and other glia cells engulf neuronal perikarya (see Dellmann and Simpson 1979), which means the neuropil surrounding the neurohemal regions is CSF-dominated. *Neuronal afferents* to the SFO are projections from the nucleus triangularis and the median preoptic nucleus (Oldfield et al. 1986). Other neuronal afferents arise from the medial septal nucleus, the medial preoptic area, anterior

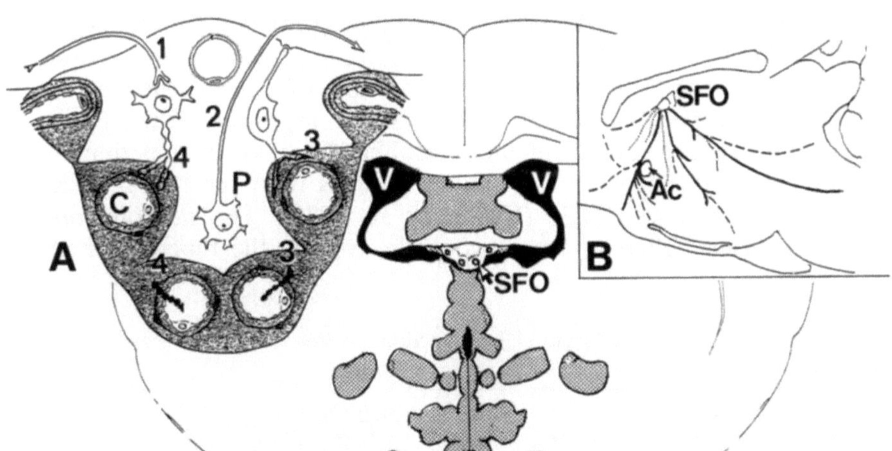

Fig. 1A,B. General organization of the neuronal connections of the subfornical organ (*SFO*), representative also for other CVOs. Coronal section through the rat brain. The *shaded areas* represent diencephalic and brain-stem nuclei in median position (*V*, ventricle) projected onto the coronal plane of the SFO. **A** Coronal section of the SFO with its limited neurohemal region (*shaded area*), fenestrated capillaries (*c*) and CSF-dominated region (*clear*) containing neuronal perikarya (*P*): *1*, neuronal afferents; *2*, neuronal efferents; *3*, humoral inputs; *4*, humoral outputs. **B** Neuronal connections of the subfornical organ (*SFO*). Midsagittal section through the rat diencephalon and brain stem. The anterior commissure (*Ac*) separates precommissural and postcommissural connections; *solid lines*, reciprocal connections; *broken lines*, efferents; *dotted lines*, afferents hypothalamus, and thalamic nuclei, including the thalamic paraventricular nucleus (see Lind and Ganten 1990). *Humoral inputs* are indicated by numerous receptors for A II; activation by blood-borne A II leads to the release of vasopressin from the hypophysial neural lobe and stimulates thirst and drinking. Binding sites for calcitonin (Rouleau et al. 1984), somatostatin (Patel et al. 1986), and atrial natriuretic peptide (Quirion et al. 1984) have been demonstrated. *Neuronal efferents* extend in a precommissural pathway towards the median preoptic nucleus, rostrally passing anterior commissure. The major bundle continues to the OVLT and to the supraoptic, the suprachiasmatic, and the arcuate nuclei (Gruber et al. 1987). In a postcommissural pathway, efferents travel with the medial corticohypothalamic tract to innervate hypothalamic areas including the paraventricular nucleus (see Lind and Ganten 1990). The lateral hypothalamus, the zona incerta, septal region, and an area of the prefrontal cortex are regions also innervated by the SFO. Its different neuronal efferents are involved in the various components of drinking behavior (see Lind and Ganten 1990). For *humoral outputs* luliberin and somatostatin are suggested to be released into the blood of the choroid plexus (Krisch and Leonhardt 1980) (Table 2).

3. The *neurohypophysis* plays a decisive role in the regulation of endocrine systems by release of neurohormones (Bargmann 1971). It comprises two neurohemal regions, the median eminence (proximal neurohypophysis) and the infundibular process or posterior pituitary (distal neurohypophysis) (Scharrer and Scharrer 1954). The extension of its neurohemal region includes the complete ventricular floor as far as it is invaded by portal vessels. The lateral borders of the neurohemal region are functional ones isolating the CSF milieu of the arcuate nucleus from the hemal milieu of the median eminence (ME) (Krisch et al. 1978). The neurohypophysis receives *neuronal afferents* from two pathways: (1) the tuberoinfundibular

Table 2. Afferent (A) and efferent (E) connections of the subfornical organ (McKinley and Oldfield 1990)

Septal and limbic nuclei A, medial septal nucleus A, triangular nucleus E, lateral septal nucleus, intermediate part E, bed nucleus of the stria terminalis E, substantia innominata *Precommissural pathway* A + E organum vasculosum laminae terminalis A + E median preoptic nucleus E suprachiasmatic nucleus E supraoptic nucleus E medial preoptic nucleus	*Postcommissural pathway* *Hypothalamic areas* A + E, lateral hypothalamic area A + E, dorsomedial nucleus A + E, zona incerta A, anterior hypothalamic area E, paraventricular nucleus E, arcuate nucleus *Thalamic and raphe nuclei* A + E, nucleus reuniens thalami A + E, median raphe nucleus E, paraventricular thalamic nucleus E, dorsal raphe nucleus E, central medial thalamic nucleus

Table 3. Afferent connections of the median eminence (McKinley and Oldfield 1990)

Supraoptic nucleus
Medial preoptic nuclei
Paraventricular nucleus
Ventromedial nucleus
Arcuate nucleus

tract and (2) the supraopticohypophyseal tract. The ME is composed of an internal and external zone where the axons of the tuberoinfundibular tract terminate. The neurons of the tuberoinfundibular tract are located outside the neurohypophysis in CSF-dominated parvocellular hypothalamic regions, including the arcuate nucleus, the paraventricular and the periventricular region, and the ventromedial hypothalamus (see Krisch 1980). The supraopticohypophyseal tract has its origin in the magnocellular perikarya of the supraoptic and paraventricular nuclei. Their axons transporting vasopressin and oxytocin pass the ME to end in the distal part of the neurohypophysis (see Bargmann 1971). *Humoral inputs* and *neuronal efferents* of the neurohypophysis are not known for certain. *Humoral outputs* are established by the fibers of the above-mentioned tracts releasing neurohormones from the ME into the portal vessels for regulation of adenohypophyseal functions, as well releasing vasopressin and oxytocin from the posterior pituitary into the general circulation (Table 3).

4. By melatonin synthesis, the *pineal gland* conveys information about light–dark/day–night cycles and in some species plays a role in regulation of reproductive activities. Although the pineal is included in the group of CVOs, it is not one of its typical representatives (see Vollrath 1981). In phylogenetic terms, there are considerable species differences (Oksche 1965; Oksche et al. 1971). The neurohemal region is represented by wide perivascular spaces (Vollrath 1981) organized in a complex manner. Similar to the situation in the SFO and ME, the number of nerve terminals and pinealocyte endings is apparently subject to functional changes. Distinct species differences exist with regard to the leakiness of pineal endothelia (Møller et al. 1978). *Neuronal afferents* to the pineal arise from the superior cervical ganglion. These sympathetic fibers enter the pineal at its posterior pole travelling either perivascularly or in association with the nervus conarii (Kenny 1985). Afferents from the retina are indirectly transmitted via the retinohypothalamic tract to the suprachiasmatic nucleus and finally via the brain stem to the sympathetic centers of the spinal cord. Additional central neuronal afferents originate from the habenula and the hypothalamus (Korf and Møller 1985). *Humoral inputs* are suggested but not definitely proven. *Neuronal efferents* are unknown in mammals. *Humoral output* consists in secretion of melatonin into the blood. Primarily, melatonin acts on hypothalamic structures (Cardinali 1974). Interactions between the pineal and the magnocellular neurosecretory nuclei have been described (Lind and

Table 4. Afferent connections of the pineal organ (McKinley and Oldfield 1990)

Superior cervical ganglion
Medial habenular nucleus
Lateral habenular nucleus
Paraventricular hypothalamic nucleus

Ganten 1990) (Table 4). Vasopressin-immunoreactive and other peptidergic nerve fibers have been observed in the pineal organ (see Leonhardt 1980).

5. The *area postrema* (AP) of most mammalian species (except rat and rabbit) is located on both sides of the fourth ventricle where it merges into the central canal. It plays an important role in mediating nausea and vomiting in response to emetic substances. Via the AP's neurohemal region vasopressin gets access to structures participating in the baroreflex (Phillips et al. 1988). The AP is part of the circuitry for cardiovascular regulation and water and energy balance. In rats, lesions of the organ reduce appetite and consequently body weight (see Lind and Ganten 1990). The neurohemal region of the AP, bordered by tanycytes, extends into wide perivascular spaces of capillaries and sinusoidal vessels. Between the circumscribed neurohemal regions neuronal somata and astrocytic cells are situated in a CSF-dominated neuropil (see Leonhardt 1980). *Neuronal afferents* to the AP originate in the lateral parabrachial nucleus and in the nucleus of the solitary tract. There is evidence of diencephalic projections to the AP (Shapiro and Miselis 1985). Direct peripheral afferents arise from the vagus (signals from thoracic and abdominal viscera), glossopharyngeal (baroreceptive information), and trigeminal nerves (Miselis et al. 1987). *Humoral inputs*: Binding sites for a great number of bioactive substances have been demonstrated in the AP, including insulin (Unger et al. 1991). Sensitivities exist to emetic substances, glucose, and sodium (Adachi and Kobashi 1985). *Neuronal efferents* project to the nucleus of the solitary tract, the ambiguus nucleus, to noradrenergic neurons of the caudal ventrolateral medulla and spinal trigeminal nucleus (Shapiro and Miselis 1985). *Humoral outputs* are unknown (Table 5).

Table 5. Afferent (A) and efferent (E) connections of the area postrema (McKinley and Oldfield 1990)

Hypothalamic nuclei	*Medullary nuclei*
A, paraventricular nucleus	A + E, mesencephalic trigeminal tract
A, dorsomedial nucleus	A + E, medial parabrachial nucleus
A, perifornical nucleus	A + E, lateral parabrachial nucleus
	E, ambiguus nucleus
	E, caudoventral lateral reticular nucleus
	E, nuclei of the solitary tract
	E, spinal trigeminal tract

Conclusions

To emphasize the high degree of connections of the CVOs it should finally be noted that the SFO is directly connected to at least seventeen, the OVLT to at least eight, the AP to seven, the ME to three, and the pineal organ to several nuclei or complex areas. Most of these brain areas are located in the midsagittal plane of the diencephalon and the brain stem and become visible in horizontal sections. A central role is played by the paraventricular nucleus; it is connected to all CVOs, partly by afferents, partly by efferents. In conclusion, in those areas of the brain connected to the CVOs autonomic and species-preserving functions are represented, i.e., respiration, control of blood pressure, circulation, food intake, water balance, reproduction, and biorhythms. Via afferent signals from the CVOs the centers located in the midsagittal plane can be influenced by hemal factors available in the neurohemal regions and, vice versa, through efferent projections into a neurohemal region the CVOs can exert influence via blood circulation.

References

Adachi A, Kobashi M (1985) Chemosensitive neurons within the area postrema of the rat. Neurosci Lett 55:137–140

Bargmann W (1971) Die funktionelle Morphologie des endokrinen Regulationssystems. In: Altmann HW, Büchner F, Cottier H (eds) Handbuch der allgemeinen Pathologie, vol VIII, part 1. Springer, Berlin Heidelberg New York, pp 1–106

Blatteis CM, Bealer SL, Hunter WS, Llanos QJ, Ahokas RA, Mashburn TA (1983) Suppression of fever after lesions in the anteroventral third ventricle in guinea pigs. Brain Res Bull 11:519–526

Cardinali DP (1974) Melatonin and the endocrine role of the pineal organ. In: James VHT, Martin L (eds) Current topics in experimental endocrinology, vol 2. Academic, New York, pp 107–127

Dellmann HD, Simpson JB (1979) The subfornical organ. Int Rev Cytol 58:333–421

Gruber K, McRae-Deguerce A, Wilkin LD, Mitchell LD, Johnson AK (1987) Forebrain and brainstem afferents to the arcuate nucleus in the rat: potential pathways to the modulation of hypophyseal secretions. Neurosci Lett 75:1–5

Kenny GCT (1985) Structural and ultrastructural analysis of the pineal in primates; its innervation. In: Mess B, Rúzsás C, Tima L, Pévet P (eds) The pineal gland: current state of pineal research. Akadémiai Kiadó, Budapest, pp 341–345

Korf HW, Møller M (1985) The central innervation of the mammalian pineal gland. In: Mess B, Rúzsás C, Tima L, Pévet P (eds) The pineal gland: current state of pineal research. Akadémiai Kiadó, Budapest, pp 47–49

Krisch B (1980) Immunocytochemistry of neuroendocrine systems. Prog Histochem Cytochem 13(2):1–163

Krisch B, Leonhardt H (1980) Luliberin and somatostatin fiber terminals in the subfornical organ of the rat. Cell Tissue Res 210:33–45

Krisch B, Leonhardt H, Buchheim W (1978) The structural and functional border between the CSF- and blood-milieu in the circumventricular organs (organum vasculosum laminae terminalis, subfornical organ, area postrema) of the rat. Cell Tissue Res 195:485–497

Leonhardt H (1980) Ependym und Circumventrikuläre Organe. In: Oksche A, Vollrath L (eds)Handbuch der mikroskopischen Anatomie des Menschen, vol IV, part 10, Springer, Berlin Heidelberg New York, pp 177–666

Lind RW, Ganten D (1990) Angiotensin. In: Björklund A, Hökfelt T, Kuhar MJ (eds) Handbook of chemical neuroanatomy, vol 9: neuropeptides in the CNS, part II. Elsevier, Amsterdam, pp 165–286

Luiten PGK, ter Horst GJ, Steffens AB (1987) The hypothalamus, intrinsic connections and outflow pathways to the endocrine system in relation to the control of feeding and metabolism. Prog Neurobiol 28:1–54

McKinley MJ, Oldfield BJ (1990) Circumventricular organs. In: Paxinos G (ed) The human nervous system. Academic, San Diego, pp 415–438

Miselis RR, Shapiro RE, Hyde TM (1987) The area postrema. In: Gross PM (ed) Circumventricular organs and body fluids, vol II. CRC Press, Boca Raton, pp 185–207

Møller M, Van Deurs B, Westergaard E (1978) Vascular permeability to proteins and peptides in the mouse pineal gland. Cell Tissue Res 195:1–15

Oksche A (1965) Survey of the development and comparative morphology of the pineal organ. Prog Brain Res 10:3–29

Oksche A, Ueck M, Rüdeberg C (1971) Comparative ultrastructural studies of sensory and secretory elements in pineal organs. Mem Soc Endocr 19:7–25

Oldfield BJ, Clevers J, McKinley M (1986) A light and electron microscopic study of the projections of the nucleus medianus with special reference to inputs to vasopressin neurones. Soc Neurosci Abstr 12:445

Patel YC, Baquiran G, Srikant CB, Posner BI (1986) Quantitative in vivo autoradiographic localization of [^{125}I-Tyr]somatostatin-14 and [Leu8,D-Trp22-^{125}I-Tyr25]somatostatin-28 binding sites in rat brain. Endocrinology 119:2262–2269

Phillips PA, Abrahams JM, Kelly J, Paxinos G, Grzonka Z, Mendelsohn FAO, Johnston CI (1988) Localization of vasopressin binding in rat brain by in vitro autoradiography using a radioiodinated V_1 receptor antagonist. Neuroscience 27:749–775

Piva F, Limonta P, Martini L (1982) Role of the organum vasculosum laminae terminalis in the control of gonadotrophin secretion in rats. J Endocrinol 93:355–364

Quirion R, Dalpe M, DeLean A, Gutkowska J, Cantin M, Genest J (1984) Atrial natriuretic factor binding sites in brain and related structures. Peptides 5:1167–1172

Rouleau MF, Warshawsky H, Goltzman D (1984) Specific receptors for calcitonin in the subfornical organ of the brain. Brain 107:107–114

Scharrer E, Scharrer B (1954) Neurosecretion. In: von Möllendorff W, Bargmann W (eds) Handbuch der mikroskopischen Anatomie des Menschen, vol VI, part 5. Springer, Berlin Göttingen Heidelberg, pp 953–1066

Shapiro RE, Miselis RR (1985) The central neural connections of the area postrema of the rat. J Comp Neurol 234:344–364

Unger JW, Moss AM, Livingston JN (1991) Immunohistochemical localization of insulin receptors and phosphotyrosine in the brain stem of the adult rat. Neuroscience 42:853–861

Vollrath L (1981) The pineal organ. In: Vollrath L, Oksche A (eds) Handbuch der mikroskopischen Anatomie des Menschen, vol VI, part 7. Springer, Berlin Heidelberg New York

Blood- and Cerebrospinal Fluid-Dominated Compartments of the Rat Brain

B. Krisch

General Aspects

Structural integrity and proper function of the brain depend on the special environment of cerebrospinal fluid (CSF). It is partly produced by the choroid plexus, partly an ultrafiltrate resulting from transport processes across brain capillary endothelia and a product of brain metabolism. Like a mosaic, the CSF compartment is composed of innumerable microcompartments (Chesler 1990), structurally supported by glial processes. The special milieu of the brain is secured by the blood–brain barrier (BBB) and the blood–CSF barriers at the border of the neurohemal regions of circumventricular organs (CVOs). Devoid of a BBB, they are dominated by a hemal environment. Being early derivatives of the brain vesicle, all CVOs are located in the embryonic midsagittal plane. This origin from structures of the midsagittal plane within a bilaterally symmetric general morphologic pattern of the brain is reflected by the persistent thin ventricular wall in areas occupied by CVOs. These areas are only secondarily augmented by ingrowing, newly developing commissures. Due to this particular structure of the ventricular wall all CVOs and their neurohemal regions are interposed between inner (ventricular) and outer (leptomeningeal) CSF compartments (see Leonhardt 1980).

These general features of CVOs induce three main complexes of questions:

1. How is the integrity of CSF compartments secured against the adjacent hemal milieu of a CVO, and vice versa?
2. Are different hemal compartments to be taken into account within a particular neurohemal region, in analogy to the various existing CSF microcompartments, and which structures establish the integrity of the former?
3. Which structures or factors make the endothelium in the neurohemal area of a CVO appear leaky, and is it possible to find structural details which might explain the lacking BBB in neurohemal regions of CVOs?

The subcommissural organ (SCO) occupies a particular position among the CVOs. Among other pecularities, it is devoid of a typical neurohemal region. Therefore, in order to answer the above questions, the other CVOs should be examined first and then compared with the observations on the SCO.

The integrity of CSF compartments surrounding the neurohemal region of CVOs is provided by different mechanisms. Partly these are common to all CVOs possessing neurohemal regions; partly they differ considerably among the CVOs, depending on their position with respect to the adjacent outer CSF compartments. Common to all CVOs is the separation of their neurohemal area from the ventricular CSF by multiply stranded tight junctions (see Leonhardt 1980). The borderline between the neurohemal regions of the CVOs and the surrounding neuropil is marked by phagocytosing glia cells (Krisch and Leonhardt 1980) and by intercellular contacts between the perivascular terminals of tanycytes (Krisch and Leonhardt 1989; Krisch et al. 1978a,b). Thus, spreading of substances is controlled (a) in the intercellular clefts adjoining the neurohemal region and (b) in the perivascular spaces of vessels common to the neurohemal region and to CSF-dominated neuropil.

Concerning the leptomeningeal CSF compartment in CVOs endowed with neurohemal regions, the mechanisms of opposition between the hemal milieu and the outer CSF spaces may differ considerably. In the case of an abrupt confrontation of both compartments, pia mater is transformed into neurothelium. In the case of a "staggered" juxtaposition and the existence of "buffer" space between the hemal and the CSF compartment, however, large amounts of basal lamina-like material form labyrinthine complexes. In these structures acidic sulfated ground substance is stainable with alcian blue, pH 1 (see Krisch and Leonhardt 1989).

Circumventricular Organs: Comparative Considerations

In three of the CVOs, the *choroid plexus (PC), the pineal organ*, and the *area postrema (AP)*, the hemal milieu and the leptomeningeal CSF are separated by just one layer of leptomeningeal cells, which consequently are transformed into neurothelium (Krisch 1986). In the *organum vasculosum laminae terminalis (OVLT)* and the *subfornical organ (SFO)*, the relationship between the organ as a whole and the leptomeningeal CSF compartments resembles that of any blood vessel invading the brain: funnel-like pia indentations penetrate these organs. However, a basic structural pattern is retained: the SFO, according to the development of pallial structures, does not have any relationship with leptomeningeal CSF compartments beyond the level of the pia mater accompanying the blood vessels, which invade the organ from the lateral aspect and interconnect it with the PC. Pia funnels and large amounts of basal lamina-like material are exclusively found in the lateral borderline area towards the PC. In contrast, in the OVLT well-developed pia funnels and excessive amounts of basal lamina-like material as well as manifold layers of leptomeningeal cells mirror the gradual confrontation of the neurohemal region with the leptomeningeal CSF compartment (Krisch et al. 1987). The *median eminence (ME)* occupies a particular position among these CVOs, in that in its basal portion the neurohemal region borders on "peripheral" hemal milieu instead of the leptomeningeal CSF compartment. In the lateral parts of the ME neurothelial layers are developed (Krisch et al. 1983), resembling the situation in the PC and the AP.

Common to all CVOs mentioned is that their neurohemal regions at least comprise the perivascular space. Concerning its fine-structural organization, leptomeningeal cells form a more or less uninterrupted layer around the fenestrated capillaries. Again, there are differences between the neurohemal regions of these CVOs, concerning the degree of completeness of the leptomeningeal layer, the amount of basal lamina-like material, and the presence of acid glycosaminoglycans.

In the PC that has the most extensive neurohemal region, in the OVLT, the pineal organ and the AP, an almost uninterrupted perivascular sheath of leptomeningeal cells encircles the capillaries. In the SFO and ME, this layer of leptomeningeal cells is penetrated by protruding neuropil elements gaining direct access to the capillary and showing function-dependent variations in their arrangement.

The PC shows the most intense staining for acid glycosaminoglycans. They are intensely stainable also in the pineal organ (particularly in its posterior portion), the ME and the OVLT, and virtually missing in the SFO. Deposits of perivascular basal lamina-like material are most extensive in the OVLT and PC, followed by the AP and the pineal organ. The OVLT occupies a particular position among the CVOs, in that its fenestrated capillaries differ locally in their permeability for horseradish peroxidase (HRP). Highly permeable capillaries in the central part of the organ are less extensively invested with basal lamina-like material in comparison to capillaries which are less permeable for HRP. In the PC, basal lamina-like material, as an exception, is located between the leptomeningeal cells and the plexus epithelia; as a rule it is exclusively deposited between leptomeningeal cells and capillary endothelium. In the SFO and the ME, basal lamina-like material is hardly developed and restricted to the lateral parts of these CVOs.

Obviously, the fine organization of the perivascular space differs in the individual CVOs; the structural arrangement found therein favors the existence of microcompartments within the neurohemal region. It can be assumed that leptomeningeal cells, which are endowed with a number of catabolic enzymes, together with acid glycosaminoglycans might influence the exchange between brain tissue and circulating blood in both directions. This may lead to an alteration of agents, which after release from axon terminals or passage via the endothelium finally reach the capillary lumen or the adjacent neuropil.

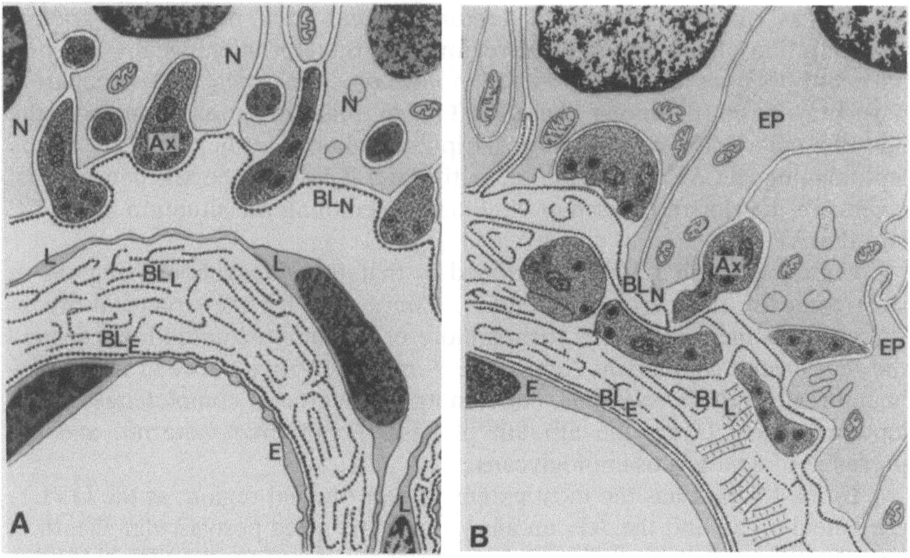

Fig. 1. Comparison between the organization of the perivascular spaces in a typical neurohemal region of a CVO (**A**) and in the SCO (**B**) without consideration of interspecies differences. **A** Axons (*Ax*) in the pericapillary spaces of fenestrated endothelia (*E*); BL_E, endothelial basal lamina; BL_L, basal lamina-like material in the pericapillary spaces; BL_N, basal lamina of the neuropil; *L*, leptomeningeal cells encircling the capillaries; *N*, neuropil = neuronal, glial or tanycytic processes. **B** Axons (*Ax*) in the pericapillary spaces of continuous endothelia (*E*); BL_E, endothelial basal lamina; BL_L, basal lamina-like material or striated bodies in the pericapillary spaces; BL_N, basal lamina of the neuropil; *EP*, secretory and nonsecretory ependymal and glial cells

Subcommissural Organ

The compiled data should now be compared with corresponding details in the SCO (Fig. 1A,B). Like the other CVOs, the SCO faces the ventricular and the leptomeningeal CSF compartment. Its tall ependymal cells show high secretory activity; the tightness of their intercellular contacts obviously displays species-specific differences (Gotow and Hashimoto 1982a,b; Rodríguez et al. 1986). The organ is free of neuronal somata and exclusively composed of actively secreting ependymal and hypependymal cells. Thus, it possesses a very particular intercellular milieu, demarcated by the BBB against the bloodstream and by a variable blood–CSF barrier agaist the ventricular CSF. However, the most obvious difference is the lacking neurohemal region in the SCO. Consequently, it seems reasonable to analyze the organization of the perivascular space of SCO capillaries. In addition to data from the literature, we have conducted a preliminary study analyzing the SCO of several rats and a rabbit. In both species the organization of the perivascular space varies between the different parts of the organ.

In the SCO, alcian blue (pH 1) does not stain any of the pericapillary spaces. Thus, they clearly differ from those in characteristic neurohemal regions but also from those of BBB capillaries. Like BBB capillaries all SCO capillaries exhibit continuous endothelia, but in contrast to BBB capillaries separate basal laminae are present for the endothelia and the neuropil. Only in the most rostral and dorsal parts of the organ, where the SCO contacts the leptomeninges, do the perivascular spaces contain sheaths of meningeal cells. These vessels might represent postcapillary venules already associated with the dead end of a pia funnel. With regard to their perivascular spaces, the remainder of vessels positioned between the secretory ependymal cells can be subdivided into two groups. One of them has narrow perivascular spaces which, except for the two separate basal laminae, do not differ significantly from BBB capillaries. The other group is more conspicuous; although meningeal cells are lacking, the perivascular spaces contain, at least in the rabbit, basal lamina-like material arranged in whorls and labyrinths like those in the neurohemal regions. Such perivascular spaces may be rather wide and may display small bundles of nerve fibers containing clear vesicles and/or small dense-core vesicles. In the rat, striated bodies are found, which may be continuous with copious basal lamina-like material. In both species, the parenchyma limiting the perivascular spaces is composed of secretory hypependymal and ependymal cells, of neuronal and probably also glial terminals. In both species this extracellular material helps to indentify extensions and arborizations of the perivascular space, which may contact vacuole-like spaces between hypependymal secretory cells, probably being in open communication with the ventricular lumen.

Conclusions

Obviously, not only the relationships between the leptomeningeal space and the neurohemal regions but also the fine organization of their capillaries and the perivascular space may show considerable differences among the individual CVOs. This fine-structural pattern also comprises the amount of glycosaminoglycans and the extent of the basal lamina-like material. However, when the BBB is lacking, the capillaries are always surrounded by a nearly uninterrupted sheath of leptomeningeal cells. It is supposed that the negatively charged glycosaminoglycans and the basal lamina-like material function like a molecular sieve in the perivascular space, thereby influencing the substance transfer in either direction between the neuropil and the fenestrated capillary.

With regard to the third question concerning the leakiness of the endothelium, the leptomeningeal cells are the most likely candidates involved in developmental processes responsible for the absent BBB. The significance of the leptomeningeal sheath is a challenging question. An answer might be found in a series of previous studies, mainly concerning the in vitro differentiation of brain endothelial and glial cells (Arthur et al. 1987; Grinspan et al. 1987; Janzer and Raff 1987). They indicate a direct interaction between endothelial elements and astrocytes with respect to the development of the BBB. Coculture with leptomeningeal cells induces leaky endothelia, possibly by impeding the above-mentioned mechanism. In the SCO, a perivascular leptomeningeal sheath is missing. In the adult rat, mature astrocytes cannot be labeled with antibodies against glial fibrillary acidic protein (GFAP) within the SCO proper, and its ependymal and hypependymal cells are obviously too highly specialized to react with any of the astrocyte markers (Didier et al. 1986). In other species, however, SCO cells react with antisera against GFAP (Redecker 1989). Hence, it can be postulated that the ependymal and hypependymal cells are derivatives of the glial-cell lineage. They may have exerted their influence on capillary endothelial differentiation before becoming specialized secretory cells.

The hypothesis of the developmental influence of leptomeninges in suppressing the formation of a regular BBB should be tested in vitro. Even though leptomeningeal cells are capable of lining the basal laminae, the origin of acid glycosaminoglycans and of basal lamina-like material is still an open question. The assumption that they are products of leptomeningeal cells is dubious in the light of the present analysis of the SCO.

References

Arthur FE, Shivers RR, Bowman PD (1987) Astrocyte-mediated induction of tight junctions in brain capillary endothelium: an efficient in vitro model. Dev Brain Res 36:155–159

Chesler M (1990) The regulation and modulation of pH in the nervous system. Prog Neurobiol 34:401–427

Didier M, Harandi M, Aguera M, Bancel B, Tardy M, Fages C, Calas A, Stagaard M, Møllgård K, Belin MF (1986) Differential immunocytochemical staining for glia fibrillary acidic (GFA) protein, S-100 protein and glutamine synthetase in the rat subcommissural organ, nonspecialized ventricular ependyma and adjacent neuropil. Cell Tissue Res 245:343–351

Gotow T, Hashimoto PH (1982a) Intercellular junctions between specialized ependymal cells in the subcommissural organ of the rat. J Neurocytol 11:363–379

Gotow T, Hashimoto PH (1982b) Fine structural studies on ependymal paracellular and capillary transcellular permeability in the subcommissural organ of the guinea pig. J Neurocytol 11:447–462

Grinspan JB, Lieb M, Stern J, Rupnick M, William S, Pleasure D (1987) Rat brain microvessel extracellular matrix modulates the phenotype of cultured rat type 1 astroglia. Dev Brain Res 33:291–295

Janzer RC, Raff MC (1987) Astrocytes induce blood-brain barrier properties in endothelial cells. Nature 325:253–257

Krisch B (1986) The functional and structural borders between the CSF- and blood-

dominated milieus in the choroid plexuses and the area postrema of the rat. Cell Tissue Res 245:101–115
Krisch B, Leonhardt H (1980) Neurohormones in the intercellular clefts and in glia-like cells of the rat brain. Cell Tissue Res 211:251–268
Krisch B, Leonhardt H (1989) Relations between leptomeningeal compartments and the neurohemal regions of circumventricular organs. Biomedical Res 10(3):155–168
Krisch B, Leonhardt H, Buchheim W (1978a) The functional and structural border of the neurohemal region of the median eminence. Cell Tissue Res 192:327–339
Krisch B, Leonhardt H, Buchheim W (1978b) The functional and structural border between the CSF- and blood milieu in the circumventricular organs (organum vasculosum laminae terminalis, subfornical organ, area postrema) of the rat. Cell Tissue Res 195:485–497
Krisch B, Leonhardt H, Oksche A (1983) The meningeal compartments of the median eminence and the cortex. A comparative analysis in the rat. Cell Tissue Res 228:597–640
Krisch B, Leonhardt H, Oksche A (1987) Compartments in the organum vasculosum laminae terminalis of the rat and their delineation against the outer cerebrospinal fluid-containing space. Cell Tissue Res 250:331–347
Leonhardt H (1980) Ependym und circumventrikuläre Organe. In: Oksche A, Vollrath L (eds) Handbuch der mikroskopischen Anatomie des Menschen, vol IV, part 10. Springer, Berlin Heidelberg New York, pp 177–666
Redecker P (1989) Immunohistochemical localization of glial fibrillary acidic protein (GFAP) and vimentin in the subcommissural organ of the Mongolian gerbil (*Meriones unguiculatus*). Cell Tissue Res 255:395–412
Rodríguez EM, Oksche A, Rodríguez S, Hein S, Peruzzo B, Schoebitz K, Herrera H (1986) The subcommissural organ – Reissner's fiber unit. In: Gross PM (ed) Circumventricular organs and body fluids, vol II. CRC Press, Boca Raton, pp 3–42

VI. Neural Inputs

The Subcommissural Organ of the Rat: An In Vivo Model of Neuron–Glia Interactions

M. Didier-Bazes, L. Chouaf, P. Lepetit, M. Aguera, and M.F. Belin

Introduction

It is established that neurotransmitters in general and serotonin (5HT) in particular have a trophic role in the maturation of the central nervous system (CNS) (Lauder 1983). This action could be mediated in part via regulation of glia since neuron–glia interactions are necessary for the maturation of both neurons and glia (Hatten and Mason 1986; Holton and Weston 1982; Le Prince et al. 1990) and since glial cells in the CNS possess receptors for neurotransmitters (Whitaker-Azmitia 1988).

The rat subcommissural organ (SCO) is formed by pseudostratified ependymocytes which have a particular phenotype: they transitorily synthesize glial fibrillary acidic protein (GFAP) during ontogeny (Chouaf et al. 1991), which indicates their glial origin, but do not express, in adults, this marker or two other glial markers: protein S100 (PS100) and glutamine synthetase (GS) (Didier et al. 1986). They accumulate gamma aminobutyric acid (GABA) by specific uptake mechanisms (Gamrani et al. 1981) and receive a synaptic serotoninergic (5HT) input (Bouchaud and Arluison 1977); a 5HT innervation of glial cells has been described in the early stages of ontogeny (Lauder et al. 1982).

Thus, the SCO provides an interesting model system to study neuron–glia interactions, in particular the role of 5HT innervation in the development and the activity of the ependymocytes.

Action of 5HT Terminals on SCO Ependymocyte Differentiation

In order to investigate the putative actions of 5HT terminals on the differentiation of SCO ependymocytes, we used a phylogenetic approach. Actually, there is great disparity in the 5HT innervation of the SCO depending on species: rat, gerbil, dog, and cat SCO possess a plexus of 5HT terminals while mouse, monkey and rabbit SCO do not (for references, see Chouaf et al. 1989).

Immunocytochemical studies of the reactivity of the SCO ependymocytes to glial markers (GFAP, GS, PS100), neuronal markers (neurofilament) and antibodies against vimentin, and autoradiographic analysis of [^{3}H]-GABA uptake have been performed in rat and cat (which display 5HT terminals in the SCO) as well as in mouse and rabbit (lacking them) (Chouaf et al. 1989). Two phenotypic characteristics appeared different according to the groups: PS100 was not detected in SCO ependymocytes of rat and cat, whereas it was observed in those of mouse and rabbit (Fig. 1b,d). Rat and cat SCO ependymocytes accumulate [^{3}H]-GABA (revealed by a strong autoradiographic labeling) while those of mouse and rabbit do not (Fig. 1a–c).

In order to define the role of 5HT terminals in the appearance of GABA carriers and the regulation of the expression of the glial markers in the SCO ependymocytes, we have studied the differentiation of these cells (1) during ontogeny in control rats, (2) after the transplantation, at birth, to a foreign environment (the fourth ventricle), and (3) after preventing the 5HT innervation of the SCO, at birth, by the administration of 5–7 dihydroxytryptamine (5–7 DHT), a neurotoxin specific to the serotoninergic system.

5HT Innervation and GABA Carriers

The capacity of a specific GABA uptake mechanism, shown by a progressive increase in the autoradiographic labeling of the SCO ependymocytes by [^{3}H]-GABA, became visible during the first two postnatal weeks. This coincides with the time of the onset of the 5HT innervation (Didier-Bazes et al. 1989; Marcinkiewicz and Bouchaud 1986). The SCO ependymocytes of newborn rats transplanted to the fourth ventricle of adult rats showed no such uptake even as long as 3 months after transplantation (Didier-Bazes et al. 1991) (Fig. 1g). This suggests the necessity of an extrinsic factor required by the SCO ependymocytes for the development of a GABA uptake mechanism. The absence of 5HT terminals in the grafts led us to examine

Fig. 1. Autoradiography of [^{3}H]-GABA uptake (**A**, **C**, **E**, **G**) and immunocytochemistry with antibodies against PS100 (**B**, **D**, **F**, **H**) in SCO ependymocytes of the mouse (**A**, **B**), control rats (**C**, **D**), rats treated with 5–7 DHT at birth (**E**, **F**), and newborn-rat SCO grafted to the fourth ventricle of an adult rat after 3-month survival (**G**, **H**). [^{3}H]-GABA autoradiography shows a high accumulation of silver grains on the SCO ependymocytes (*curved arrow*) of the control rat (**C**) while the SCO ependymocytes of the mouse (**A**), the rat pretreated with 5–7 DHT at birth (**E**), and the ependymocytes of the grafted SCO (**G**), are covered by a diffuse, nonspecific precipitate. Immunoreactivity pattern with antibodies against PS100 in SCO ependymocytes (*curved arrow*): control rat SCO cells (**D**) are immunonegative to PS100 while some SCO ependymocytes in the mouse (**B**), in a rat treated with 5–7 DHT at birth (**F**), and in the grafted SCO (**H**) are immunopositive. **A**, **B** ×200; **C**, **E** ×160; **D** ×300; **F**, **G**, **H** ×380

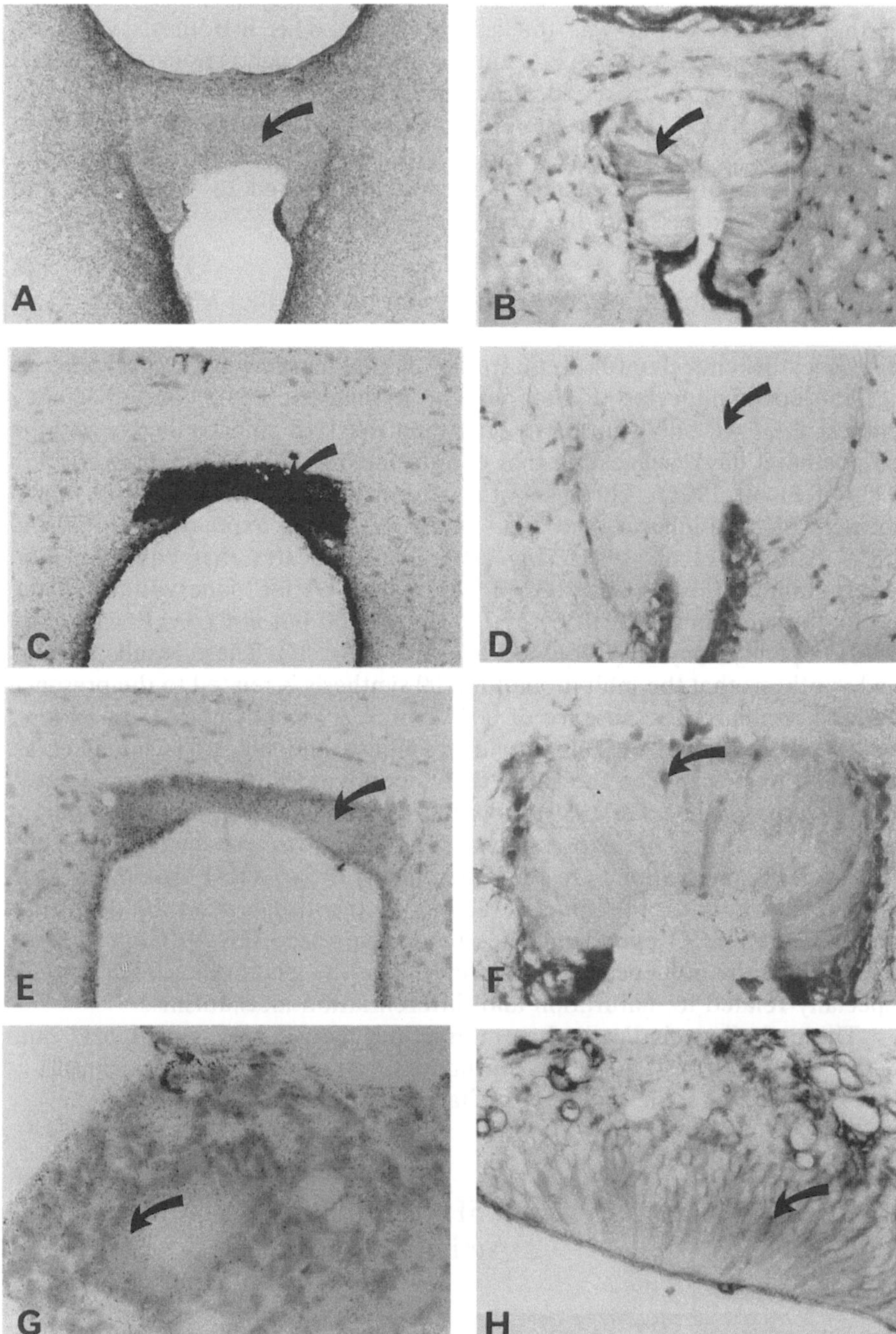
A
B
C
D
E
F
G
H

whether the 5HT input was the critical factor. This hypothesis was confirmed by the absence of GABA uptake into SCO ependymocytes deprived of 5HT innervation by the destruction of the serotoninergic neurons at birth (Didier-Bazes et al. 1991) (Fig. 1e). However, 5HT denervation in adults did not suppress GABA uptake (see below). This indicates that the mechanisms involved in this phenomenon are related to maturation and differentiation processes rather than to regulation.

5HT Innervation and Expression of the Glial Markers

Immunocytochemical ontogenetic studies of glial markers in the SCO showed that rat ependymocytes do not appear to express PS100 and that they synthesize GFAP only transiently (between the 19th embryonic day and the 3rd postnatal day), suggesting that certain factors inhibit their transcription (Chouaf et al. 1991). Three months after transplantation into the fourth ventricle of an adult rat, the SCO of newborn rats expressed GFAP and PS100 (Chouaf et al. 1991) (Fig. 1h). This indicates that environmental factors inhibit GFAP and PS100 expression. After denervation of the serotoninergic input at birth by 5–7 DHT, PS100 but not GFAP was found in SCO ependymocytes (Chouaf et al. 1991) (Fig. 1f). These results support the hypothesis that the inhibition of PS100 synthesis is related to the presence of 5HT terminals. The absence of GFAP in SCO ependymocytes of animals genetically devoid of 5HT innervation (mouse and rabbit) (Chouaf et al. 1989) and its presence in the gerbil SCO ependymocytes that are innervated by 5HT axons (Redecker 1989) show that, in the SCO, there is an apparent correlation between the expression of GFAP and the presence of a certain type of 5HT innervation. On the other hand, if 5–7 DHT was injected in adult rats, whatever the time of action of the drug (from 10 days to 3 months), their SCO ependymocytes never expressed PS100 (Chouaf et al. 1991). Thus, the influence of 5HT afferents on the glial marker expression is especially related to maturation and differentiation mechanisms.

These findings indicate that interactions between neuronal 5HT and SCO ependymocytes take place during maturation, and serve the control of the expression of some phenotypic characteristics.

Neuromodulating Effects of 5HT on the SCO Ependymocyte Activity

The presence of 5HT receptors on glial cells (Whitaker-Azmitia 1988) suggests that this neuromediator might be capable of modulating some aspects of glial activity. Thus we investigated the effect of 5HT on the rates of protein synthesis and [^{3}H]-GABA accumulation in SCO ependymocytes.

Action of 5HT on Protein Metabolism

The rate of protein synthesis can be estimated by means of quantitative autoradiographic measurement of the incorporation of ^{35}S methionine into newly synthesized proteins (Lestage et al. 1987). In a preliminary study, the action of 5HT on protein synthesis has been evaluated by increasing its concentration by a systemic administration of 5-hydroxytryptophan (5HTP), the immediate precursor of 5HT (Lepetit et al. 1991). This treatment induced a 50% decrease in the quantity of ^{35}S methionine incorporated into the SCO ependymocyte protein (Lepetit et al. 1991). This suggests that 5HT may inhibit general protein synthesis in SCO ependymocytes. The accumulation of glycoproteins observed by Léger et al. (1983) after the destruction of the 5HT terminals could thus be explained by an increase in their synthesis. However, an effect of 5HT on their release is by no means excluded.

Action of 5HT on GABA Accumulation

Rat SCO ependymocytes accumulate [^{3}H]-GABA by a highly specific uptake mechanism (Gamrani et al. 1981). Destruction of 5HT terminals by 5–7 DHT or inhibition of 5HT synthesis by parachlorophenylalanine (PCPA) increased the rate of [^{3}H]-GABA accumulation (Didier-Bazes et al. 1989; Gamrani et al. 1981) (Table 1). Action of PCPA was not due to side effects since the simultaneous injection of 5HTP which restores the 5HT content prevented the increase in [^{3}H]-GABA accumulation. However, an increase in 5HT content of the SCO by pargyline, an inhibitor of its degradation, appeared to have no effect on [^{3}H]-GABA uptake (Didier-Bazes et al. 1989) (Table 1). This suggests that the control of [^{3}H]-GABA accumulation in SCO ependymocytes by 5HT is sensitive to an absence but not to an excess of 5HT. The functional role of GABA uptake into the SCO ependymocytes

Table 1. Modification of [^{3}H]-GABA accumulation, GABA content, and 5HT content in the SCO of untreated, pCPA−, pCPA+, 5-HTP- and pargyline-treated rats (From Didier-Bazes et al. 1989)

	Control rats	pCPA-treated rats	pCPA + 5HTP-treated rats	Pargyline-treated rats
[3H]-GABA accumulated in the SCO (Bq)	130 ± 24	364 ± 48*	151 ± 39	125 ± 41
GABA content (nmol/mg tissue)	1.26 ± 0.24	1.19 ± 0.22	1.07 ± 0.12 (ns)	1.10 ± 0.19 (ns)
5-HT content (nmol/mg tissue)	0.45 ± 0.03	Not detectable	0.49 ± 0.06	Not measured

* $P < 0.01$; *ns*, not significant

is unknown; its penetration into the nucleus of the SCO ependymocytes suggests that it may act on genomic transcription. In a preliminary study, we have labeled newly synthesized SCO protein by intraventricular injection of ^{35}S cysteine and analyzed, by sodium dodecyl sulfate polyacrylamide gel electrophoresis, the concanavalin A (Con A)-binding to SCO glycoproteins. Our results showed that following the inhibition of GABA synthesis in vivo with allylglycine, there was a general increase in the synthesis of ^{35}S ConA-binding glycoprotein (Fig. 2b). Conspicuous changes were observed in the level of incorporation into five bands in particular (Fig. 2c); there was a 44% decrease in incorporation to a 90-kDa band and increases of 955%, 221%, 157%, and 222% to 83-, 77-, 71-, and 46-kDa bands, respectively. These preliminary results suggest that some modifications in the metabolism of

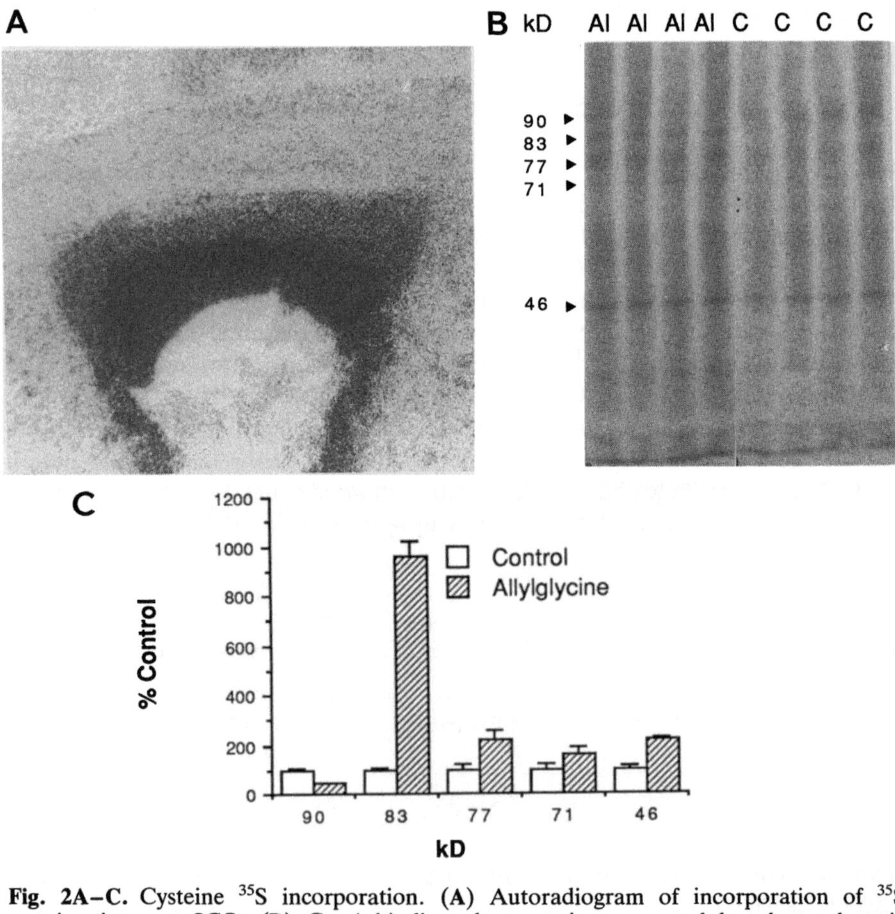

Fig. 2A–C. Cysteine ^{35}S incorporation. (**A**) Autoradiogram of incorporation of ^{35}S cysteine into rat SCO. (**B**) ConA-binding glycoproteins separated by electrophoretic migration on SDS-PAGE, in allylglycine (*Al*)-treated rats and control (**C**) rats, after incorporation of ^{35}S cysteine. (**C**) Changes in incorporation of ^{35}S cysteine into five ConA-binding glycoproteins after treatment with allylglycine

ConA-binding SCO glycoproteins are mediated by GABA and that 5HT may thus control SCO secretory activity not only via classical transducing factors but also by regulating GABA uptake mechanisms.

In conclusion, these findings indicate that interactions between neuronal 5HT and SCO ependymocytes take place (1) during the maturation of the SCO ependymocytes leading to a control of the expression of some of their phenotypic characteristics, and (2) in differentiated ependymocytes by modulating their activity. Although the SCO ependymocytes represent a distinct population of glial cells with regard to their phenotype and their persistent 5HT input, it is noteworthy that other glial cells, such as oligodendrocytes and astrocytes, possess 5HT receptors (Whitaker-Azmitia 1988) and thus have the capacity to take up GABA (Reynolds and Herschkowitz 1987; Schousboe et al. 1986), and that transitory serotoninergic contacts are observed on other glial cells during ontogeny (Lauder et al. 1982).

Acknowledgments. This work was supported by INSERM CJF 90.10 and CNRS URA 1195. The authors thank E. Derrington for reading the proofs.

References

Bouchaud C, Arluison M (1977) Serotoninergic innervation of ependymal cells in the rat subcommissural organ; a fluorescence electron microscopic and radioautographic study. Biol Cell 30:65–72

Chouaf L, Didier-Bazes M, Aguera M, Tardy M, Sallanon M, Kitahama K, Belin MF (1989) Comparative marker analysis of the ependymocytes of the subcommissural organ in four different mammalian species. Cell Tissue Res 257:255–262

Chouaf L, Didier-Bazes M, Hardin H, Aguera M, Fèvre-Montange M, Voutsinos B, Belin MF (1991) Developmental expression of glial markers in subcommissural organ ependymocytes of the rat: role of the environment. Cell Tissue Res 266:553–561

Didier M, Harandi M, Aguera M, Bancel B, Tardy M, Fages C, Calas A, Stagaard M, Møllgård DK, Belin MF (1986) Differential immunocytochemical staining for glial fibrillary acidic (GFA) protein, S100 protein and glutamine synthetase in the rat subcommissural organ and non specialized ventricular ependyma and adjacent neuropil. Cell Tissue Res 245:243–351

Didier-Bazes M, Aguera M, Chouaf L, Harandi M, Calas A, Meiniel A, Belin MF (1989) Neuronal control of [^{3}H] GABA uptake in the ependymocytes of the subcommissural organ: an in vivo model of neuron-glia interaction. Brain Res 489:137–145

Didier-Bazes M, Chouaf L, Hardin H, Aguera M, Fèvre-Montange M, Belin MF (1991) Developmental neuron-glia interaction: role of the serotonin innervation upon the onset of GABA uptake into the ependymocytes of the rat subcommissural organ. Dev Brain Res 63:135–139

Gamrani H, Belin MF, Aguera M, Calas A, Pujol JF (1981) Radioautographic evidence for an innervation of the subcommissural organ by GABA-containing nerve fibres. J Neurocytol 10:411–424

Hatten ME, Mason CA (1986) Neuron-astroglia interactions in vitro and in vivo. Trends Neurosci 9:168–174

Holton B, Weston J (1982) Analysis of glial cell differentiation in peripheral nervous tissue. II. Neurons promote S-100 synthesis by purified glial precursor cell populations. Dev Biol 89:72–81

Lauder JM, Wallace JA, Krebs H, Petruz P, Mc Carthy K (1982) In vivo and in vitro development of serotonergic neurons. Brain Res Bull 9:605–625

Lauder JM (1983) Hormonal and humoral influences on brain development. Psychoneuroendocrinology 8:121–155

Léger L, Degueurce A, Lundberg JJ, Pujol JF, Møllgård K (1983) Origin and influence of the serotoninergic innervation of the subcommissural organ in the rat. Neuroscience 2:411–423

Lepetit P, Touret M, Grange E, Gay N, Bobillier P (1991) Inhibition of methionine incorporation into brain proteins after the systemic administration of p-chlorophenylalanine and L-5-hydroxytryptophan. Eur J Pharmacol 209:207–212

Le Prince G, Copin MC, Hardin H, Belin MF, Bouilloux JP, Tardy M (1990) Neuron-glia interactions: effect of serotonin on the astroglial expression of GFAP and of its encoding message. Dev Brain Res 51:295–298

Lestage P, Gonon M, Lepetit P, Vitte PA, Debilly G, Rossatto C, Lecestre D, Bobillier P (1987) An in vivo kinetic model with L-^{35}S-methionine for the determination of local cerebral rates of methionine incorporation into protein in the rat. J Neurochem 48:352–363

Marcinkiewicz M, Bouchaud C (1986) Formation and maturation of axo-glandular synapses and concomitant changes in the target cells of the rat subcommissural organ. Biol Cell 56:57–65

Redecker P (1989) Immunohistochemical localisation of glial fibrillary acidic protein (GFAP) and vimentin in the subcommissural organ of the Mongolian gerbil (*Meriones unguiculatus*). Cell Tissue Res 255:595–600

Reynolds R, Herschkowitz N (1987) Oligodendroglial and astroglial heterogeneity in mouse primary central nervous system culture as demonstrated by differences in GABA and D-aspartate transport and immunocytochemistry. Dev Brain Res 36:13–25

Schousboe A, Drejer J, Meier E, Larsson OM, Schousboe I (1986) Regional heterogeneity of astrocytic transport processes for glutamate and GABA. Adv Biosci 61: 225–233

Whitaker-Azmitia PM (1988) Astroglial serotonin receptors. In: Kimelberg HK (ed) Glial cell receptors. Raven, New York, pp 107–120

Neural Inputs to the Subcommissural Organ

C. Bouchaud

General Aspects

Among the circumventricular organs of the vertebrates, the subcommissural organ (SCO), a specialized area of ependyma, is one of the most fascinating. Its modified ependymal cells arranged in a pseudostratified epithelium are elongated, high cylinders that contact apically the ventricular cerebrospinal fluid. The SCO ependymocytes appear as specialized glial cells displaying some cytochemical characteristics that differentiate them from other types of glial cells (Didier et al. 1986; Chouaf et al. 1989). In all vertebrates, the SCO is easy to recognize, but despite its constant morphology, this organ, which differentiates very early during embryonic life, shows conspicuous specific variations especially in the innervation of its ependymal and hypendymal elements. Thus, a few species of higher vertebrates possess a dense innervation of the SCO while most species are totally devoid of such innervation. The axoglandular innervation of the SCO is richest and most differentiated in the rat. Hence, the rat SCO provides a classical model for studying certain aspects of neuron–glia interactions.

The Rat Model of SCO Innervation

In 1964, Stanka identified in the rat SCO a plexus of zinc–iodide positive nerve fibers. More recently, using silver staining techniques Léger et al. (1983) described a complex network of nerve fibers in the hypendyma, generally a mix of very fine and somewhat thicker fibers. From this plexus innervating the ependyma thin fibers extend to the nuclear level of the SCO ependymocytes.

Ultrastructural study of the SCO of adult rats has mainly shown the presence of numerous terminal or preterminal boutons synapsing on the SCO cells, especially on the ependymocytes but also on some hypendymal cells (Møllgård and Wiklund 1979). To date, the rat SCO constitutes the classical example of true synapses on ependymal cells in the mammalian central nervous system (Fig. 1a). In this type of synapselike contact, however, the postsynaptic cell is a non-neuronal element.

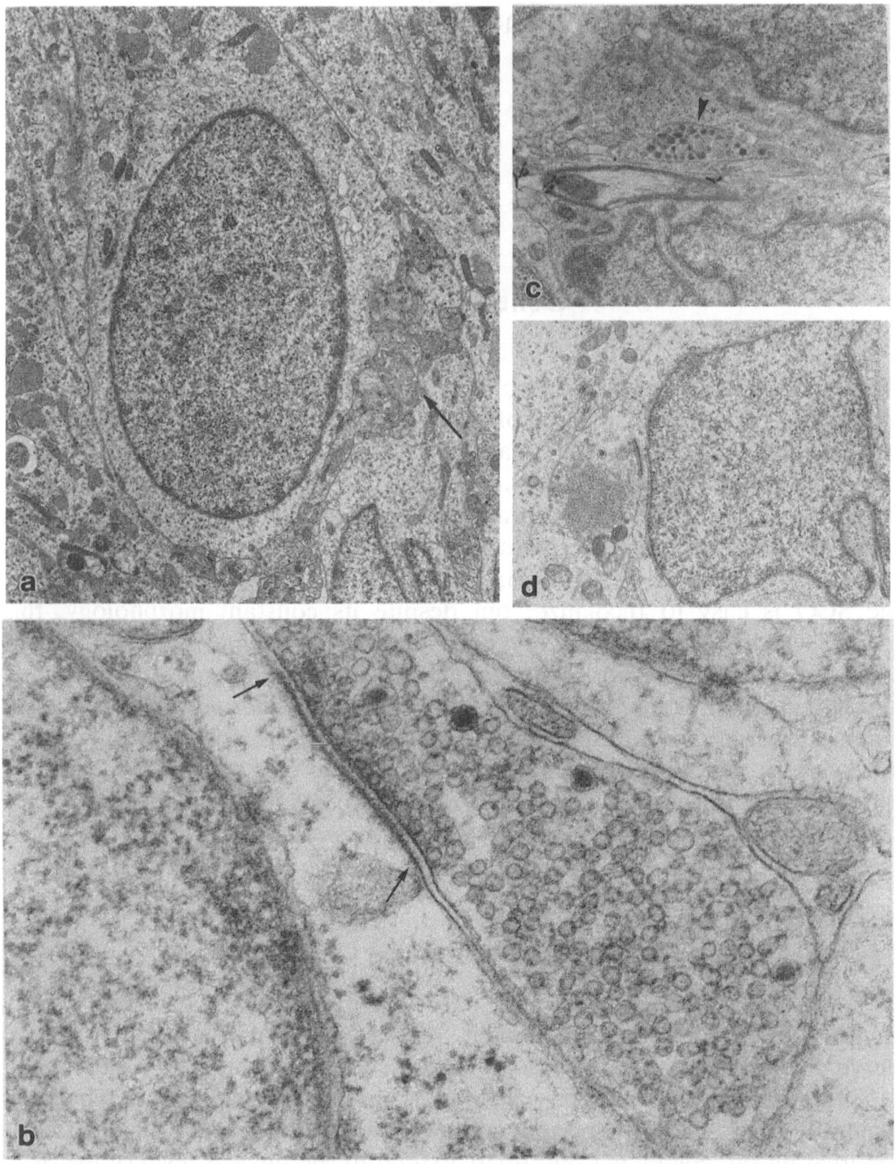

Fig. 1a–d. Synapses on target cells in the rat subcommissural organ. **a** A dense plexus (*arrow*) of unmyelinated nerve fibers and terminals contacts the laterobasal part of an ependymocyte. ×6600. **b** Typical axoglandular synapse (*arrows*) near the cell nucleus of an ependymocyte (*left*). Note the small dots in the clear synaptic vesicles. ×41 000. **c** A rare, probably peptidergic fiber (*arrowhead*) located between two hypendymal cells. ×6000. **d** Axon terminal containing numerous synaptic vesicles in a characteristic paracrystalline arrangement. ×5300

Most of the synaptic contacts arise from fine, varicose unmyelinated axons terminating on SCO ependymal cells in the immediate vicinity of their nuclei (about 0.1–0.4 μm) or basal processes. The varicosities issued from small fiber bundles located in the deep hypendyma beneath the posterior commissure form the presynaptic element of an axoglandular synapse (Fig. 1b). These boutons have an average diameter of 0.4–0.7 μm; successive boutons are separated from each other by intervaricose segments as in the case of supraependymal nerve fibers at the surface of the ventricles (Arluison et al. 1982). Some very rare varicosities belonging to fibers "en passant" contain only large granular vesicles (peptidergic fibers?) (Fig. 1c). Generally, the boutons exhibit an abundant population of clear vesicles (55 nm in diameter), dispersed in the axoplasm or more rarely grouped in the central part in a paracrystalline pattern (Fig. 1d), and a few rare large granular vesicles (100 nm in diameter) representing less than 0.6% of the vesicles contained in the boutons (Bouchaud and Arluison 1977). The synaptic membrane specializations are characteristic of Gray's type-I synapses: the postsynaptic membrane is thicker than the presynaptic membrane (asymmetrical synaptic contact). The junctional cleft contains an electron-dense material.

As a general rule, the innervated ependymocytes receive several inputs and frequently a bouton innervates several SCO ependymocytes (Bouchaud and Arluison 1977; Møllgård and Wiklund 1979; Bouchaud and Bosler 1986). Synapses of the same type are also present at other postsynaptic sites such as hypendymal cells and basal processes of SCO ependymocytes (Møllgård and Wiklund 1979).

An unanswered question is whether all SCO ependymocytes receive innervation. Møllgård and Wiklund (1979) suggest that all SCO ependymocytes are innervated. However, we have frequently noticed that in transverse sections through the SCO containing 200–300 ependymocytes, almost no synaptic varicosities can be found. This observation suggests that many cells are free of innervation. In the adult rat, all SCO ependymocytes display morphofunctional signs indicating synthesis, storage, and release of secretory products (Marcinkiewicz and Bouchaud 1983); a direct communication between contiguous ependymocytes is suspected. To our knowledge, however, no gap junctions have been described in this special ependyma (Madsen and Møllgård 1979), in contrast to typical ependymal cells (Brightman and Reese 1969). Nevertheless, confronting hypolemmal associations involving mitochondria, similar to the subsurface-cistern mitochondria observed in various excitable cells (Bouchaud 1987), could play the role of gap junctions and be involved in the coupling between contiguous ependymocytes.

The Innervation of the Rat SCO is Predominantly of the Serotoninergic Type

General characteristics

The characteristics of the SCO innervation have largely been elucidated by both anatomical and cytochemical techniques associated with pharmacological treatments. The first histochemical technique applied to visualize the innervation of the rat SCO was the Falk-Hillarp fluorescence method (formaldehyde-induced fluorescence). This method showed that the basal portion of the organ is innervated by a dense monoaminergic plexus (Dahlström and Fuxe 1964) (Fig. 2a). In this connection, it has been impossible to demonstrate the presence of a catecholaminergic input (Bosler and Descarries 1983), but in contrast a massive serotoninergic innervation has been detected by several investigators (Bouchaud and Arluison 1977; Møllgård and Wiklund 1979). The distribution of the projections of serotoninergic neurons has been confirmed by immunocytochemistry using antibodies to biosynthetic enzymes such as tryptophan hydroxylase or to serotonin linked to a protein carrier (see reviews, see Geffard and al. 1988, and Takeuchi 1988; see Fig. 2b) and by radioautography using ventricular perfusion with tritiated serotonin (Fig. 2c).

Ultrastructural autoradiography has allowed identification especially of serotoninergic terminals constituting at least 75% of the inputs received by this organ in adult animals (Bouchaud 1979). The boutons, labeled after intraventricular perfusion of [^{3}H]-5 hydroxytryptamine (serotonin, ^{3}H-5 HT) and previous treatment with a monoamine oxidase inhibitor, contain a characteristic population of round and elliptical clear vesicles (Møllgård and Wiklund 1979) that are often pleomorphic and exhibit a dense central dot (Figs. 1b, 2d). These clear vesicles often appear to fill the varicosities, in association with less than 1% of large dense vesicles; a particular arrangement of these elements, however, has not been revealed.

The fine serotoninergic fibers end with abundant boutons contacting the perikarya at the level of the nuclei of the elongated ependymocytes or in the region of the basal processes of the ependymocytes and hypendymal cells. Occasionally, varicosities with a dystrophic appearance can be observed in contact with the perikarya of SCO ependymocytes (Bouchaud and Bosler 1986) or SCO microvessels (Fig. 2e).

Most of the boutons labeled by ventricular perfusion of [^{3}H]-5 HT contact SCO ependymocytes, which receive not only several serotoninergic inputs but also one or two unlabeled boutons indicating that these cells are polyinnervated (Bouchaud 1979; Bouchaud and Bosler 1986) (Fig. 2d). About 50% of the nonlabeled (nonserotoninergic) terminals display "paracrystalline" inclusions (Bouchaud and Bosler 1986); their neuromediator, however, remains unknown. Exceptionally, varicosities may contain a population

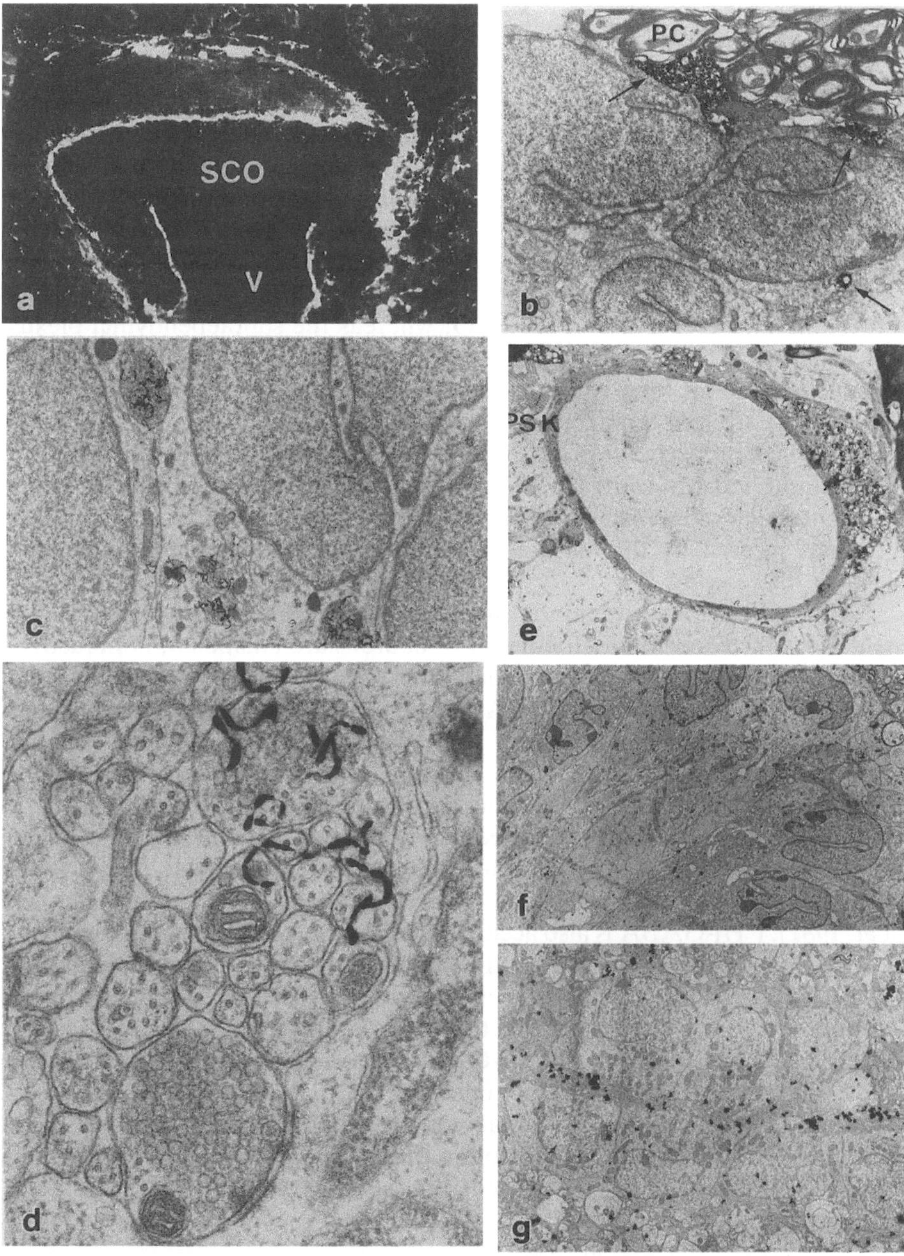

Fig. 2a–g. Serotoninergic innervation in the SCO area. **a** Fluorescent nerve fibers in a frontal section of the rat SCO. Note serotoninergic terminals in the basal region and supraependymal fibers in the immediate vicinity of the SCO, ×175; *V*, ventricle. **b** Immunocytochemical demonstration of serotonin-immunoreactive terminals (*arrows*) in SCO hypendyma; *PC*, posterior commissure. ×5600. **c–g** Radioautography after intraventricular administration of [^{3}H]5-HT. In the rat **c–e**, numerous terminals synapsing on the SCO ependymocytes are heavily labeled while another ending (*bottom*) is unlabeled (**d**). **c** ×3500; **d** ×33 500. Sometimes, dystrophic terminals contacting a SCO capillary take up the [^{3}H]5-HT; *PSK* "periodisch strukturierte Körper" (**e** ×3850). **f,g** Radioautography in the adult mouse; the SCO is devoid of labeling (**f**) while the supraependymal fibers are heavily labeled. ×1500

of large granular vesicles (Fig. 1c). This aspect which speaks in favor of peptidergic fibers is characteristic of probably less than 1/1000 boutons. These elements may correspond to the peptidergic fibers described by Pelletier et al. (1975) and Buijs and Pévet (1980); they are found immediately adjacent to the SCO. Nevertheless, Weindl and Sofroniew (1981) claimed that the presence of small amounts of arginine-vasopressin as revealed by radioimmunoassay is not supported by immunohistochemical observations.

In the mouse, radioautographic studies confirm the absence of a serotoninergic innervation of the SCO, whereas the supraependymal plexus takes up [^{3}H]-5 HT as holds true in the rat (Fig. 2f,g). Sparse data concerning a possible cholinergic innervation based on acetylcholinesterase histochemistry (Köhl 1975) are less convincing.

Specific neurotoxic destruction of serotoninergic terminals was achieved by Wiklund and Møllgård (1979) and Møllgård and Wiklund (1979), using intraventricular administration of 5,6- or 5,7-dihydroxytryptamine. Axon terminals injured by this neurotoxin can be easily identified by light and electron microscopy.

Autoradiography, immunohistochemistry, and neurotoxic destruction yield the same result: a small number of boutons are unlabeled, nonreactive, or do not show degenerative changes. These elements probably represent a minor non-monoaminergic input. A careful examination of these varicosities shows that their vesicular content may resemble that of labeled terminals (however, without pleomorphic vesicles) or display unusual features (paracrystalline arrangement). Their neurotransmitters remain unknown; a possible colocalization with neuropeptides is suggested.

In the rat, SCO ependymocytes accumulate extracellular [^{3}H]-GABA (tritiated γ-aminobutyric acid) by a specific uptake mechanism (Gamrani et al. 1981). [^{3}H]-GABA is also accumulated in numerous boutons innervating the secretory cells (Harandi et al. 1986). It appears that the uptake of GABA into SCO ependymocytes is controlled by the serotoninergic innervation since in the mouse and in the newborn rat (where the SCO is not yet innervated), no uptake of [^{3}H]-GABA occurs (Didier-Bazes et al. 1989). The 5-HT innervation is probably involved not only in the activation/stimulation of the GABA carriers but also in the inhibition of the expression in SCO ependymocytes of S100, a protein that is characteristic of various types of glial cells, including typical ciliated ependymocytes, astrocytes, and tanycytes (Chouaf et al. 1989).

Origin of the Serotoninergic Innervation

In mammals, most serotoninergic neurons are localized in the brain stem in certain raphe nuclei (Dahlström and Fuxe 1964; Steinbusch 1981; Takeuchi et al. 1983). A well known fact is that these neurons branch extensively; this was very clearly demonstrated for the supraependymal plexus along the walls of the cerebral ventricles (Arluison et al. 1982). Björklund et al. (1973) described the ascending serotoninergic fiber bundles issuing from the mesencephalic and pontine raphe nuclei and innervating the SCO area. This was confirmed later by Parent et al. (1981) who studied the organization of ascending serotoninergic systems innervating the SCO and other cerebral structures.

The most data concerning the source of the SCO innervation are documented for the rat. Searching for the origin by tract-tracing methods using, for instance, retrograde transport is difficult because the SCO borders the ventricles that are covered by serotoninergic supraependymal nerve fibers. The work of Léger et al. (1983) using electrolytic lesioning of different raphe nuclei enabled the cell bodies and the pathways innervating the SCO to be determined. It was suggested that the SCO innervation originates in three raphe nuclei, each contributing about one third of the input: (1) nucleus raphe centralis superior, (2) nucleus raphe dorsalis, and, probably, (3) nucleus raphe pontis. The perikaryon of the SCO ependymal cells appears to be approached by the finest fibers arising from the nucleus raphe centralis superior. The fibers originating from the nucleus raphe dorsalis may innervate the hypendymal region.

The examination of sagittal sections immunohistochemically treated for detecting serotoninergic cell bodies and their projections shows that the innervation reaches the SCO via its rostral part, and passes then under the lamina intercalaris, which contains a group of immunoreactive pinealocytelike cells (Møllgård and Wiklund 1979; Bouchaud and Bosler 1986) and the posterior commissure. This dense plexus extends throughout the organ. The density of the plexus decreases in the rostrocaudal direction since the fibers are distributed at all levels in the organ.

Innervation of the SCO in Other Vertebrate Species

As stated above, there are marked specific variations. Generally, lower vertebrates and sauropsids lack an innervation of their SCO. However, a few reports indicate the existence of a sparse peptidergic innervation (see, e.g., neurophysinergic innervation in the snake, *Natrix maura*; Fernandez-Llebrez et al. 1987).

In the following, results obtained in other rodents and some carnivores are reviewed.

In the Mongolian gerbil (Wiklund et al. 1977) and in the guinea pig (Tramu et al. 1983), a sparse serotoninergic innervation of the SCO has been described. No innervation exists in the mouse or in the rabbit, even if in the latter a few subcommissural organ-associated neurons were found in fetuses and neonates (Kimble and Møllgård 1975).

In the cat, sparse 5-HT-immunoreactive varicose fibers and terminals were found in the basal portion of the SCO (Sakumoto et al. 1984). In the dog, a pattern very different from that representing the characteristic innervation in rodents was described by Matsuura and Sano (1987). The fibers form a conspicuous basal plexus but also run apically between the SCO ependymocytes, sometimes protruding into the ventricular lumen and returning to the basal aspect of the SCO after having formed loops. This innervation is predominantly serotoninergic but moderate numbers of peptidergic fibers also exist (mainly vasopressin-immunoreactive axons). True synapses between the ependymal cells and axonal varicosities have not been described in carnivores.

Ontogenetic Development of the SCO Innervation in the Rat

The SCO differentiates early in the fetus (embryonic day 13 in the mouse; Castañeyra-Perdomo et al. 1983) and the differentiation of the secretory ependymocytes is independent of serotoninergic influence. Using formaldehyde-induced fluorescence, Wiklund et al. (1977) showed that the serotoninergic innervation which is absent at birth progressively develops and appears completed at the end of the first postnatal month. By means of immunohistochemistry employing 5-HT antibodies, we were able to identify some serotoninergic fibers in the rostral portion of the SCO from postnatal day 3 (Marcinkiewicz and Bouchaud 1986). On day 6, the immunoreaction was markedly increased in the rostral part and on day 9, the entire SCO displayed a positive 5-HT reaction. Ultrastructural studies revealed on postnatal day 1 large immature varicosities (swollen bulbs) in contact with SCO ependymocytes, but the first synaptic contacts were observed not earlier than postnatal day 3. The cytological character of the innervation was considered analogous to that in adults after postnatal day 21 (Marcinkiewicz and Bouchaud 1986).

Deafferentation After Intraventricular Administration of Neurotoxic Monoamines

Neurotoxic indolamines such as 5,6- or 5,7-dihydroxytryptamine can be used for destroying central serotoninergic fibers (Baumgarten et al. 1972). They induce a selective degeneration of serotoninergic terminals and can serve as tools for analysis of the role of serotoninergic innervation. As indicated by

Møllgård and Wiklund (1979), the rat SCO represents a unique model for the study of the serotonin-dependent neuron–secretory cell interaction. After various survival times (6 h to 3 days) changes in the ultrastructure of axons and synaptic profiles are evident. The selective chemical destruction associated with histochemical and ultrastructural study of the SCO secretory cells reveals conspicuous changes in the target cells (accumulation of secretory granules, increased release of secretory material; Møllgård and Wiklund 1979).

One month after denervation it was possible to observe reinnervation of SCO by newly formed synapses (Wiklund and Møllgård 1979). The reinnervation of the denervated SCO fully replaces the serotoninergic synapses at the original sites (e.g., close to the SCO perikarya). They fail, however, to exert inhibitory influence on the secretion. The newly formed boutons are not serotoninergic and originate from sprouting non-monoaminergic neurons. This result is in apparent contrast to the data obtained after chemical axotomy in the hypothalamus and certain circumventricular organs such as OVLT, showing that the serotoninergic neurons have capacities for axonal sprouting and regeneration (Saïdi and Bosler 1990). Given that the true synapses in the SCO represent specific features (usually serotoninergic fibers establish "en passant" contacts; Soghomonian et al. 1988), it might be that only nonsynaptic serotoninergic connections can be reestablished after neurotoxic injury; the nonjunctional fibers may have inherent capacities for continued remodeling (Wiklund and Møllgård 1979).

In a recent report, Ueda et al. (1988) described an original experiment in the rat in which the SCO had been previously chemically denervated. Three months after implantation of the raphe region from a normal rat fetus near the SCO into the extension of the third ventricle, numerous serotonin-containing fibers were found in the SCO. If the grafts were implanted 2.5 mm rostral to the SCO, there was no sign of reinnervation.

While no microglial cells were identified in the SCO hypendyma of control rats, they appeared on day 3 following intraventricular administration or 5,6 dihydroxytryptamine (Stagaard et al. 1987). These microglia elements phagocytosed debris originating from degenerated presynaptic profiles.

Conclusions

Among the numerous mammalian species in which the SCO represents a constant feature of the adult brain, the organ displays an innervation only in some rodents (gerbil, guinea pig, rat) and carnivores (cat, dog). The innervated cells are essentially the secretory ependymocytes, but also some hypendymal cells may establish neural contacts.

The SCO of the rat constitutes the best model for studying neuron–secretory cell interaction, the latter releasing their secretory products into the ventricular cerebrospinal fluid. The numerous axoglandular synapses, unique in the central nervous system, are predominantly serotoninergic. Morphological and histochemical studies of the SCO ependymocytes during postnatal development and after selective denervation in the adult clearly show that the serotoninergic innervation exerts a powerful inhibition on the synthetic machinery of the SCO ependymocytes producing and releasing glycoproteins.

The role of non-monoaminergic fibers, whose transmitter is unknown, and of the very rare peptidergic fibers remains open to discussion.

References

Arluison M, Bouchaud C, Derer P, Tramu G (1982) Etude immunocytochimique de l'arborisation des axones sérotoninergiques du rat. CR Acad Sci (Paris) 294:875–880

Baumgarten HG, Björklund A, Holstein AF, Nobin A (1972) Chemical degeneration of indolamine axons in the rat brain by 5,6-dihydroxytryptamine. Z Zellforsch 129:256–271

Björklund A, Nobin A, Stenevi U (1973) Effects of 5,6-hydroxytryptamine on nerve terminal serotonin and serotonin uptake in the rat brain. Brain Res 53:117–127

Bosler O, Descarries L (1983) Uptake and retention of [^{3}H] adrenaline by central monoaminergic neurons: a light- and electron-microscope radioautographic study after intraventricular administration in the rat. Neuroscience 8:561–581

Bouchaud C (1979) Evidence for a multiple innervation of subcommissural ependymocytes in the rat. Neurosci Letters 12:253–258

Bouchaud C (1987) Confronting hypolemmal associations involving mitochondria in contiguous ependymocytes of the rat subcommissural organ. A site of cellular interaction? Wiss Z Karl-Marx-Univ 36:114–115

Bouchaud C, Arluison M (1977) Serotoninergic innervation of ependymal cells in the rat subcommissural organ. A fluorescence electron microscopic and radioautographic study. Biol Cell 30:61–64

Bouchaud C, Bosler O (1986) The circumventricular organs of the mammalian brain with special reference to monoaminergic innervation. Int Rev Cytol 105:283–327

Brightman MW, Reese TS (1969) Junctions between intimely apposed cell membranes in the vertebrate brain. J Cell Biol 40:648–677

Buijs RM, Pévet P (1980) Vasopressin- and oxytocin-containing fibers in the pineal gland and the subcommissural organ of the rat. Cell Tissue Res 205:11–17

Castañeyra-Perdomo A, Meyer G, Ferres-Torres R (1983) Development of the subcommissural organ in the albino mouse (a Golgi study). J Hirnforsch 24:363–370

Chouaf L, Didier-Bazes M, Aguera M, Tardy M, Sallanon M, Kitahama K, Belin MF (1989) Comparative marker analysis of the ependymocytes of the subcommissural organ in four different mammalian species. Cell Tissue Res 257:255–262

Dahlström A, Fuxe K (1964) Evidence for the existence of monoamine-containing neurons in the central nervous system. I. Demonstration of monoamines in the cell bodies of brain stem neurons. Acta Physiol Scand 62[Suppl 232]:1–55

Didier M, Harandi M, Aguera M, Bancel B, Tardy M, Fages C, Calas A, Stagaard M, Møllgård K, Belin MF (1986) Differential immunocytochemical staining for glial fibrillary acidic protein, S-100 protein and glutamine synthetase in the rat subcommissural organ, non specialized ventricular ependyma and adjacent neuropil. Cell Tissue Res 245:343–351

Didier-Bazes M, Aguera M, Chouaf L, Harandi M, Calas A, Meiniel A, Belin MF (1989) Neuronal control of [^{3}H] GABA uptake in the ependymocytes of the subcommissural organ: an in-vivo model of neuron-glia interaction. Brain Res 489:137–145

Fernández-Llebrez P, Perez J, Cifuentes M, Alvial G, Rodríguez EM (1987) Immunocytochemical and ultrastructural evidence for a neurophysinergic innervation of the subcommissural organ of the snake *Natrix maura*. Cell Tissue Res 248:473–478

Gamrani H, Belin MF, Aguera M, Calas A, Pujol JF (1981) Radioautographic evidence for an innervation of the subcommissural organ by GABA-containing nerve fibres. J Neurocytol 10:411–424

Geffard M, Tuffet S, Peuble L, Patel S (1988) Production of antisera to serotonin and their use in immunocytochemistry. In: Osborne NN, Hamon M (eds) Neuronal serotonin. Wiley, Chichester, pp 1–23

Harandi M, Didier M, Aguera M, Calas A, Belin MF (1986) GABA and serotonin (5-HT) pattern in the supraependymal fibres of the rat epithalamus: combined radioautographic and immunocytochemical studies. Effect of 5-HT content on [^{3}H] GABA accumulation. Brain Res 370:241–249

Kimble JE, Møllgård K (1975) Subcommissural organ-associated neurons in fetal and neonatal rabbit. Cell Tissue Res 159:195–204

Köhl N (1975) Enzymatic organization of the subcommissural organ. J Histochem Cytochem 7:1–50

Léger L, Degueurce A, Lundberg JJ, Pujol JF, Møllgård K (1983) Origin and influence of the serotoninergic innervation of the subcommissural organ in the rat. Neuroscience 10:411–423

Madsen JK, Møllgård K (1979) The tight epithelium of the Mongolian gerbil subcommissural organ as revealed by freeze-fracturing. J Neurocytol 8:481–491

Marcinkiewicz M, Bouchaud C (1983) The ependymal secretion of the fetal and adult rat subcommissural organ. Morphological aspects linked to the synthesis, storage and release of the secretory products. Biol Cell 48:47–52

Marcinkiewicz M, Bouchaud C (1986) Formation and maturation of axo-glandular synapses and concomitant changes in the target cells of the rat subcommissural organ. Biol Cell 56:57–65

Matsuura T, Sano Y (1987) Immunohistochemical demonstration of serotoninergic and peptidergic nerve fibres in the subcommissural organ of the dog. Cell Tissue Res 248:287–295

Møllgård K, Wiklund L (1979) Serotoninergic synapses on ependymal and hypendymal cells of the rat subcommissural organ. J Neurocytol 8:445–467

Parent A, Descarries L, Beaudet A (1981) Organization of ascending serotonin systems in the adult rat brain. A radioautographic study after intraventricular administration of [^{3}H]5-hydroxytryptamine. Neuroscience 6:115–138

Pelletier G, Leclerc D, Dube F, Labrie F, Puviani R, Arimura A, Schally AV (1975) Localization of growth hormone-releasing-inhibiting hormone (somatostatin) in the rat brain. Am J Anat 142:397–400

Rakic P, Sidman RL (1968) Subcommissural organ and adjacent ependyma: autoradiographic study of their origin in the mouse brain. Amer J Anat 122:317–335

Saïdi H, Bosler O (1990) Serotonin reinnervation of the rat organum vasculosum laminae terminalis (OVLT) after 5,7-dihydroxytryptamine deafferentation. Brain Res 530: 151–155

Sakumoto T, Sakai K, Salvert D, Sasaki H, Kimura H, Maeda T, Jouvet M (1984) Possible role of serotonin in the secretory activity of the subcommissural organ of the cat. Neurosc Res 1:191–197

Soghomonian JJ, Beaudet A, Descarries L (1988) Ultrastructural relationship of central serotonin neurons. In: Osborne NN, Hamon M (eds) Neuronal serotonin. Wiley, Chichester, pp 57–92
Stagaard, M, Baslev Y, Lundberg JJ, Møllgård K (1987) Microglia in the hypendyma of the rat subcommissural organ following brain lesion with serotonin neurotoxin. J Neurocytol 16:131–142
Stanka P (1964) Untersuchungen über eine Innervation des Subkommissuralorgans der Ratte. Z Zellforsch Mikrosk Anat Forsch 71:1–9
Steinbusch HWM (1981) Distribution of serotonin-immunoreactivity in the central nervous system of the rat. Cell bodies and terminals. Neuroscience 6:557–618
Takeuchi Y (1988) Distribution of serotonin neurons in the mammalian brain. In: Osborne NN, Hamon M (ed) Neuronal serotonin. Wiley, Chichester, pp 25–56
Takeuchi Y, Sano Y (1983) Serotonin distribution in the circumventricular organs of the rat. An immunohistochemical study. Anat Embryol 167:311–319
Takeuchi Y, Kojima M, Matsuura T, Sano Y (1983) Distribution of serotoninergic neurons in the central nervous system. A peroxidase-antiperoxidase study with anti-serotonin antibodies. J Histochem Cytochem 31:181–185
Tramu G, Pillez A, Leonardelli J (1983) Serotonin axons of the ependyma and circumventricular organs in the forebrain of the guinea pig. Cell Tissue Res 228:297–311
Ueda S, Ihara N, Tanabe T, Sano Y (1988) Reinnervation of serotonin fibres in the denervated rat subcommissural organ by fetal raphe transplants. An immunohistochemical study. Brain Res 444:361–365
Weindl A, Sofroniew MV (1981) Relation of neuropeptides to mammalian circumventricular organs. In: Martin JB, Reichlin S, Bick KL (eds) Neurosecretion and brain peptides. Raven, New York, pp 303–319
Wiklund L, Møllgård K (1979) Neurotoxic destruction of the serotoninergic innervation of the rat subcommissural organ is followed by reinnervation through collateral sprouting of non-monoaminergic neurons. J Neurocytol 8:469–480
Wiklund L, Lundberg JJ, Møllgård K (1977) Species differences in serotoninergic innervation and secretory activity of rat, gerbil, mouse and rabbit subcommissural organ. Acta Physiol Scand [Suppl 452]:27–30

Impairment of the Serotoninergic Innervation of the Rat Subcommissural Organ After Early Postnatal X-Irradiation of the Brain Stem

N. DELHAYE-BOUCHAUD and C. BOUCHAUD

Introduction

The rat subcommissural organ (SCO) is remarkable for the glandular activity of its ependymocytes, which release their secretory products in the cerebrospinal fluid. Great attention has been paid to the nervous input to the SCO and marked species differences have been reported (see review by Bouchaud, this volume). In all mammals, in contrast to many other regions of the ventricular lining, the apical surface of the SCO is devoid of a supraependymal serotoninergic plexus; however, in the rat, the basal part of the organ is innervated by a dense plexus of fine serotoninergic fibers arising from different raphe nuclei (Léger et al. 1983). The serotoninergic fibers run between the SCO ependymocytes and their axonal varicosities form true synaptic contacts, mainly with the laterobasal portion of the ependymocytes (Bouchaud and Arluison 1977).

Evidence from developmental studies and denervation experiments has shown that the serotoninergic innervation exerts an inhibitory control on the secretory activity of the SCO. Indeed, the investment of the SCO by serotoninergic fibers begins on postnatal day 3 and is accompanied by a progressive decrease in the secretory function (Marcinkiewicz and Bouchaud 1986). Furthermore, in the adult, chemical deafferentation by means of serotonin (5-HT) neurotoxins leads to a marked increase in the synthesis and release of secretory products (Møllgård and Wiklund 1979).

The serotoninergic fibers which innervate virtually all regions of the mammalian central nervous system constitute a diffusely organized projection system (Parent et al. 1981). Most of these fibers do not establish synapselike junctions, and junctional 5-HT varicosities as in the rat SCO are exceptional (Chan-Palay 1975; Soghomonian et al. 1988).

Beaudet and Sotelo (1981) asked whether this dual mode of innervation is inherent to serotoninergic neurons or dependent on the local environment. Using the cerebellum as a model, they showed that a modification of the target area, namely, the deletion of the granule cell population by early postnatal X-irradiation, induces a profound alteration of the diffuse serotoninergic system which evolves from a "mostly non junctional into an entirely junctional input." They interpreted their findings as proof of the determining influence of the target on the synaptic modeling of the incoming serotoninergic nerve fibers. However, a possible direct effect of ionizing radiations on the developing serotoninergic axons was not considered or appraised.

The rat SCO is a convenient model to reinvestigate this issue, since, (a) as in the cerebellum, the serotoninergic fibers develop gradually during the postnatal period (Lidov and Molliver 1982) and establish exclusively synapselike junctions with their target, and (b) unlike the cerebellum, the SCO is already differentiated at birth and should not be modified by the radiation procedure which affects only proliferating cells (Hicks and d'Amato 1963).

Materials and Methods

The study was performed on normal adult Wistar rats and on rats which had received repeated X-irradiations on their brain stem during the early postnatal period (200 R at birth and 150 R on days 3, 4, 5, 6, and 7; total dose 950 R). The animals were allowed to survive for 2–3 months. They were pretreated with a monoamine-oxidase inhibitor (pheniprazine, 10 mg/kg i. p., 1 h preop.) and, under anesthesia, administered stereotaxically with 100 µl of [^{3}H]5-HT (10^{-4} *M*) injected into the lateral ventricle. One hour later, the rats were perfused with buffered glutaraldehyde via a vascular route. Tissue blocks containing the SCO were sectioned and the sections were processed for autoradiography or immunohistochemistry with specific 5-HT antibodies. Some sections were postfixed in osmium tetroxide and processed for electron microscopy.

Results

Effect of X-Irradiation on Serotoninergic Fibers

Radioautography. In control rats, the SCO area displays two locations of labeling after intraventricular administration of tritiated serotonin, i.e., in the basal portion of the SCO ependyma and in the supraependymal plexus (Fig. 1a). Under the "typical" ependyma, many parenchymal fibers are also labeled. In irradiated rats, the labeling of the serotoninergic input to the SCO is absent while the labeling of the supraependymal fibers is similar to that in control rats (Figs. 1b–d, 2b).

Immunohistochemistry. After incubation with specific 5-HT antibodies, in irradiated animals a faint reactivity is consistently observed at the basal pole

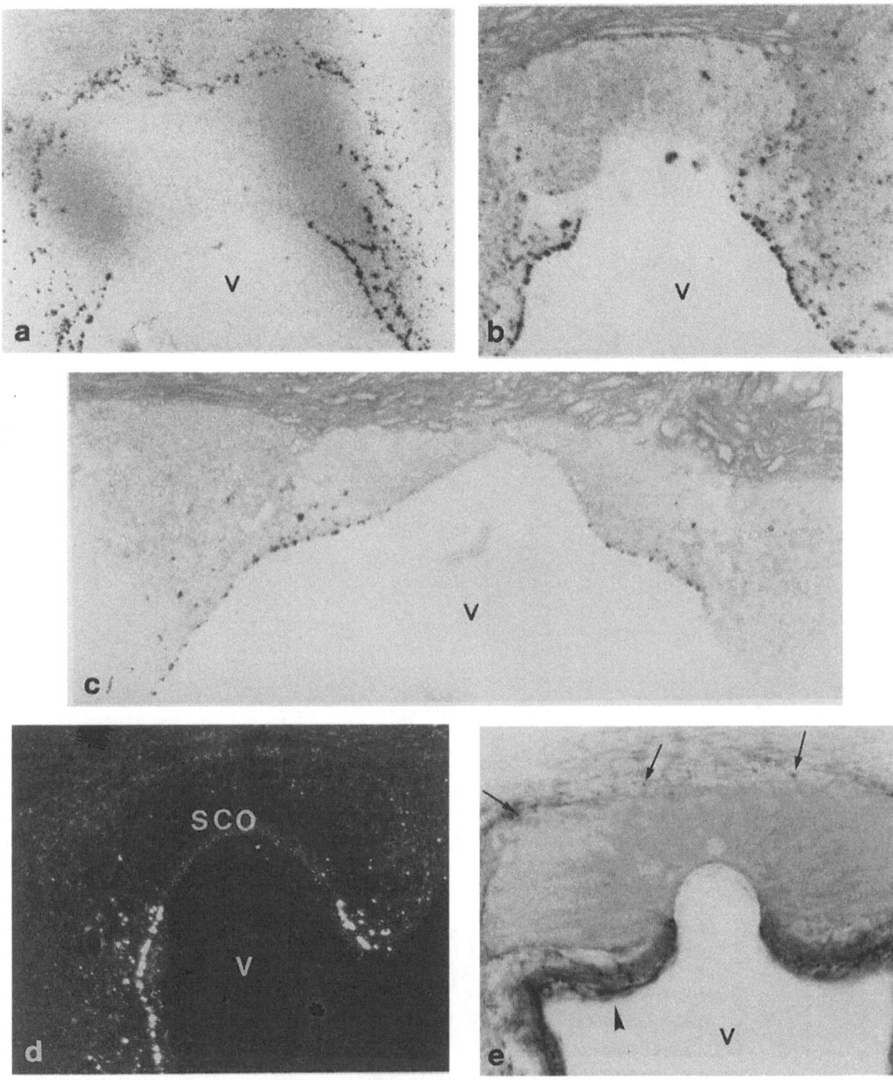

Fig. 1a–d. Light-microscopic [^{3}H]5-HT labeling in paraffin sections through the SCO of a control rat (**a**; ×190) and of X-irradiated rats (**b–d**; ×160; **d**, dark field). Note the almost total disappearance of axonal labeling in the irradiated rats SCO, while the supraependymal fibers appear normally labeled. *V*, Ventricle. **e** Rat SCO stained by means of 5-HT immunohistochemistry (×250). After irradiation, the supraependymal plexus is strongly immunoreactive (*arrowhead*), but the 5-HT fibers at the base of the SCO are only faintly reactive (*arrows*)

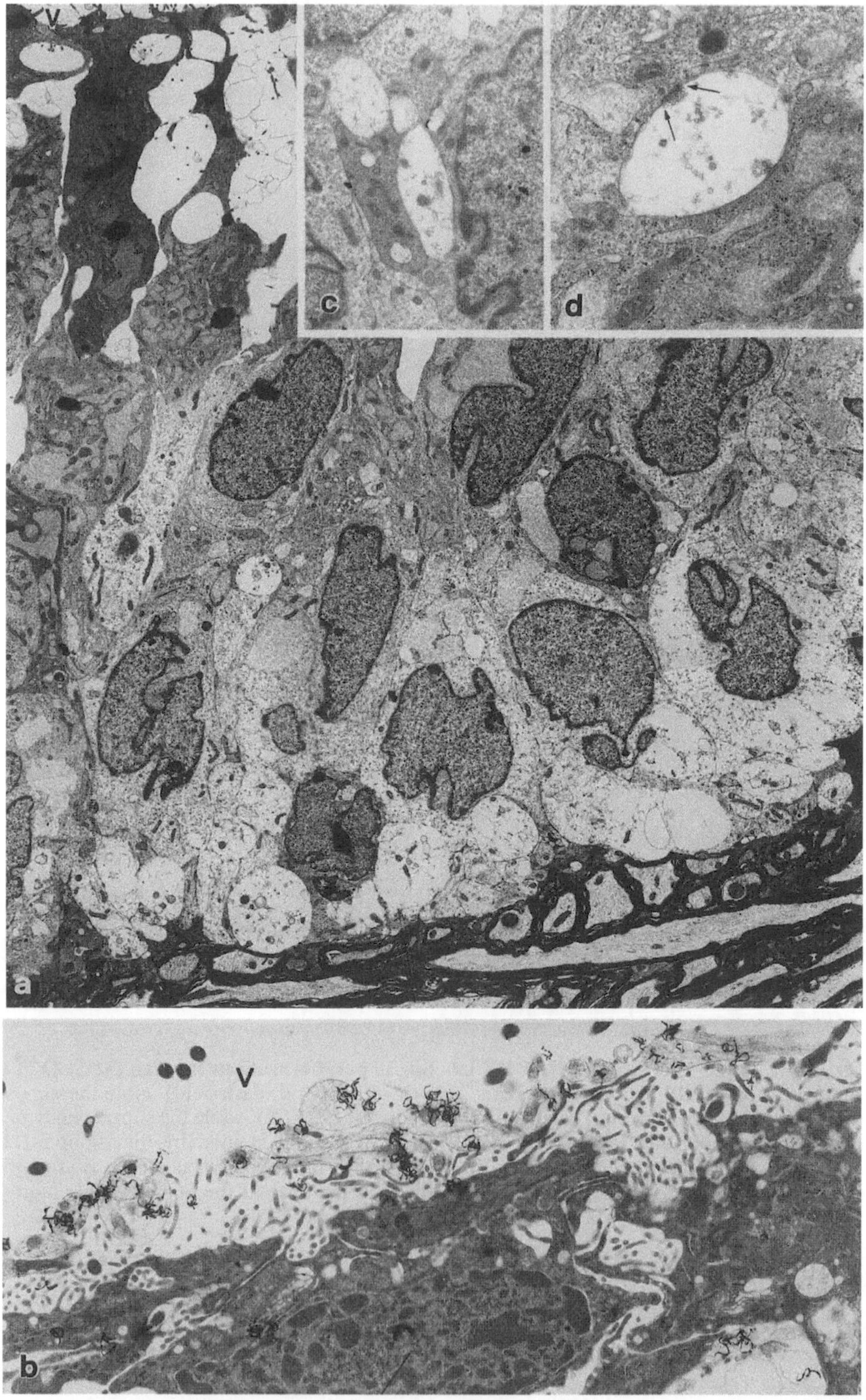
V
c
d
a
V
b

of the SCO, under the posterior commissure. The supraependymal plexus on the surface of the "typical" ependyma is always more reactive (Fig. 1e).

Ultrastructural Data. In the SCO of irradiated rats, the synaptic contacts are very rare, and the unlabeled varicosities are swollen and almost totally devoid of synaptic vesicles (Fig. 2a,c,d). In contrast, the varicosities of the supraependymal plexus have a normal appearance and are capable of specific uptake and retention of [^{3}H]5-HT (Fig. 2b).

Discussion

Two techniques have been used to visualize serotoninergic axons: (1) a radioautographic detection using specific uptake and retention of [^{3}H]5-HT, and (2) an immunohistochemical method using specific antibodies.

In the irradiated rats, 5-HT accumulating nerve fibers could no longer be observed in contact with the SCO. However, the faint serotonin immunoreactivity still present in the SCO indicates that the radiation procedure did not fully prevent the investment of the organ by serotoninergic axon terminals, a process which normally takes place postnatally (Wiklund et al. 1977; Marcinkiewicz and Bouchaud 1986). Although axonal guidance and cell recognition have occurred, electron-microscopic studies show that the varicosities contacting the SCO ependymocytes have an atypical, swollen appearance and contain only rare vesicles. In addition, the presynaptic 5-HT bouton and the postsynaptic glandular ependymocyte are almost devoid of synaptic differentiations. Therefore, there is a marked change in the mode of innervation, the serotoninergic synaptic input being replaced by a mostly nonjunctional contact. In contrast, the supraependymal serotoninergic fibers retain their normal features and appear to have preserved their uptake/retention capacity.

In normal control rats, the major difference between serotoninergic SCO axons and supraependymal nerve fibers lies in the presence of true synapses established on the SCO ependymocytes, whereas supraependymal fibers never form synaptic specializations with the ependymal cells.

In the rat SCO, it has been shown that serotoninergic innervation is directly or indirectly involved in the onset of the GABA-uptake carriers on

Fig. 2a–d. Rat SCO 3 months after irradiation (**a**; ×4800). Radioautography after administration of [^{3}H]5-HT shows no labeling in the SCO, in contrast to the heavily labeled supraependymal plexus (**b**; ×7200). The apical portion of the SCO ependymocytes and the "typical" ependymocytes display marked shrinkage with enlarged intercellular spaces. In the SCO, the presynaptic elements are swollen and almost devoid of synaptic vesicles (**c**,**d**; ×9600). A circumscribed synaptic differentiation is shown in **d** (*arrows*) the glandular ependymocytes (Didier-Bazes et al. 1989). In the same line, we can speculate that, during development, the SCO ependymocytes may interact with the contiguous axonal varicosities and control the uptake and/or the retention of 5-HT, as well as the synaptic differentiation. In our experiments, the irradiation procedure altered both the "typical" ependymal cells and the secretory SCO ependymocytes and thus possibly modified the interaction between the target cells and the afferent 5-HT fibers. The subsequent impairment of the 5-HT uptake mechanism would lead to a decreased synthesis and release of serotonin. This "functional" denervation might account for the hypersecretory aspect of the SCO ependymocytes, similar to that described after chemical denervation (Møllgård and Wiklund 1979).

Thus, our irradiation procedure did not prevent the proliferation of the raphe neurons innervating the SCO since these neurons are formed during the embryonic life (Wallace and Lauder 1983); neither did it prevent the axonal outgrowth and arborization nor the process of recognition of the target cell. On the other hand, it exerted an effect on the secretory ependymocyte, although it was already differentiated, possibly by modifying the interaction between the target cell and the afferent 5-HT fiber. Our results are consistent with previous findings on the cerebellar serotoninergic system (Beaudet and Sotelo 1981) and confirm the role of the target on the modeling of the serotoninergic input.

References

Beaudet A, Sotelo C (1981) Synaptic remodeling of serotonin axon terminals in rat agranular cerebellum. Brain Res 206:305–329

Bouchaud C, Arluison M (1977) Serotoninergic innervation of ependymal cells in the rat subcommissural organ. A fluorescence, electron microscopic and radioautographic study. Biol Cell 30:61–64

Chan-Palay V (1975) Fine structure of labelled axons in the cerebellar cortex and nuclei of rodents and primates after intraventricular infusions with tritiated serotonin. Anat Embryol 148:235–265

Didier-Bazes M, Aguera M, Chouaf L, Harandi M, Calas A, Meiniel A, Belin MF (1989) Neuronal control of [^{3}H] GABA uptake in the ependymocytes of the subcommissural organ: an in-vivo model of neuron-glia interaction. Brain Res 489:137–145

Hicks SP, d'Amato CJ (1963) Low dose radiation of developing brain. Science 141:903–905

Leger L, Deguerce A, Lundberg JJ, Pujol JF, Møllgård K (1983) Origin and influence of serotonergic innervation of the subcommissural organ in the rat. Neuroscience 10:411–423

Lidov HGW, Molliver ME (1982) An immunohistochemical study of serotonin neuron development in the rat: ascending pathways and terminal fields. Brain Res Bull 8:389–430

Marcinkiewicz M, Bouchaud C (1986) Formation and maturation of axo-glandular synapses and concomitant changes in the target cells of the rat subcommissural organ. Biol Cell 56:57–65

Møllgård K, Wiklund L (1979) Serotoninergic synapses on ependymal and hypendymal cells of the rat subcommissural organ. J Neurocytol 8:445–467
Parent A, Descarries L, Beaudet A (1981) Organization of ascending serotonin systems in the adult rat brain. A radioautographic study after intraventricular administration of [^{3}H]5-hydroxytryptamine. Neuroscience 6:115–138
Soghomonian JJ, Beaudet A, Descarries L (1988) Ultrastructural relationship of central serotonin neurons. In: Osborne NN, Hamon M (eds) Neuronal serotonin. Wiley, Chichester, pp 57–92
Wallace JA, Lauder JM (1983) Development of the serotonergic system in the rat embryo: an immunocytochemical study. Brain Res Bull 10:459–479
Wiklund L, Lundberg JJ, Møllgård K (1977) Species differences in serotoninergic innervation and secretory activity of rat, gerbil, mouse and rabbit subcommissural organ. Acta Physiol Scand [Suppl 452]:27–30

Immunocytochemistry of Neuropeptides and Neuropeptide Receptors in the Subcommissural Organ of the Rat

K.M. Knigge and D. Schock

Introduction

Seven small circumventricular organs (CVOs) of the brain have modifications in the anatomy of their neuropil, ependyma, or vasculature, which suggest that they may be involved in communication between brain and body. This communication may take the form of signals from brain to body via the systemic circulation (neurohemal) or from body into the brain (hemoneural). Of these windows of the brain, evidence for and the anatomical substrate for neurohemal communication is well established in the case of median eminence (ME), neurohypophysis (NP), and the pineal gland (PG); hemoneural communications may take place in the subfornical organ (SFO), the organum vasculosum laminae terminalis (OVLT), and the area postrema (AP). The subcommissural organ (SCO) is the least well understood with regard to its participation in this mode of brain–body communication.

Of the known bioactive peptides which leave the brain, specific sites of action and specific recognition mechanisms (receptors) in the body have been identified in almost all cases. Simply conceived, hemoneural communication involves blood-borne signals which originate in body (or in the brain) and are received in the CVOs for some specific action there or in some further locus in the brain. Implicit in the premise of specific actions is the presence of appropriate receptors. Although some peptides have been identified in the SFO, OVLT, AP, and SCO, substantial evidence for sites and mechanisms of action and receptors is limited.

In this report, using the SCO as a model, we review the CVOs as sites of hemoneural communication and report on immunocytochemical techniques which may identify receptors at a level of resolution which permits a more precise localization and inference with regard to sites of action.

Methods

Brains and pituitary glands of Sprague-Dawley rats (some pretreated with intraventricular infusion of 10–50 μg of colchicine) were fixed by whole body perfusion (at 100 mmHg) using a variety of fixatives, adjusted to 320–440 mosm. To a base solution of 4% paraformaldehyde and picric acid, glutaraldehyde or acrolein were added in various proportions, based on trial and error experiments to establish optimal fixatives for immunostaining of each peptide or receptor. Brains were Vibratome-sectioned at 20–40 μm; pituitary glands were embedded in gelatin before sectioning. Immunocytochemistry (ICC) was performed routinely with successive incubations in primary, link, and PAP antibody (Abs) with final color development using nickel-intensified diaminobenzidine (DAB). For EM, appropriate areas of stained sections were flat-embedded in Epon and reviewed with a Zeiss EM10 microscope.

Primary polyclonal or monoclonal *idiotypic* Abs were used to survey the distribution of serotonin (5HT) and several peptides including LRF, SRIF, CRF, TRF, ACTH, α-MSH, oxytocin (OXY), substance P (SP), neurotensin (NT), β-endorphin (βE), leu-enkephalin (l-ENK), met-enkephalin (m-ENK), dynorphin A (DYN-A), dynorphin B (DYN-B), neuropeptide Y (NPY), gastrin (GA), atrial natriuretic peptide (ANP), cholecystokinin (CCK), and vasoactive intestinal peptide (VIP). Polyclonal antisera containing *anti-idiotypic* antibodies (AIAs) available for receptor ICC include those for AVP, OXY, CRF, SRIF, ACTH, LRF, α-MSH, and l-ENK. Abs generated against peptide mRNA complementary peptides (CPAs; see "Discussion") include those for AVP and OXY. NADPH-diaphorase activity was revealed by the direct method descibed by Scherer-Singler et al. (1983); sections were incubated in a 50 m*M* TRIS-HCL solution, pH 8.0, with 0.5 m*M* nitroblue tetrazolium chloride (NBT) and 0.2 m*M* NADPH at 37°C for 30–60 min. For dual staining, ICC for CRF or CRF-AIA was performed first, followed by enzyme histochemistry.

Results

Peptides in the Central Gray (CG) and SCO. The central gray, particularly the dorsal and rostral regions around the SCO, contains perikarya and/or fibers of virtually all the peptides examined in this study (Fig. 1). The boundary between neuropil and ependyma of the SCO is demarcated clearly and in general indicates entrance of immunostained fibers into the area of the SCO (Fig. 2). In contrast to the adjacent neuropil and to other CVOs, the SCO is remarkably sparse in immunocytochemically demonstrable perikarya or processes of peptidergic neurons; AVP, CRF, and CCK are the only peptides which we could demonstrate consistently. AVP and CRF were seen in the form of beaded processes located primarily in the hypendymal region and in the nuclear zone of the ependyma (Fig. 3). Although only one to two processes were seen in any given section, they were present fairly consistently throughout the rostrocaudal extent of the SCO. Immunostaining

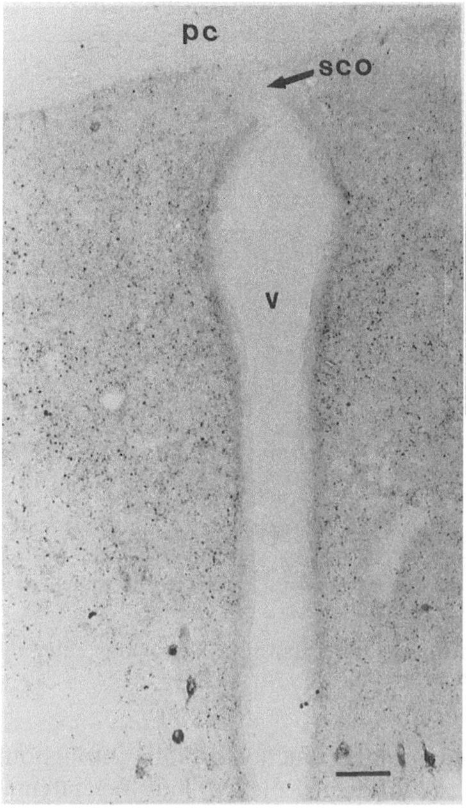

Fig. 1

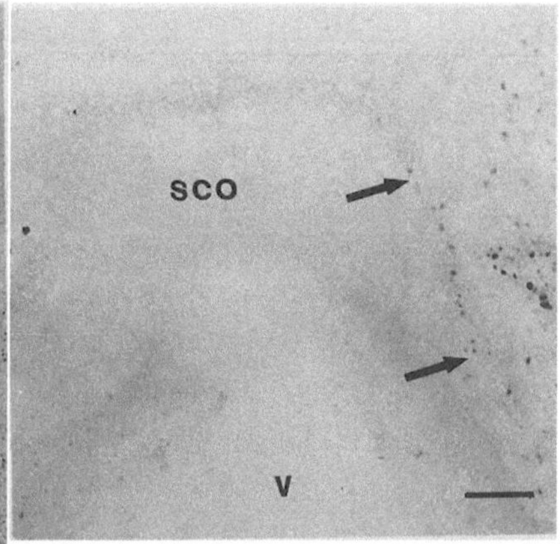

Fig. 2

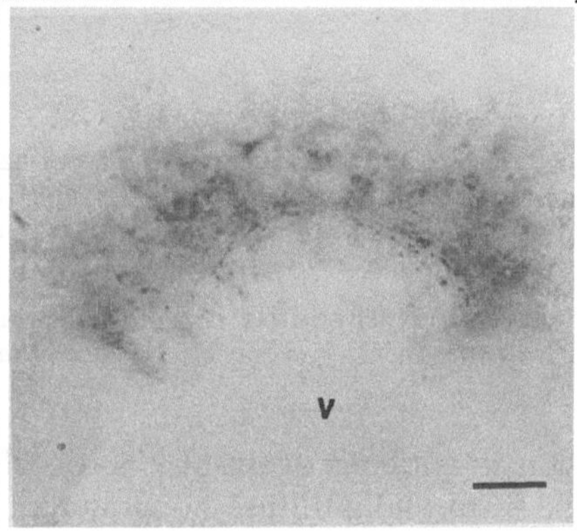

Fig. 4

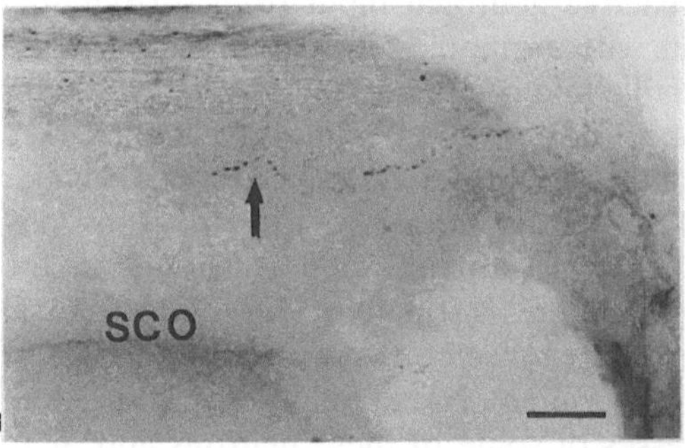

Fig. 3

Fig. 1. Midbrain dorsal CG and *SCO*, immunostained for SP, illustrating the heavy concentration of perikarya and fibers of these peptidergic neurons in the CG. *PC*, posterior commissure; *V*, ventricle III. *Scale bar* = 100 μm

Fig. 2. SCO of Fig. 1 at higher magnification. Clear-cut demarcation between the SCO and adjacent neuropil (*arrows*) for peptidergic axons not entering the SCO. *V*, ventricle III. *Scale bar* = 25 μm

Fig. 3. Peptidergic axons (immunoreactive for AVP, CRF) enter the SCO from adjacent neuropil (*arrow*); they are present sparsely throughout the rostrocaudal extent of the SCO. *Scale bar* = 50 μm

Fig. 4. SCO immunostained for CCK. The immunoreactive material is present in an atypical granular and diffuse form. *V*, ventricle III. *Scale bar* = 25 μm

Fig. 5

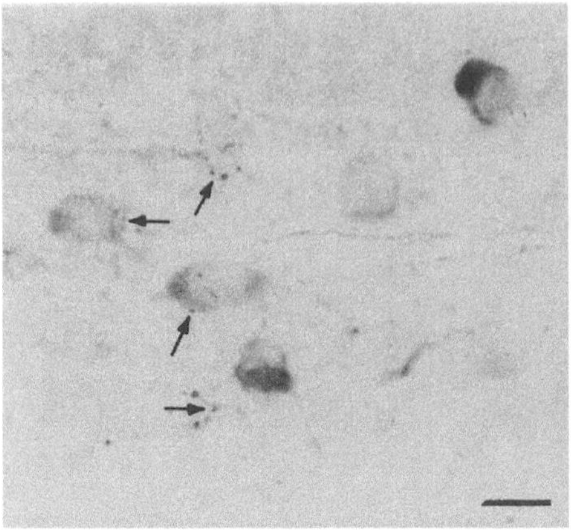

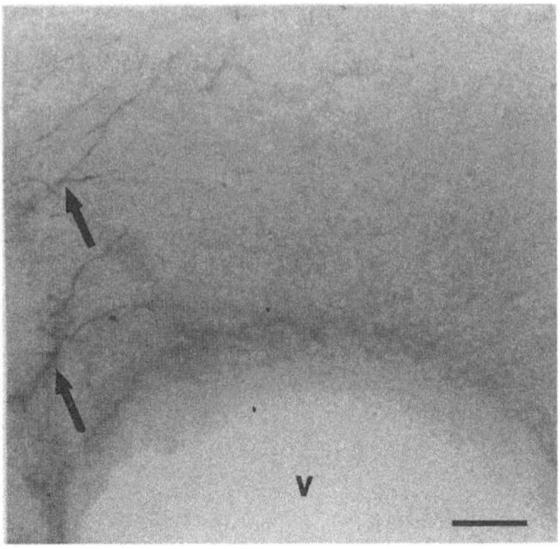

Fig. 6

Fig. 5. NADPH-positive neuronal perikarya in neuropil adjacent to the SCO; CRF-immunoreactive terminals (*arrows*) abut on both perikarya and fibers. In the original color photograph, NADPH (perikarya and fibers) are blue and CRF terminals are brown. *Scale bar* = 10 µm. Preparation of G. Michael

Fig. 6. NADPH-positive nerve fibers from the neurons shown in Fig. 5 may enter the area of the SCO. *V*, ventricle III. *Scale bar* = 50 µm

with an antibody against CCK revealed granular reaction product throughout the SCO and a diffuse nongranular reaction in the nuclear and hypendymal zones (Fig. 4). No CCK-immunoreactive neuronal input from surrounding neuropil could be found. The results with antibodies against three peptides – m-ENK, ANP, and NT – were variable and require further study; treatment of the animals with colchicine, specifically dose and time of action before sacrifice, may be critical in revealing these peptides in the SCO. Of particular interest are large numbers of NADPH-diaphorase-positive neurons distributed throughout the dorsal CG, with particular concentrations in the dorsolateral aspects of the CG and in the rostral portions of the dorsal raphe nucleus. Double-stain preparations revealed that virtually all of these neurons have CRF-immunoreactive terminal (synaptic) contacts on their soma and processes (Fig. 5). Processes of some of these neurons appear to enter the area of the SCO (Fig. 6).

Receptor ICC. Of the AIAs and CPAs used for receptor immunocytochemistry, only the CRF *anti-idiotype* revealed receptor images in the SCO. The "images" consist of dark, punctate dots 1–1.5 µm in diameter; they are distributed fairly uniformly throughout the SCO. It was not possible to establish their relationship to ependyma or blood vessels; in many instances, several images appeared to surround or be associated with small capillaries (Figs. 7, 8). Staining of these images is eliminated by preincubation of the antisera with preparations of whole-brain neural membranes or coincubation with excess CRF. Using adjacent sections to establish colocalization as

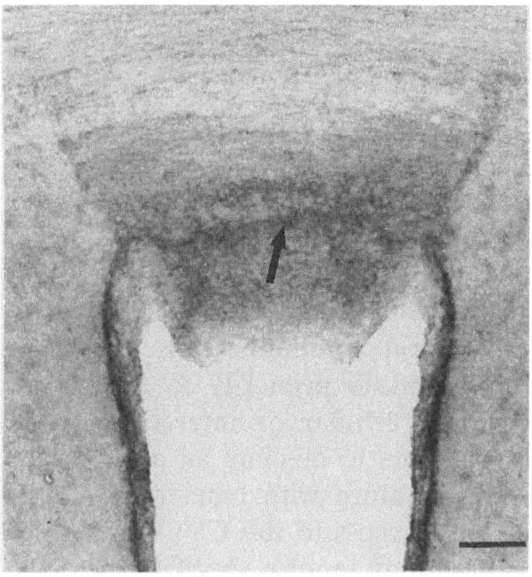

Fig. 7

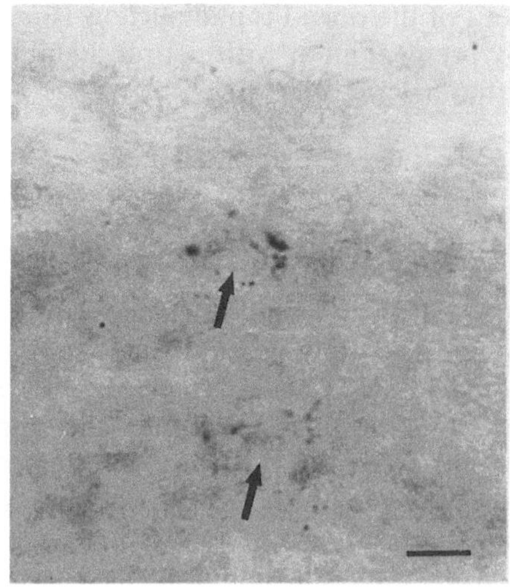

Fig. 8

Fig. 7. SCO immunostained with the CRF-AIA, at low magnification; receptor sites (*arrow*) are barely apparent in a diffuse, light brown background. *Scale bar* = 50 μm

Fig. 8. At higher magnification, CRF receptor sites (*arrows*) are randomly distributed throughout the SCO; in some instances they appear to encircle capillaries. *Scale bar* = 10 μm

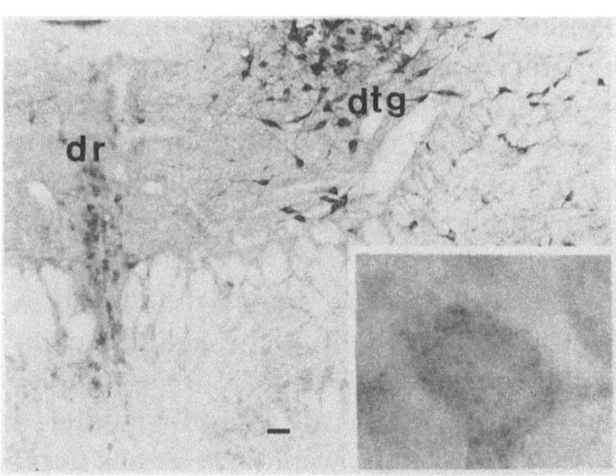

Fig. 9

Fig. 10

Fig. 9. Immunostaining of neurons in the dorsal raphe (*dr*) and dorsal tegmental nuclei (*dtg*) with the use of CRF-AIA. *Scale bar* = 40 μm. *Insert*, higher magnification of CRF–AIA-positive neurons showing the punctate nature of this staining

Fig. 10. An adjacent control section to Fig. 9 in which the CRF–AIA was preincubated with rat neural membranes (1.5 mg/ml dilute antiserum) overnight before staining. *Scale bar* = 50 μm

well as double-stain preparations to identify coexistence, it was shown that the CRF *anti-idiotype* stained also the NADPH-diaphorase-positive cells in the CG described above. The very distinct punctate nature of the immunostaining with the CRF *anti-idiotype* is revealed particularly well in cells of the dorsal raphe nucleus (Fig. 9). Absorption of this immunostaining by preincubation with neural membranes is demonstrated easily in this cell group (Fig. 10).

Discussion

Several anatomical considerations must be entertained in considering the hypothesis that receptor-mediated hemoneural communication occurs in some of the CVOs. These considerations pertain primarily to the access of blood-borne signals into the appropriate cellular or interstitial space of the CVO and the presence there of receptors to mediate an appropriate response. In those CVOs containing a vasculature with fenestrated capillaries, blood-borne bioactive molecules have access to the CVO interstitial space. A specific action of these signals here may require specific receptors, with the most likely sites of action being the specialized ependyma or neuronal perikarya or processes. Implicit in various schemata of "long" and "short" feedback for the control in the median eminence of hypothalamo-adenohypophyseal peptides, for example, is the requirement for peptide-specific recognition on receptors. Vascular blood flow may be regulated by specific signals with receptors in the capillary endothelium. The milieu of the interstitial space in CVOs where peptide–receptor interactions occur may have certain requirements such as composition and flow. Normally, the neuropil of the CNS is irrigated by a substantial flow into the ventricular cavities. It is of interest to speculate that the ependymal tight junctions in CVOs serve the purpose of restricting this flow and maintaining a composition physiologically more compatible with the plasma from which the peptide signals come.

Evidence for receptors in the CVOs challenges the morphologist to develop techniques with sufficient resolution to identify the cellular components involved. This level of resolution becomes critical, for example, in identifying the sites (glial, ependymal, neural) of receptor-mediated hemoneural communication in the CVOs. Autoradiography (AR) has heretofore been the method of choice for localizing receptors. At the light-microscope level, resolution is limited to brain regions; although the method can be brought to the EM level, the techniques are tedious and time-consuming and used infrequently. A further difficulty lies in the inconsistencies between AR and ICC and biochemical assay, a problem due most likely to different techniques and radiolabeled ligands as well as to receptor subtypes with different binding characteristics. ICC using antireceptor Abs offers a convenient method with resolution which can be brought easily to the EM level.

The several methods of creating anti-receptor antibodies include the generation of (a) poly- or monoclonal Abs against purified preparations of receptor, (b) sequencing the receptor, generating oligonucleotides that can be used to probe cDNA libraries for clones expressing the receptor, (c) anti-idiotypic Abs, and (d) Abs against "complementary" peptides. The latter two methods are being explored in our work. Methods of generating AIAs and their characterization and applications have been reviewed by us extensively (Knigge et al. 1986, 1987, 1988, 1989; Piekut et al. 1987; Abood et al. 1989a,b; Piekut and Knigge 1989). A complementary peptide represents one translated from sense mRNA although evidence for such translation is limited at present. As early as 1981, Biro (Biro 1981a,b,c) hypothesized that amino acids encoded by sense mRNA bind to corresponding amino acids encoded by antisense RNA. This sense–antisense complementarity in protein–protein interaction has been explored by Blalock and colleagues (Blalock and Smith 1984; Bost et al. 1985a,b; Blalock and Bost 1986; Weigent et al. 1986; Elton et al. 1988; Bost et al. 1988; Pascual et al. 1989) with consistent evidence that the sense–antisense relationship may form the molecular basis for protein recognition, e.g., hormone–receptor and enzyme–substrate interactions. Theoretical aspects of G-protein-coupled receptors (SRIF, LRF, α-2- and β-2-adrenergic, muscarinic, opiate, substance K, and angiotensin) have been discussed by Slootstra and Roubos (Slootstra and Roubos 1990, 1991); these peptides exhibit sense–antisense complementarity with all four extracellular regions of their receptors.

Results of the present study suggest that the SCO receives a limited input from brain peptidergic neurons. AVP- and CRF-immunoreactive fibers were identified. Significant immunostaining with our CCK antibody was observed, but in an atypical granular and diffuse form; the interpretation of this awaits further study. Putative CRF receptors were identified using a CRF *anti-idiotypic* Ab. This Ab is unusual in that it recognizes NADPH-associated nitric oxide synthetase (Hope et al. 1991). This enzyme is involved in the generation of nitric oxide, a novel transmitter (Garthwaite 1990). It is suggested that our CRF anti-idiotypic Ab recognizes a subunit of the CRF receptor comprised of this NADPH–nitric oxide synthetase complex.

The relationship of CRF to the SCO may be hypothesized as follows. CRF neurons, the location of whose perikarya has not been determined, send axonal processes into the area of the SCO. These fibers may have receptor contacts on some postsynaptic element whose identity is not yet established and must be proven at the ultrastructural level. Unusually, the receptor images revealed by the CRF-AIA appear not to be NADPH-positive; this may indicate that in some instances the receptor/NADPH-NO synthetase complex may be separated. Additional CRF input into the SCO may be present via an intermediate NADPH-positive neuron in the adjacent neuropil. Our results, together with considerations of the anatomical organization of the SCO, suggest that the activities of the SCO are influenced by neural input (CRF, AVP) from the surrounding brain, and that this CVO is not a window of extensive hemoneural communication between brain and body.

References

Abood LG, Michael GJ, Xin L, Knigge KM (1989a) Interaction of putative vasopressin receptor proteins of rat brain and bovine pituitary gland with an antibody against a nanopeptide encoded by the reverse message of the complementary mRNA to vasopressin. J Recept Res 9:19–25

Abood LG, Xin L, Michael G, Knigge KM (1989b) Interaction of putative vasopressin receptors in rat brain and bovine pituitary gland with a vasopressin anti-idiotype antibody as revealed by immunoblotting. Peptides 9:1407–1409

Biro J (1981a) Comparative analysis of specificity in protein-protein interactions, part I: a theoretical and mathematical approach to specificity in protein-protein interaction. Med Hypotheses 7:969–979

Biro J (1981b) Comparative analysis of specificity in protein-protein interactions, part II: the complementary coding of some proteins as the possible source of specificity in protein-protein interactions. Med Hypotheses 7:981–993

Biro J (1981c) Comparative analysis of specificity in protein-protein interactions, part III: models of the gene expression based on the sequential complementary coding of some pituitary proteins. Med Hypotheses 7:995–1007

Blalock JE, Bost KL (1986) Binding of peptides that are specified by complementary RNA's. Biochem J 234:679–683

Blalock JE, Smith EM (1984) Hydropathic anti-complementarity of amino acids based on the genetic code. Biochem Biophy Res Commun 121:203–207

Bost KL, Blalock JE (1989) Preparation and use of complementary peptides. Methods Enzymol 168:16–28

Bost KL, Smith EM, Blalock JE (1985a) Similarity between the corticotropin (ACTH) receptor and a peptide encoded by an RNA that is complementary to ACTH mRNA. Proc Natl Acad Sci USA 82:1372–1375

Bost KL, Smith EM, Blalock JE (1985b) Regions of complementarity between the messenger RNAs for epidermal growth factor, transferrin, interleukin-2 and their respective receptors. Biochem Biophys Res Commun 128:1373–1380

Bost KL, Smith LR, Blalock JE (1988) A molecular recognition code: its use for the purification of ACTH, endorphin, and LHRH receptors. In: Linthicum S, Farid A (eds) Antiidiotypies, receptors and molecular mimicry. Springer, Berlin Heidelberg New York, pp 35–44

Elton TS, Dion LD, Bost KL, Oparil S, Blalock JE (1988) Purification of an angiotensin II binding protein by using antibodies to a peptide encoded by angiotensin II complementary RNA. Proc Natl Acad Sci USA 85:2518–2522

Garthwaite J (1990) Nitric oxide synthesis linked to activation of excitatory neurotransmitter receptors in the brain. In: Moncada S, Higgs EA (eds) Nitric oxide from L-arginine: a bioregulatory system. Excerpta Medica, Amsterdam, pp 115–138

Hope BT, Michael GJ, Knigge KM, Vincent SR (1991) Neuronal NADPH-diaphorase is a nitric oxide synthetase. Proc Natl Acad Sci USA 88:2811–2814

Knigge KM, Piekut DT, Berlove D (1986) Immunocytochemistry of magnocellular neurons of supraoptic and paraventricular nuclei of normal and Brattleboro rats using vasopressin anti-idiotype antibody. Cell Tissue Res 246:509–513

Knigge KM, Piekut DT, Berlove D, Junig J, Melrose PA (1987) Staining of magnocellular neurons of the supraoptic and paraventricular nuclei with vasopressin anti-idiotype antibody: a potential method for receptor immunocytochemistry. Mol Brain Res 2:69–78

Knigge KM, Piekut DT, Berlove D (1988) Immunocytochemistry of a vasopressin (AVP) receptor with anti-idiotype antibody: inhibition of staining with a peptide (PVA) encoded by an RNA that is complementary to AVP mRNA. Neurosci Lett 86:269–271

Knigge KM, Piekut DT, Abood LG, Joseph SA, Berlove DJ, Michael GJ (1989) Use of anti-idiotype antibodies in receptor immunocytochemistry. Methods Enzymol 178:212–221
Pascual DW, Blalock JE, Bost KL (1989) Antipeptide antibodies that recognize a lymphocyte substance P receptor. J Immunol 143:3697–3702
Piekut DT, Knigge KM (1989) Immunocytochemistry of putative CRF receptors in rat forebrain using a CRF anti-idiotypic antibody. Neurosci Res Commun 4:167–173
Piekut DT, Knigge KM, Berlove D, Junig J, Melrose PA (1987) Application of a vasopressin anti-idiotype antibody for studies on vasopressin receptors. In: Goheen SC (ed) Membrane proteins, proceedings of the membrane protein symposium. Bio-Rad Laboratories, pp 587–600
Scherer-Singler U, Vincent SR, Kimura H, McGeer EG (1983) Demonstration of a unique population of neurons with NADPH-diaphorase histochemistry. J Neurosci Methods 9:229–234
Slootstra JW, Roubos EW (1990) Sense-antisense complementarity of hormone-receptor interaction sites. Trends Biotech 8:279–281
Slootstra JW, Roubos EW (1991) Sense-antisense complementarity in protein-protein interaction sites. In: Mol JNM, van Der Krol AR (eds) Antisense nucleic acids and proteins. Dekker, New York pp 205–222
Weigent DA, Hoedrich PD, Bost KL, Brunck TK, Reiher WE, Blalock JE (1986) The HTLV-III envelope protein contains a hexapeptide homologous to a region of interleukin-2 that binds to the interleukin-2 receptor. Biochem Biophys Res Commun 139:367–374

VII. Connections with Other Brain Structures

Relationships Between the Subcommissural Organ and the Pineal Complex

A. Oksche and H.-W. Korf

General Aspects

The close spatial relationship between the pineal body and the subcommissural organ (SCO) has been noted by all investigators of the diencephalic roof. The interest in this aspect of circumventricular organs has been further emphasized by developmental studies. In comparative terms, the SCO is a very ancient and persistent structure of the vertebrate brain, in contrast to the pineal complex that displays striking evolutionary changes with respect to its sensory and secretory structures and functions. In spite of this remarkable comparative variability in the organization of the roof of the third ventricle, the spatial contact between both areas is always very intimate.

Ontogenetic Development

In the ontogenetic development, no clear-cut borderline is visible between the primordia of the pineal and the SCO. Thus, the question of a possible shift of matrix material in one direction or the other has repeatedly been a matter of debate (see Bargmann 1943; Leonhardt 1980). This situation is properly illustrated by a developmental study in early anuran larvae (*Rana temporaria*) (see Oksche 1961; Fig. 1A). Under these premises the question appears justified whether neural and/or vascular connections of functional importance exist between the pineal and subcommissural areas.

Neural Connections

In a number of lower vertebrates (anamniotes), especially frogs, the final neural pathway (pineal tract) of the photosensory pineal complex, on its way to the pretectal area and to the reticular formation of the brain stem, runs along the basal portion of the SCO (Fig. 1B) and is encompassed by processes of SCO cells rich in secretory material (see Oksche 1954, 1955,

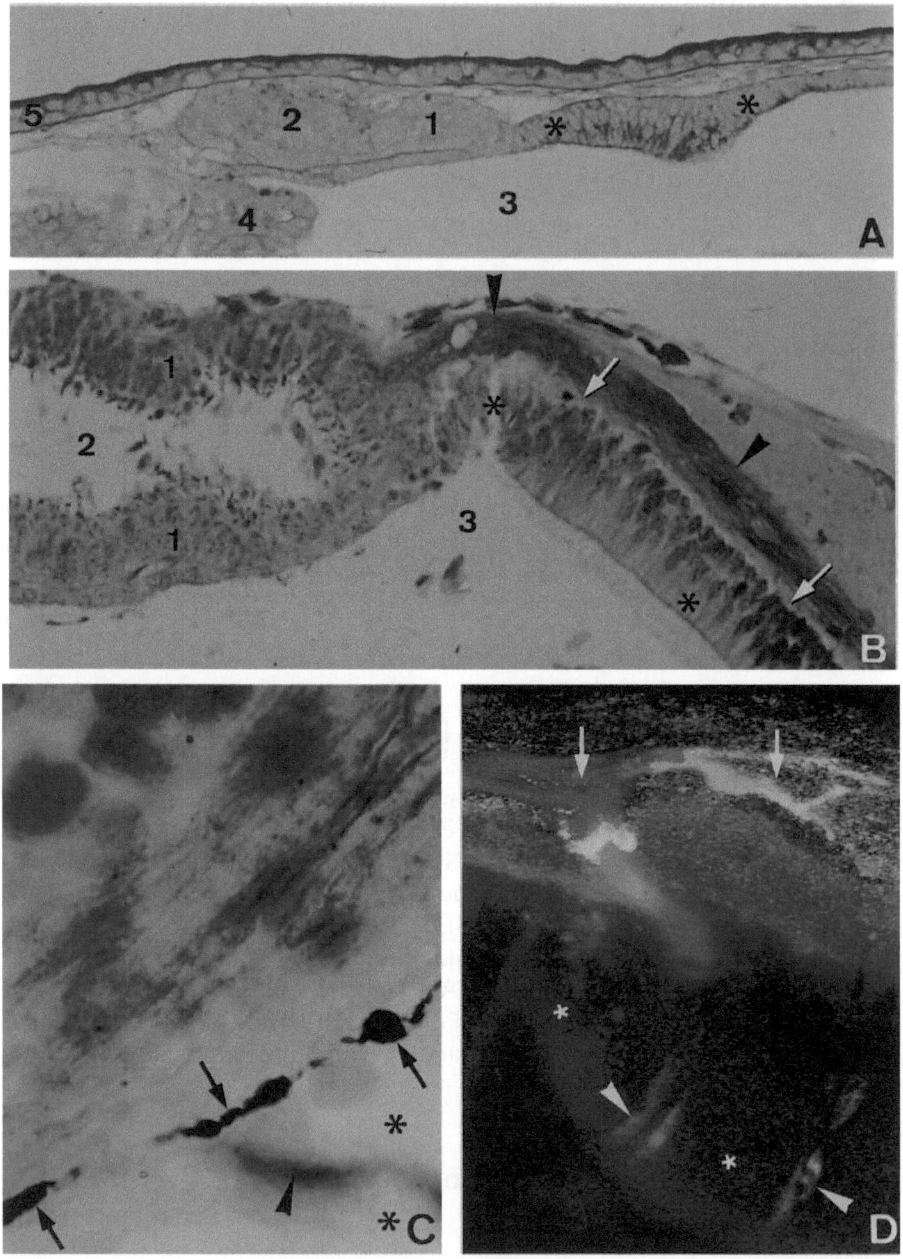

Fig. 1. A Primordium of the subcommissural organ (*asterisks*) and the pineal complex (*1*, pineal organ proper; *2*, frontal organ) in a larva of *Rana temporaria*. Third ventricle (*3*); choroid plexus of the third ventricle (*4*); epidermis (*5*). Paraldehyde fuchsin (Halmi-Dawson). ×200 (see Oksche 1961) **B** Close spatial relationships (*arrows*) between the pineal tract (*arrowheads*) and the subcommissural organ (*asterisks*) in adult *Rana esculenta*. *1*, pineal organ; note outer segments of photoreceptor cells extending into the

1956; Mautner 1965). Of course, spatial contacts of this type may be merely accidental and without any functional importance. Therefore, in order to deal with this problem in an adequate way, a closer look at the anatomical situation is needed.

In silver impregnations of the frog brain, some axons leave the pineal tract; they appear to penetrate into the SCO and to contact the basal aspect of its secretory cells with a knoblike swelling (Oksche 1954, 1955). Such light-microscopic images, however, do not prove the existence of synaptic (synaptoid) junctions. In electron-microscopic investigations of the frog pineal–SCO borderline area, Oksche and Vaupel-von Harnack (1965) were able to observe presynaptic axon terminals on basal structures of the SCO cells. Similar findings were reported by Rodríguez (1970) in toads (cf. Fig. 2C).

However, in both cases the exact proof that these vesicle-rich terminals indeed belong to axons of the pineal tract is missing. Too, they might originate from the very prominent posterior commissure containing fiber systems of different origin and functional significance. After transection of the pineal tract, Mautner (1965) and Paul et al. (1971) could clearly demonstrate degenerating axonal structures in intimate contact with the SCO (Fig. 1C). Both studies, however, were conducted exclusively at the light-microscopic level and thus do not provide evidence for the synaptic character of the described "endings." In addition, a population of acetylcholinesterase-reactive neurons resembling pineal second-order nerve cells accompanies the pineal tract in close contact with the basal aspect of the SCO (Wake et al. 1974). Similar findings were also obtained in the rainbow trout (Korf 1974). A reinvestigation of this region and the structures mentioned is needed by means of an anterograde tracer labeling the pineal tract and subsequent electron-microscopic study of the presumed contact area. (See p. 209 for recent developments.)

In the early period of investigations, Oksche (1954, 1955) proposed a working hypothesis regarding the pineal organ as a photic sensor and the SCO as a secretory effector. It was speculated that this complex might serve color-change mechanisms, since after longer periods on black background (followed by pigment expansion), the SCO of frogs (*Rana temporaria*) contained increased amounts of selectively stained secretory material. Similar

←

pineal lumen (*2*); *3*, third ventricle. Iron hematoxylin (Heidenhain), ×250. **C** Degenerating nerve fiber (*arrows*) of the pineal tract (20 days after transection) in close proximity to the subcommissural organ (*asterisks*), *Rana temporaria*. Nauta-Fink-Heimer. Capillary (*arrowhead*). ×500 (see Paul et al. 1971). **D** Relationship between the pineal tract (*arrows*) and the subcommissural organ (*asterisks*) of the rainbow trout as demonstrated by laser scanning microscopy of an 80-µm-thick frontal (coronal) Vibratome section. The brain was fixed with 4% paraformaldehyde. The pineal tract tract is labeled by DiI, which was placed into the pineal end-vesicle and allowed to difuse along the axonal membranes for 14 days. Different *colors* correspond to different planes within the section. Note a few DiI-labeled cells (*arrowheads*) in the subcommissural organ. ×320 changes in secretory activity had already been observed by Legait (1942), who used conventional staining procedures. All these studies were performed before the discovery of melatonin (Lerner et al. 1958) and the neurophysiological recordings revealing direct reponses of the pineal organ of lower vertebrates to light (Heerd and Dodt 1961; Dodt and Heerd 1962).

So far there is no conclusive proof of the above-mentioned systemic hypothesis. Shortly after his experiments with frogs, Oksche (unpublished results) realized that in some species of teleosts, which possess a distinct pineal tract, this pathway hardly establishes an anatomical connection to the SCO. These considerations are in accord with recent observations by Pérez-Fígares et al. (this volume) that the SCO of the goldfish is not innervated by fibers of the pineal tract. This statement is based on tracer studies (HRP) conducted at the electron-microscopic level. However, the fact that considerable differences among teleost species may exist should not be overlooked. Thus, Ekström and van Veen (1983) observed axons of the pineal tract penetrating deeply into the SCO of the three-spined stickleback; they used anterograde tracer techniques at the light-microscopic level. On the other hand, in several species of anamniotes and also in mammals lacking a pineal organ, the SCO is a spectacular, highly active secretory structure (see Oksche, this volume). Thus, the originally suggested neural connection cannot be regarded as a general property of the vertebrate brain. However, a reinvestigation of the pineal tract–SCO relationships in anurans (and other species of poikilothermic vertebrates) by means of combined tracer–ultrastructural studies is needed (see Fig. 2C and p. 209).

In these future studies it should also be examined whether a link between the SCO and the pineal organ can be established by cellular processes extending from the SCO into the pineal organ. The existence of such projections may be inferred from the labeling of scattered perikarya of SCO cells after tracer application to the pineal organ of the goldfish, rainbow trout (Fig. 1D), and golden hamster. In teleosts, these results were obtained by in vitro application of Di I (a lipophilic tracer) into the pineal organ (H.-W. Korf, H. Wicht and A.J. Jiménez, unpublished observations); in the golden hamster, the SCO perikarya were labeled after microiontophoretic injection of horseradish peroxidase into the pineal organ (Møller and Korf 1987).

Vascular Connections

A systemic link between the pineal organ and the SCO might also be established by vascular connections. In anurans, of particular interest because of their distinct pineal tracts, a careful study of the vascular

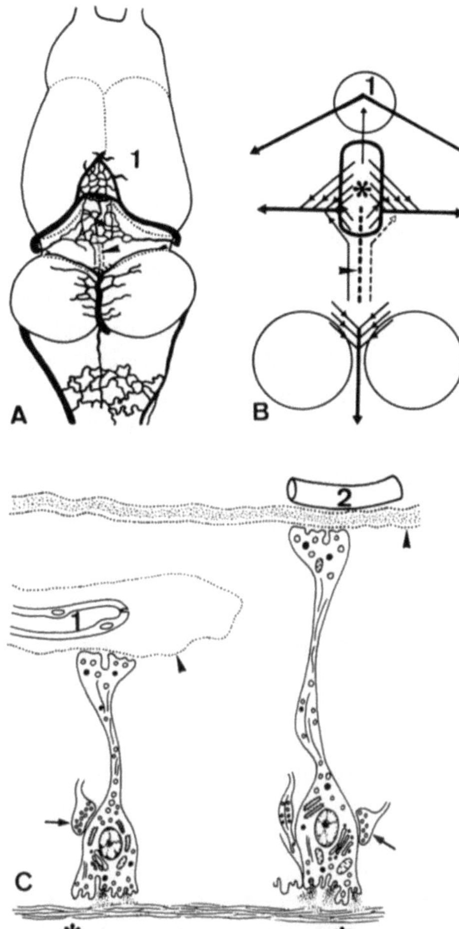

Fig. 2A–C. A,B Vascular system at the level of the diencephalic roof of frogs. For details, see text. **A** Specimen injected with India ink. Pineal organ (*asterisk*); pineal tract (*arrowhead*) marking the position of the subcommissural organ; paraphysis + choroid plexus (*1*) of the third ventricle. Semidiagrammatic representation. **B** The exact direction of the blood flow (*arrows*) in these vessels has been verified by observations in live, anesthetized animals (see Mautner 1965). Diagrammatic representation. **C** Secretory ependymocytes in the subcommissural organ of an anuran displaying contact with intrinsic (*1*) and leptomeningeal (*2*) blood vessels; basal lamina (*arrowheads*). Reissner's fiber (*asterisks*). Note presynaptic terminals (*arrows*) contacting the secretory ependymal cells. (Diagram modified after Rodríguez 1970; see p. 209.)

system was conducted by Mautner (1965). In whole mounts of India ink-injected frog brains, Mautner observed three different variations of a basic vascular pattern at the dorsal aspect of the brain (Fig. 2A). These specimens showed capillaries accompanying the pineal tract and penetrating into the pineal organ. Such preparations, however, are unsuitable to provide precise information concerning the blood flow in vivo; in this regard, vascular charts based on injected brains remain speculative. Functional deductions were considerably improved by Mautner's direct intravital observations in anesthetized frogs (Fig. 2B). The latter experiments clearly showed a blood flow in the capillaries extending from the posterior commissure to the pineal organ. However, there was no indication of the existence of an epithalamic portal vascular system, and some capillaries accompanying the pineal tract exhibited blood flow in the mesencephalic direction.

Does a similar body of observations exist for mammals? Oksche (1956) was puzzled by the occasional occurrence in several mammalian species (e.g., rat) of capillaries ascending from the region of the posterior commissure or SCO and penetrating into the pineal organ. In very careful and extensive studies with India-ink injections Duvernoy and Koritké (1964) could show that, in the cat, a system of connecting capillaries links the capillary beds of the SCO and the pineal organ. Again, there was no evidence for the existence of a portal system.

Possible Exchange of Matrix Material

Is there any recent evidence indicating an embryonic invasion of cellular elements from the SCO into the pineal body or vice versa? The particular developmental situation has already been discussed. In this context, a new avenue of thought was opened by the use of antisera against secretory glycoproteins of the SCO with the brains of a representative number of different vertebrates. These polyclonal antisera were shown to react (cross-react), in addition to the SCO cells, with pinealocytes of the receptor line in lamprey larvae, coho salmon, a toad, two species of lizards, and the domestic fowl (Rodríguez et al. 1988). By use of double labeling the coexistence of the reactive epitope with photoreceptor markers (rhodopsin and retinal S-antigen immunoreactions) could be verified.

Immunoreactive pinealocytes were observed also in the pineal organs of the rat and bovine, in the latter, however, only when an antiserum raised against an extract of the entire SCO was used (antisera against Reissner's fiber were, in this case, without any effect). After preadsorption with a pineal extract the effective antiserum no longer stained any cell of the pineal gland but continued to stain the secretory material in the SCO. These results appear to indicate the existence of a peculiar proteinaceous material in pinealocytes. These observations, however, leave the question unanswered of whether the pinealocytes immunoreactive with antisera against a certain component of the SCO secretion should be regarded as scattered derivatives of the SCO primordium (matrix).

New information available may, in addition, indicate another type of similarity by demonstrating a pineal-specific compound, at least temporarily, in the SCO as well. J. Bennett (personal communication) investigated rhodopsin promotor-regulated gene expression in pinealocytes of transgenic mice. These studies revealed common elements controlling gene expression in the pineal organ and in the retina. Surprisingly, during a certain period of the development a positive reaction was observed in the SCO too. So far, there is no answer concerning the participation of the SCO in this reaction. These results show, however, that important new impulses can be expected from molecular studies.

Conclusions

In view of the present evidence, it is uncertain whether the pineal and the SCO can be regarded as a functional complex (see Oksche, this volume). In the literature, the existence of an epithalamo-epiphysial complex has been repeatedly postulated (see Roussy and Mosinger 1938). This complex, in a wider sense, would include not only the habenular ganglia with their numerous neural connections, but also the choroid plexus of the third ventricle and the vascular connections of the epithalamic region. In functional terms, all these relationships are so far hypothetical. With respect to the SCO, one should keep in mind that a highly secretory active SCO also exists in species devoid of a pineal organ (e.g., *Myxine glutinosa*, all crocodilian species so far examined, armadillo). Even if a discrete, disseminated population of pinealocytes could be demonstrated in these species by molecular methods, a typical pineal body as a component of an anatomically defined epithalamic system would still be missing. This leaves us with a number of unanswered questions concerning relationships between the SCO and the pineal complex.

Pineal serotonin and/or melatonin may act on the SCO in a paracrine or CSF-mediated fashion, thus complementing the inhibitory influence of serotoninergic projections ascending from the lower brain stem in some (but not all) mammalian species (see Rodríguez et al. 1992). On the other hand, specific agents of the SCO might be conveyed by the circulating CSF to the pineal organ. In anurans possessing a pineal lumen in open communication with the third ventricle, this lumen may contain fibrous and reticular material containing (partly metachromatic) glycoproteins of enigmatic origin (see Oksche and Hartwig 1975). The latter may exhibit strong fluorescence for a serotonin precursor (5-HTP) occurring in sensory pinealocytes. Similar luminal structures can be observed in the pineal organ of turtles and lizards. All these problems in the line of pineal–SCO relationships deserve further careful investigation.

References

Bargmann W (1943) Die Epiphysis cerebri. In: Von Möllendorff W (ed) Handbuch der mikroskopischen Anatomie des Menschen, vol VI, part 4. Springer, Berlin Heidelberg New York, pp 309–502

Dodt E, Heerd E (1962) Mode of action of pineal nerve fibres in frogs. J Neurophysiol 25:405–429

Duvernoy H, Koritké JG (1964) Contribution à l'étude de l'angioarchitectonie des organes circumventriculaires. Arch Biol [Suppl] 75:693–748

Ekström P, van Veen T (1983) Central connections of the pineal organ in the three-spined stickleback, *Gasterosteus aculeatus* L. (Teleostei). Cell Tissue Res 232:141–155

Heerd E, Dodt E (1961) Wellenlängen-Diskriminatoren im Pinealorgan von *Rana temporaria*. Pflugers Arch Gesamte Physiol 274:33–34

Korf HW (1974) Acetylcholinesterase-positive neurons in the pineal and parapineal organs of the rainbow trout (with special reference to the pineal tract). Cell Tissue Res 155:475–489

Legait E (1942) Les organes épendymaires du troisième ventricule. L'organe sous-commissural. L'organe sub-fornical. L'organe para-ventriculaire. Med thesis, University of Nancy

Leonhardt H (1980) Circumventriculäre Organe. In: Oksche A, Vollrath L (eds) Handbuch der mikroskopischen Anatomie des Menschen, vol IV, part 10. Springer, Berlin Heidelberg New York, pp 362–538

Lerner AB, Case JD, Takahashi Y, Lee Y, Mori W (1958) Isolation of melatonin, the pineal gland factor that lightens melanocytes. J Am Chem Soc 80:2587

Mautner W (1965) Studien an der Epiphysis cerebri und am Subcommissuralorgan der Frösche. Mit Lebendbeobachtung des Epiphysenkreislaufs, Totalfärbung des Subcommissuralorgans und Durchtrennung des Reissnerschen Fadens. Z Zellforsch 67:234–270

Møller M, Korf HW (1987) Neural connections between the brain and the pineal gland of the golden hamster (*Mesocricetus auratus*). Cell Tissue Res 247:145–153

Oksche A (1954) Über die Art und Bedeutung sekretorischer Zelltätigkeit in der Zirbel und im Subkommissuralorgan. Anat Anz 101:88–96

Oksche A (1955) Untersuchungen über die Nervenzellen und Nervenverbindungen des Stirnorgans, der Epiphyse und des Subkommissuralorgans bei anuren Amphibien. Gegenbaurs Morphol Jahrb 95:393–425

Oksche A (1956) Funktionelle histologische Untersuchungen über die Organe des Zwischenhirndaches der Chordaten. Anat Anz 102:404–419

Oksche A (1961) Vergleichende Untersuchungen über die sekretorische Aktivität des Subkommissuralorgans und den Gliacharakter seiner Zellen. Z Zellforsch 54:549–612.

Oksche A, Hartwig HG (1975) Photoneuroendocrine systems and the third ventricle. In: Knigge KM, Scott DE, Kobayashi H, Ishii S (eds) Brain-endocrine interaction II. The ventricular system in neuroendocrine mechanisms. Karger, Basel, pp 40–53

Oksche A, Vaupel-von Harnack M (1965) Elektronenmikroskopische Untersuchungen an den Nervenbahnen des Pinealkomplexes von *Rana esculenta* L. Z Zellforsch 68:389–426

Paul E, Hartwig HG, Oksche A (1971) Neurone und zentralnervöse Verbindungen des Pinealorgans der Anuren. Z Zellforsch 112:466–493

Rodríguez EM (1970) Ependymal specializations. III. Ultrastructural aspects of the apical secretion of the toad subcommissural organ. Z Zellforsch 111:32–50

Rodríguez EM, Korf HW, Oksche A, Yulis CR, Hein S (1988) Pinealocytes immunoreactive with antisera against secretory glycoproteins of the subcommissural organ: a comparative study. Cell Tissue Res 254:469–480

Rodríguez EM, Oksche A, Hein S, Yulis R (1992) Cell biology of the subcommissural organ. Int Rev Cytol 135:39–121

Roussy G, Mosinger M (1938) Le complexe epithalamo-epiphysaire. Rev Neurol 69: 459–470
Wake K, Ueck M, Oksche A (1974) Acetylcholinesterase-containing nerve cells in the pineal complex and subcommissural area of the frogs, *Rana ridibunda* and *Rana esculenta*. Cell Tissue Res 154:423–442

Added in proof. For new details, see the following paper: Jiménez AJ, Pérez-Fígares JM, Rodríguez EM, Fernández-Llebrez P, Oksche A (1993) Synapse-like contacts between axons of the pineal tract and the subcommissural organ in *Rana perezi* (Anura) and their absence in *Carassius auratus* (Teleostei): ultrastructural tracer studies (Cell Tissue Res, in press). This report, based on experimental evidence, reveals principal differences between anuran amphibians and teleosts. In anurans, labeled axons emerging from the pineal tract establish numerous synapse-like contacts with secretory ependymocytes of the SCO, in addition to terminals of unknown source and nature.

The Goldfish Subcommissural Organ Is Not Innervated by Fibers of the Pineal Tract

J.M. PÉREZ-FÍGARES, A.J. JIMÉNEZ, and E.M. RODRÍGUEZ

Introduction

An innervation of the subcommissural organ (SCO) by nerve fibers arising from the pineal organ (PO) has been suggested by several authors (Diederen 1975; Korf 1976; Ekström and van Veen 1983; Ekström and Korf 1985). However, no conclusive evidence has so far been obtained (see Oksche and Korf, this volume).

Neuroanatomical tracing studies in teleosts have revealed a prominent bundle of nerve fibers arising from the pineal tract and running toward midbrain areas in close apposition to the basal portion of the SCO cells (Sathyanesan and Sastry 1982; Ekström and van Veen 1983, 1984; Puzdrowski and Northcutt 1989). Furthermore, ultrastructural studies in some species have shown presynaptic terminals of different origin on the SCO cells, in amphibians (Murakami and Tanizaki 1963; Oksche and Vaupel von-Harnack 1965; Altner 1968; Diederen 1970; Rodríguez 1970) and mammals (Kimble and Møllgård 1973; Bouchaud and Arluison 1977; Møllgård et al. 1978; Bouchaud 1979; Møllgård and Wiklund 1979). According to Møllgård and Wiklund (1979), the predominant neural input to the SCO is serotoninergic. Bouchaud (1979) has estimated that about 75% of the nerve endings contain serotonin, while the remaining 25% are endowed with an unknown transmitter. Gamrani et al. (1981) and Weissman-Nanopoulus et al. (1983) have presented evidence that a small population of nerve terminals on SCO cells contains GABA.

Regardless the existence of a possible neural connection between these two circumventricular organs, other forms of communication such as endocrine and paracrine, via the cerebrospinal fluid, may exist in lower vertebrates, in which the pineal lumen freely communicates with the third ventricle just rostral to the SCO.

Tracer methods have been used recently at the light-microscopic level, whereas the earlier ultrastructural studies of the synaptic terminals on the SCO cells did not employ tracers. Thus, the existence of axons of pineal origin establishing synaptic contacts with the SCO cells has not been definitely investigated. The aim of the present study was to prove whether the SCO of *Carassius auratus* is innervated by nerve fibers arising from the PO. For this purpose a neuroanatomical tracing method was applied at the light- and electron-microscopic levels.

Materials and Methods

Animals. Sixty adult goldfish (3–15 cm length), *Carassius auratus*, of both sexes (2–12 g body weight) were used.

Light Microscopy. Fifty animals were maintained under light anesthesia (3-aminobenzoic acid ethyl ester, methanosulfonate salt, 0.06 g/l, Sigma) for 30–60 min at 5°C, with the cranial roof over the water surface. A small incision was made just over the pineal stalk and pineal vesicle. Then 0.1–0.5 µl of 10%–30% HRP (Sigma, Type VI) solution diluted in 0.9% ClNa was applied either onto the intact or cut pineal vesicle and pineal tract. After 1 h, the opening was sealed, and the fish were placed back into fresh water.

Eight, 16, 24 and 48 h after surgery, the fish were again anesthetized (see above) and the brains dissected out and fixed in Bouin's fluid (Baker 1946) for 24 h. Embedding was in paraffin. Sections, 10 µm thick, were treated with xylene, dehydrated, and immunostained according to the PAP method of Sternberger (1986) with an anti-HRP antiserum raised in rabbit (dilution 1:6000). Finally, two intensifying methods were used, i.e., double PAP method (Vacca et al. 1980) and silver-methenamine (Rodríguez et al. 1984). In some cases, after silver-methenamine intensification, the sections were processed for a second immunostaining using an antiserum against the bovine Reissner's fiber (AFRU, see Rodríguez et al. 1987). After this double immunostaining procedure, the nerve fibers of the pineal tract appeared black and the secretion of the SCO was brown.

Electron Microscopy. In 10 animals the HRP solution was applied as described above for regular light microscopy. After a survival time of 24 h the animals were anesthetized and perfused transcardially with 0.1 *M* phosphate buffer, followed by fresh Karnovsky's fixative (1% paraformaldehyde and 1% glutaraldehyde in 0.1 *M* phosphate buffer, pH 7.4). The brains were dissected out and immersed in the same fixative for 3 h at 4°C.

Vibratome and cryostat sections were processed for the demonstration of the enzymatic activity of peroxidase using 3,3′-diaminobenzidine and hydrogen peroxide as described by Graham and Karnovsky (1966). Some of the sections through the SCO displaying peroxidase-labeled nerve fibers were treated with 1% OsO_4 for 2 h, dehydrated in graded series of ethanol and acetone and embedded in Araldite. From each animal 3–5 sections, 20–30-µm thick, were processed for electron microscopy. Each of these slices was cut on the ultramicrotome, and ultrathin sections were collected from approximately four different planes throughout the 20 to 30-µm-thick section. Sections were either unstained or stained with lead citrate for 2 min. Seventy-five grids corresponding to four animals were carefully studied under the electron microscope and 240 electron micrographs were obtained.

Results

Light Microscopy. Forty animals showed an uptake of HRP, and the course of the fibers of the pineal tract was clearly visualized. The best results were obtained in specimens killed 24 h after HRP administration. The nerve fibers of the pineal tract appeared heavily labeled with HRP; at variance, the fibers of the posterior commissure were free of the label. In almost all animals, the ventricular lumen and the tissues adjacent to the pineal fibers were free of HRP. Occasionally, SCO cells close to the pineal tract showed an intracellular labeling.

The goldfish pineal tract penetrates the brain at two levels: (1) the habenular commissure and (2) rostral to the posterior commissure, just in front of the SCO. This latter tract bifurcates into two bundles located at both sides of the midsagittal plane and extending toward the SCO ependymal cells (Fig. 1). It contains two types of fibers: (1) thin unmyelinated axons, displaying swellings, and (2) thick myelinated elements.

At rostral levels of the SCO, pineal fibers were shown to leave the two compact bundles and project laterally forming a dense plexus at the SCO–posterior commissure interface (Fig. 1). Basal processes of the SCO cells established close spatial contacts with fibers of the pineal tract (Fig. 1). At caudal levels of the SCO, the pineal tracts were exclusively formed by myelinated fibers, which proceeded caudad toward thalamic areas. At this level, a neural plexus over the SCO was missing.

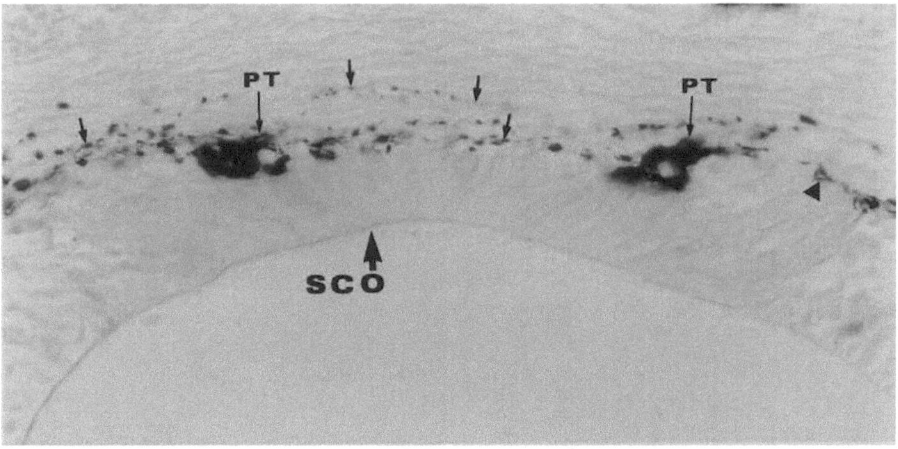

Fig. 1. Transverse paraffin section through the rostral region of the SCO of a goldfish killed 24 h after administration of HRP into the pineal organ and tract. Immunostaining with HRP antiserum and subequent intensification with silver methenamine. The two main bundles of the pineal tract appear transversally cut in a bilateral position (*PT*). Fine beaded fibers (*small arrows*) extend along the base of the subcommissural organ (*SCO*). *Arrowheads*, close contact between ependymal processes and pineal fibers; *V*, third ventricle. ×240

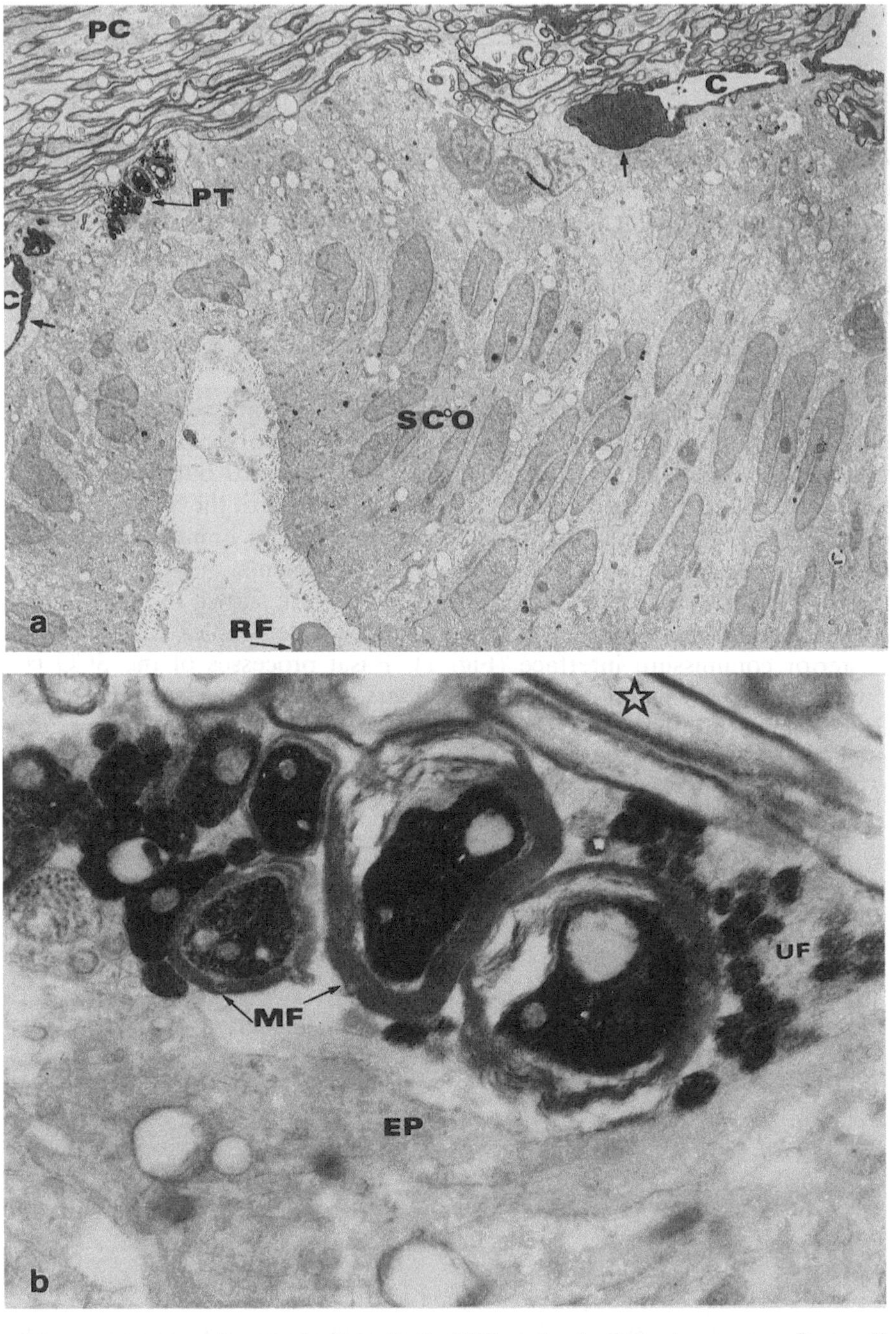

Fig. 2a–e. Electron micrographs through the SCO region in *Carassius auratus* after application of HRP into the pineal tract. Enzymatic activity of HRP was revealed in cryostat and Vibratome sections. **a** Survey of the rostral portion of the subcommissural organ (*SCO*). Endogenous peroxidase occurs in endothelial cells (*arrows*) of blood capillaries (*C*) and in peroxisomes. Exogenous peroxidase is found exclusively in fibers of the pineal tract (*PT*). *PC*, posterior commissure; *RF*, Reissners fiber. ×1600. **b** Selective labeling of myelinated (*MF*) and unmyelinated (*UF*) pineal axons. *Star*, unlabeled myelinated fiber of the posterior commissure; *EP*, ependymal processes of SCO cells. ×13 500. **c** Detailed

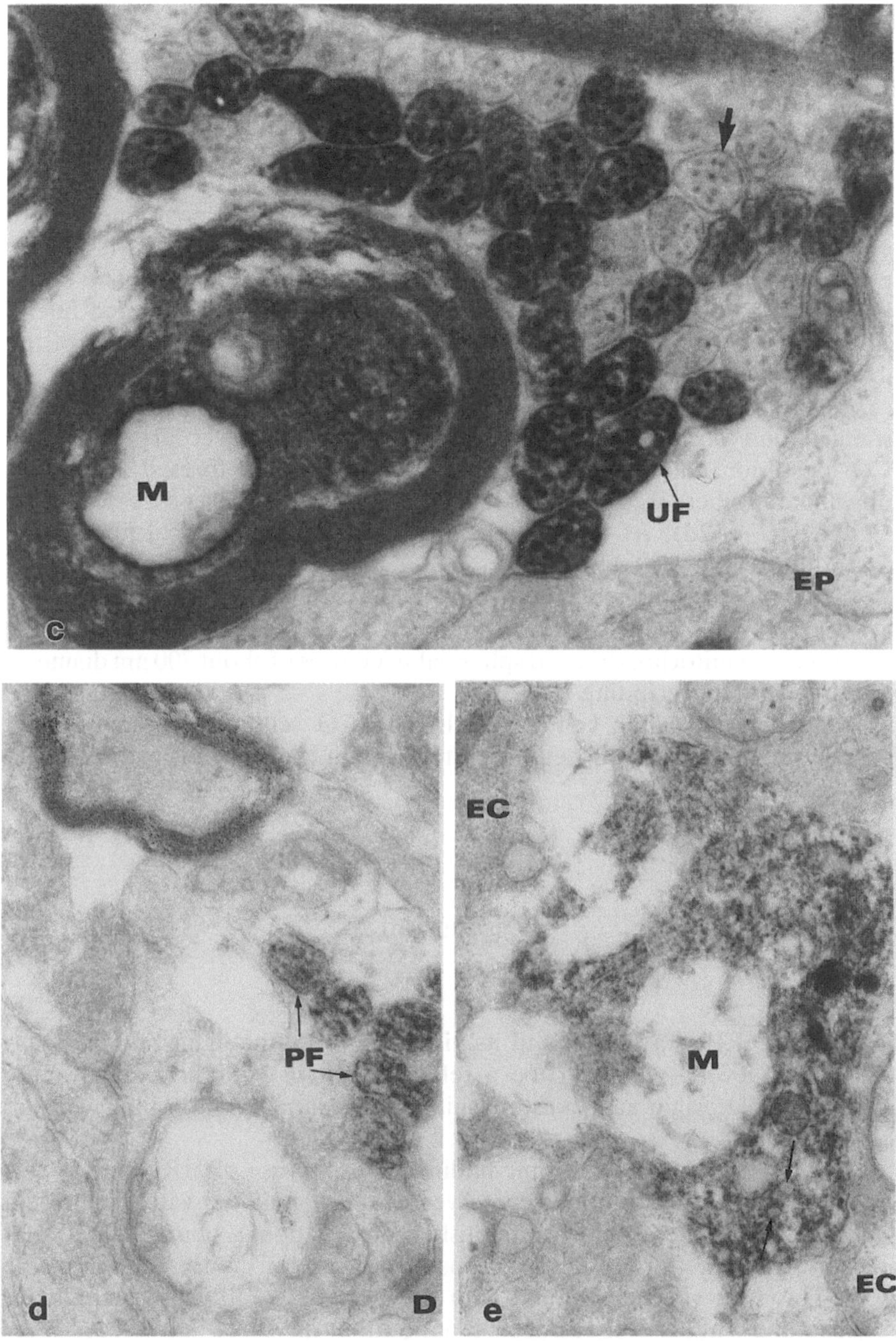

magnification of the pineal tract, showing labeled myelinated and unmyelinated nerve fibers (*UF*). HRP appears to bind preferentially to microtubules; this location is more apparent in weakly labeled fibers (*arrow*). *M*, swollen mitochondrion; *EP*, ependymal processes of SCO cells. ×38 500. **d** Cluster of labeled pineal nerve fibers (*PF*) located between SCO cells; *D*, desmosome. ×20 600. **e** Labeled axon terminal profile located between somata of SCO cells (*EC*). *Arrows*, unlabeled "synaptic" vesicles; *M*, swollen mitochondrion within a terminal. ×25 200

Electron Microscopy. The HRP reaction product appeared as a material of high electron density that clearly constrasted with the unstained background tissue (see Fig. 2b–e). Although this material filled the axoplasm it appeared particularly associated with microtubules and neurofilaments (see Fig. 2c).

In the rostralmost part of the SCO, the labeled pineal fibers were densely packed in two tracts located in a midline position at the border between the SCO cells and the posterior commissure (Fig. 2a). The entire pineal tract consisted of 50–60 unmyelinated and 12–16 myelinated fibers. Labeled myelinated fibers had a diameter of 0.7–3 μm; the diameter of the labeled unmyelinated fibers ranged between 0.15 and 0.5 μm (Fig. 2b,c).

As the pineal tract proceeded caudally, its fibers became more loosely arranged and occupied a more lateral position, however always at the SCO–posterior commissure interface. Small bundles formed by a few unmyelinated fibers were found between cell bodies or cell processes of the SCO cells (Fig. 2d). At the caudal third of the SCO no pineal fibers were seen in the region of the SCO.

In all the sections examined, only two labeled profiles corresponding to axon terminals were seen to contact SCO cells. The label filled these structures, but mitochondria and spherical structures of about 700 nm diameter were free of the precipitate (Fig. 2e).

Nonlabeled synaptic contacts with the SCO cells were not observed. Fibers containing classical neurosecretory granules occurred in the vicinity of the SCO.

Discussion

The use of HRP as a tracer, and its visualization at the light- and electron-microscopic levels, proved to be a reliable tool for identification of nerve fibers of pineal origin and in establishing their spatial relationships with the SCO.

The present light-microscopic findings with the use of HRP as a tracer, and those obtained with fluorescent dyes, carbocyanine derivatives (DiI and D16) applied to fixed brains (unpublished observations), demonstrated that pinealofugal fibers form a dense plexus at the base of the rostral half of the SCO. This is in good agreement with previous observations in amphibians (Oksche and Vaupel-von Harnack 1965; Paul et al. 1971) and teleost fishes (Sathyanesan and Sastry 1982; Ekström and van Veen 1983, 1984).

The ratio of unmyelinated to myelinated fibers of approximately 4:1 and the diameters of both types of fibers are quite similar to those described by Oksche and Vaupel-von Harnack in the frog (1965). This is also in accord with findings in other poikilothermic species. Depending on the species, up to 20% of the total nerve fibers are myelinated.

In teleost fishes, *Puntius sophore* (Sathyanesan and Sastry 1982) and *Gasterosteus aculeatus* (Ekström and Veen 1983), and in the lamprey, *Ichthyomyzon unicuspis* (Puzdrowski and Northcutt 1989), pineal fibers apparently penetrate the SCO ependyma and reach the third ventricle. Such transependymal pineal fibers were not found in the rainbow trout (Hafeez and Zerihun 1974) or in *Carassius carassius* (Ekström and Veen 1984). In the present study, only occasionally were HRP-labeled fibers seen among the SCO ependymal cells.

In *Rana esculenta*, synaptic structures of unknown origin are found in the area between the pineal tract and the SCO (Oksche and Vaupel-von Harnack 1965). In the present ultrastructural study, only two labeled profiles that might correspond to axon terminals were observed in this location. Thus it can be concluded that, in *Carassius auratus*, fibers of the pineal tract virtually do not establish synaptic contacts with the SCO cells. Furthermore, in contrast to other vertebrates, the SCO of *Carassius auratus* appears to be completely free of innervation. The SCO of other fish species, such us the coho salmon and *Liza aurata*, also appear to be devoid of a neural input, as shown by a conventional ultrastructural study (R. Yulis and P. Fernández-Llebrez, personal communication).

The question then arises whether in these species the SCO is under a control mechanism mediated via a neurohumoral pathway. In the fish brain, a system of CSF-contacting serotoninergic neurons has been described in the paraventricular and posterior recess organs (Meurling and Rodríguez 1990) that, by means of release of this indolamine into the ventricular CSF, could exert a control on the activity of the SCO. In addition, an influence of humoral–hormonal type from the PO upon the SCO may be considered. Indeed, in teleosts the pineal lumen communicates freely with the third ventricle just in front of the SCO. If biologically active principles present in the CSF, particularly serotonin, are to play a role in the secretory activity of the SCO, the presence of a specific receptor at the apical plasma membrane of the ependymal secretory cells would be a necessary prerequisite. The existence of such receptors, however, has not yet been investigated.

The probable innervation of the SCO by pineal fibers should be investigated by applying adequate tracers at the ultrastructural level to species displaying numerous synaptoid contacts with the SCO cells and lacking a neural serotoninergic input. Thus, frogs (Oksche and Vaupel-von Harnack 1965; Paul et al. 1971) and toads (Rodríguez 1970) can be regarded as appropriate animal models. A study along this line is in progress in our laboratory (see p. 219).

Acknowledgments. This work has been partially supported by grants PB90-0804 from DGICYT, MAR88-0751, from CICYT, Madrid, and MEC conceded to J.M.P.F., Spain; grant 0890/88 from CONICYT (Chile), and grant S-98-01 from the Dirección de Investigaciones, Universidad Austral de Chile, Chile, to E.M.R. The authors are gratefully indebted to Mr. L. Delannoy for his technical assistance.

References

Altner H (1968) Untersuchungen am Ependym und Ependymorganen im Zwischenhirn niederer Wirbeltiere (Neoceratodus, Urodelen, Anuren). Z Zellforsch 84:102–140

Baker JR (1946) Cytological technique: the principles underlying routine methods. Wiley, New York

Bouchaud C (1979) Evidence for a multiple innervation of subcommissural ependymocytes in the rat. Neurosci Lett 12:253–258

Bouchaud C, Arluison M (1977) Serotoninergic innervation of ependymal cells in the rat subcommissural organ. A fluorescence, electron microscopic and radioautographic study. Biol Cell 30:61–64

Diederen JHB (1970) The subcommissural organ of *Rana temporaria* L. A cytological, cytochemical, cytoenzymological and electronmicroscopical study. Z Zellforsch 111: 379–403

Diederen JHB (1975) A possible functional relationship between the subcommissural organ and the pineal complex and the lateral eyes in *Rana esculenta* and *Rana temporaria*. Cell Tissue Res 158:37–60

Ekström P, Korf HW (1985) Pineal neurons projecting to the brain of the rainbow trout, *Salmo gairdneri* Richardson (Teleostei). Cell Tissue Res 240:693–700

Ekström P, van Veen T (1983) Central connections of the pineal organ in the three-spined stickleback, *Gasterosteus aculeatus* L. (Teleostei). Cell Tissue Res 232:141–155

Ekström P, van Veen T (1984) Pineal neural connections with the brain in two teleosts, the crucian carp and the European eel. J Pineal Res 1:245–261

Gamrani H, Belin MF, Aguera M, Calas A, Pujol IF (1981) Radioautographic evidence for an innervation of the subcommissural organ by GABA-containing nerve fibers. J Neurocytol 10:411–424

Graham RC Jr, Karnovsky MJ (1966) The early stages of injected horseradish peroxidase in the proximal tubules of mouse kidney: ultrastructural cytochemistry by a new technique. J Histochem Cytochem 14:728

Hafeez MA, Zerihun L (1974) Studies on central projections of the pineal nerve tract in rainbow trout, *Salmo gairdneri* Richardson, using cobalt chloride iontophoresis. Cell Tissue Res 154:485–510

Kimble JE, Møllgård K (1973) Evidence for basal secretion in the subcommissural organ of the adult rabbit. Z Zellforsch 142:223–239

Korf HW (1976) Histological, histochemical and electron microscopical studies on the nervous apparatus of the pineal organ in the tiger salamander, *Ambystoma tigrinum*. Cell Tissue Res 174:475–497

Meurling P, Rodríguez E (1990) The paraventricular and posterior recess organs of elasmobranchs: a system of cerebrospinal fluid-contacting neurons containing immunoreactive serotonin and somatostatin. Cell Tissue Res 259:463–473

Møllgård K, Wiklund L (1979) Serotoninergic synapses on ependymal and hypendymal cells of the rat subcommissural organ. J Neurocytol 8:445–467

Møllgård K, Lundberg JJ, Wiklund L, Lochenmajer L, Baumgarten HG (1978) Morphological consequences of serotonin neurotoxin administration: neuron-target cell interaction in the rat subcommissural organ. Ann N Y Acad Sci 305:262–288

Murakami M, Tanizaki T (1963) An electron microscopic study on the toad subcommissural organ. Arch Histol Jpn 23:337–358

Oksche A, Vaupel von-Harnack MV (1965) Elektronenmikroskopische Untersuchungen an den Nervenbahnen des Pinealkomplexes von *Rana esculenta* L. Z Zellforsch 68:389–426

Paul E, Hartwig HG, Oksche A (1971) Neurone und zentralnervöse Verbindungen des Pinealorgans der Anuren. Z Zellforsch 112:406–493

Puzdrowski RL, Northcutt RG (1989) Central projections of the pineal complex in the silver lamprey *Ichthyomyzon unicuspis*. Cell Tissue Res 225:269–274

Rodríguez EM (1970) Ependymal specializations. III. Ultrastructural aspects of the basal secretion of the toad subcommissural organ. Z Zellforsch 111:32–50
Rodríguez EM, Yulis R, Peruzzo B, Alvial G, Andrade R (1984) Standardization of various applications of methacrylate embedding and silver methenamine for light and electron microscopy immunocytochemistry. Histochemistry 81:253–263
Rodríguez EM, Hein S, Rodríguez S, Herrera H, Peruzzo B, Nualart F, Oksche A (1987) Analysis of secretory products of the subcommissural organ. In: Scharrer B, Korf HW, Hartwig HG (eds) Functional morphology of neuroendocrine systems. Evolutionary and environmental aspects. Springer, Berlin Heidelberg New York, pp 189–201
Sathyanesan AG, Sastry VKS (1982) Pineal innervation of the third ventricle ependyma in the teleost, *Puntius sophore* (Ham.). J Neural Transm 53:187–192
Sternberger LA (1986) Immunocytochemistry. Wiley, New York
Vacca LL, Abrahams SJ, Naftchi NE (1980) A modified peroxidase-antiperoxidase procedure for improved localization of tissue antigens: localization of substance P in rat spinal cord. J Histochem Cytochem 28:297
Weissman-Nanopoulus D, Belin MF, Didier M, Aguera M, Partisani M, Pujol JF (1983) Immunohistochemical evidence for neuronal and non-neuronal synthesis of GABA in the rat subcommissural organ. Neurochem Int 5:785

Added in proof. For new evidence see Jiménez AJ, Pérez-Fígares JM, Rodríguez EM, Fernández-Llebrez P, Oksche A (1993) Synapse-like contacts between axons of the pineal tract and the subcommissural organ in *Rana perezi* (Anura) and their absence in *Carassius auratus* (Teleostei): ultrastructural tracer studies. Cell Tissue Res (in press). Compare also Oksche and Korf (this volume, p. 209)

Immunochemical Relationships Between the Subcommissural Organ and Hypothalamic Neurons

C.R. Yulis, B. Peruzzo, R.I. Muñoz, S. Hein, and E.M. Rodríguez

Introduction

Preliminary immunocytochemical observations in several vertebrate species revealed that certain hypothalamic neurons of the periventricular and arcuate nuclei and their axons projecting to the external zone of the median eminence (ME) contain material immunoreactive with antisera against Reissner's fiber (RF) (Rodríguez et al. 1987). It was also demonstrated that this immunoreactive RF-like material coexists with immunoreactive somatostatin (SOM) in the same neurons. The present study was undertaken with the aim to obtain further information on the nature of the immunoreactive RF material present in these hypothalamic structures as well as in other SOM-secreting cells such as those of the endocrine pancreas.

Materials and Methods

Light Microscopy

The brain and pancreas of adult rats (Holtzman strain) were fixed by immersion or vascular perfusion with Bouin's fluid. Tissue blocks containing the subcommissural organ (SCO), basal hypothalamus, and portions of the pancreas were dissected out and embedded in Paraplast or methacrylate (Rodríguez et al. 1984a). Serial sections (7 µm) were obtained from the Paraplast-embedded blocks and mounted on gelatin-coated slides. The methacrylate-embedded tissue was sectioned at 1 µm using an ultramicrotome; single adjacent sections were mounted on separate uncoated slides.

Electron Microscopy

Rats were fixed by vascular perfusion using a mixture of 2% paraformaldehyde and 0.5% glutaraldehyde in 0.1 *M* phosphate buffer, pH 7.4, for 20 min. Tissue blocks containing the basal hypothalamus and portions of pancreas were dissected out and postfixed by immersion in the same fixative for 2 h. Embedding was performed in Lowicryl according to a protocol described elsewhere (Peruzzo and Rodríguez 1989). Ultrathin sections were mounted on uncoated nickel grids and processed for single or double immunostaining using the protein A-gold method (pAg, Bendayan 1982).

Preparation of Adsorbents

Aqueous and ammonium bicarbonate extracts of the bovine RF were prepared as described elsewhere (Rodríguez et al. 1984b; Nualart et al. 1991). Ammonium bicarbonate and acetic acid extracts of the bovine SCO were prepared according to Nualart et al. (1991) and Jones and Pickering (1969), respectively. Protein concentration of the extracts was assessed by means of the Bradford's method (1976). Aliquots of an antiserum raised against an extract of bovine RF in EDTA–DTT–urea (AFRU, Rodríguez et al. 1984b), diluted 1:1500–1:4000, were preincubated for 24h at room temperature with the above-mentioned RF or SCO extracts at the following concentrations: aqueous extract of RF, 100 μg/ml AFRU; ammonium bicarbonate extract of RF, 5–200 μg/ml AFRU; ammonium bicarbonate extract of RF, 100–1200 μg/ml AFRU; acetic acid extract of SCO, 100 μg/ml AFRU. SOM (SOM28 15–28, kindly provided by Dr. J. Rivier, The Salk Institute, California) and SOM28 1–14 (Sigma) were also used for liquid-phase adsorptions of anti-SOM (1:2000) and AFRU (1:1500) at concentrations of 10–60 μg/ml of diluted sera.

Tissue blocks containing the SCO and the region of the ME–pituitary stalk of bovine brains were collected at the Valdivia slaughterhouse and fixed by immersion in Bouin's fluid. The blocks were sectioned using a Vibratome (100 μm thickness). After removal of the fixative by repeated washes in TRIS buffer, sections of the SCO or ME thus obtained were placed in glass vials containing diluted AFRU serum (1:1500–1:4000). Sections corresponding to 2–5 blocks of tissue were used for each milliliter of diluted antiserum. The antiserum was preincubated with free-floating sections for 24h at room temperature; then it was removed from the vials and used for immunostainings on Paraplast sections of the material described above.

Light-Microscopic Immunocytochemistry

The peroxidase–antiperoxidase (PAP) immunocytochemistry method was applied. Antisera developed against an aqueous extract of RF (AFRA, Rodríguez et al. 1984b), an EDTA–DTT–urea extract of RF (AFRU, Rodríguez et al. 1984b), an ammonium bicarbonate extract of the SCO (ASO, Rodríguez et al. 1988) and SOM (kindly provided by Dr. A. Weindl, University of Munich) were used. All the dilutions of antisera were prepared in a solution containing carrageenan (0.7%) and Triton X-100 (0.1%) in TRIS buffer, pH 7.7. Dilutions of the untreated and preadsorbed antisera were used on Paraplast sections of the rat SCO, ME, and pancreas.

Reagent concentrations and incubation times were as follows: (1) First antiserum, 1:1000–1:4000, 18 h (AFRU was also used at increasing dilutions up to 1:400 000); (2) second antiserum (goat anti-rabbit IgG), 1:25, 30 min; (3) PAP complex, 1:50, 30 min; (4) DAB, 0.2% in TRIS buffer containing 0.05% H_2O_2, 15 min. TRIS buffer washings (3 × 5 min) were performed between each incubation.

Electron-Microscopic Immunocytochemistry

Single immunostaining was carried out according to the procedure described by Peruzzo and Rodríguez (1989). Briefly, it consisted of the following steps: (1) washings in PBS, 2 × 5 min; (2) 1% BSA in PBS, 5 min; (3) first antiserum diluted 1:1000 (AFRU) or 1:1500 (anti-SOM) in PBS–BSA 1%, 18 h, 4°C; (4) washings in PBS, 3 × 5 min; (5) pAg (15-nm gold particle diameter) diluted 1:10 in PBS from a stock solution, 45 min; (6) washings in PBS, 2 × 5 min; (7) distilled H_2O, several washes.

A double-labeling technique (Bendayan 1982) was used by performing a complete immunostaining procedure on one side of the grid with anti-SOM serum (15-nm gold particles), followed by immunostaining of the opposite side using AFRU (5-nm gold particles). The sequence of steps, concentration of solutions, and time of incubations were the same as used in the single immunocytochemistry procedure. After completion of both (single and double) immunocytochemical procedures, the grids were counterstained with uranyl acetate (2% aqueous solution, 7 min)-lead citrate (2 min) and then carbon coated.

Results

Light Microscopy

Neurons of the arcuate and periventricular nuclei and a dense network of fibers projecting to the external zone of the ME displayed a strong immunoreactivity with anti-SOM and -AFRU sera (Fig. 1). Immunostainings on 1-μm-thick methacrylate sections confirmed the colocalization of both immunoreactivities in cell bodies of the arcuate nucleus. The reaction in the basal hypothalamus was observed with AFRU serum at dilutions between 1:1500 and 1:4000. At higher dilutions the ME fibers were no longer observed, but the reaction in the SCO persisted at dilutions up to 1:400 000. AFRA and ASO sera did not react with neuronal perikarya or fibers in the basal hypothalamus. On the contrary, the rat SCO and RF were strongly immunostained with the AFRU, AFRA, and ASO sera, but the anti-SOM did not react with the SCO–RF complex (see Table 1

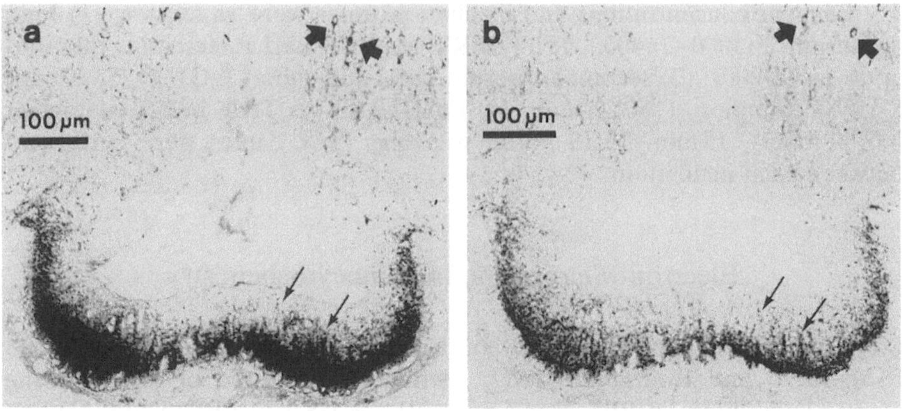

Fig. 1a,b. Immunostainings using anti-SOM (1:2000) and AFRU (1:1500) sera, respectively, of arcuate nucleus perikarya (*thick arrows*) and nerve fibers projecting to the external zone of the ME (*thin arrows*)

Table 1. Summary of immunostainings of the subcommissural organ (SCO) and median eminence (ME) using different antisera and adsorbents

Antiserum[a]	Adsorbent	SCO	ME
AFRU	None	++++	++
-SOM	None	–	++++
-SOM	SOM28 (15–28)	–	–[b]
AFRU	SOM28 (15–28)	++++	++
AFRA	None	+++	–
AFRU	Aqueous extract of RF	–	++
AFRU	Ammonium bicarbonate extract of RF	–	++
ASO	None	++	–
AFRU	Ammonium bicarbonate extract of SCO	–	++
AFRU	Acetic acid extract of SCO	++	––+
AFRU	SCO sections	+	–
AFRU	ME sections	++++	–

[a] For abbreviations concerning antisera, see "Materials and Methods" and p. 280; cf. p. 228.
[b] Intensity of immunostainings with adsorbed antisera is correlated to the immunostaining of control sections using unadsorbed antiserum at the same dilution.

for a summary of results obtained with the different immunostaining procedures).

The reaction of the anti-SOM serum was completely abolished by liquid-phase adsorption with SOM28 15–28 (biologically active peptide), whereas adsorption with the fragment SOM28 1–14 of prosomatostatin had no effect in decreasing this reactivity. Preincubation of the AFRU serum with either SOM28 1–14 or SOM28 15–28 did not result in any decrease in the immunostaining pattern of this antiserum.

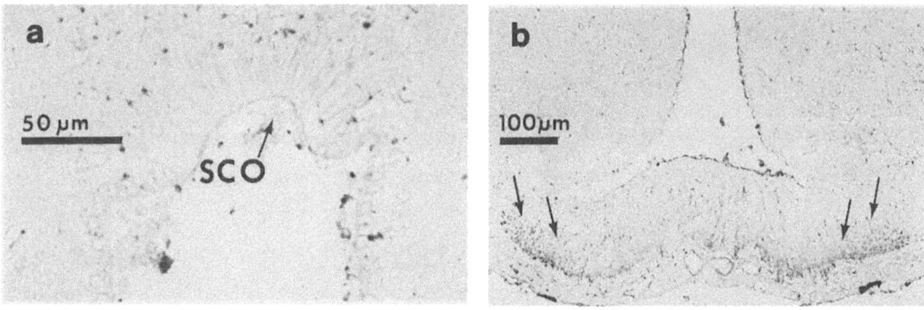

Fig. 2. Immunostainings using AFRU serum, diluted 1:3000, preadsorbed with an ammonium bicarbonate extract of the bovine SCO. Note absence of reaction in the SCO (compare with Fig. 3a), whereas ME fibers remain positive (**b**, *arrows*)

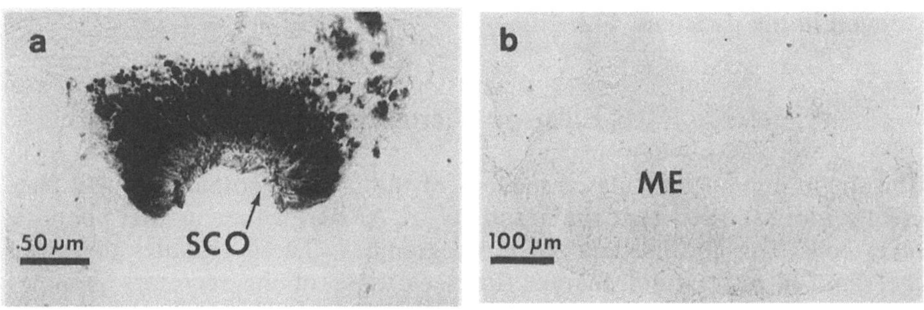

Fig. 3a,b. Immunostainings using AFRU serum (1:2000) preadsorbed with ME sections. The reaction in the SCO (**a**, *SCO*) cells is maintained, whereas fibers in the external zone of the ME (*ME*) are no longer observed (**b**)

AFRU serum preincubated with aqueous or ammonium bicarbonate extracts of bovine RF lost its reactivity with the SCO, but continued reacting with ME fibers. Preincubation with high concentrations of an ammonium bicarbonate extract of bovine SCO also blocked the reaction of AFRU in the SCO without affecting the reaction in the ME (Fig. 2). Preincubation of AFRU with an acetic acid extract of bovine SCO was the only liquid-phase preadsorption procedure that reduced the reactions in both SCO and ME. AFRU serum preincubated with Vibratome sections of the bovine ME did not immunostain the ME (Fig. 3b), but it continued to react strongly with the SCO (Fig. 3a). Preincubation of AFRU with sections of the bovine SCO also blocked the reaction in the ME and produced a marked decrease in the staining of the rat SCO.

The AFRU serum reacted strongly with somatostatinergic (D) cells of the pancreatic islets (Fig. 4a). The reactivity of these cells under the conditions depicted above (i.e., dilutions of AFRU and use of a number of adsorbents) paralleled the results obtained in structures of the rat basal hypothalamus

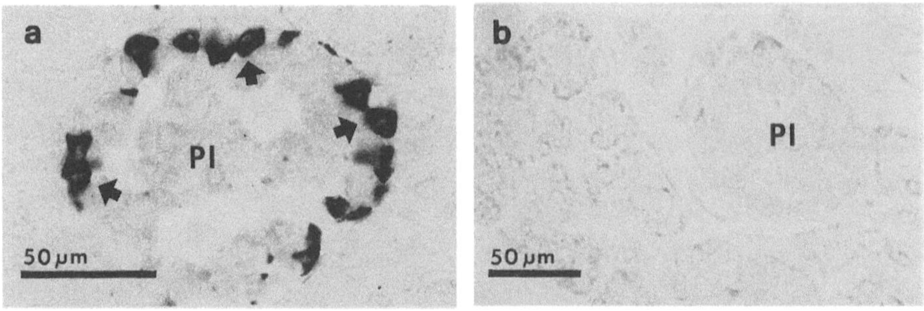

Fig. 4a,b. D cells of the pancreatic islets (*PI*) immunoreactive with AFRU (**a**, *arrows*) become negative after preadsorption of the AFRU serum with ME sections (**b**)

(i.e., conditions that abolished the reaction in the ME also abolished the reaction in the pancreas, Fig. 4b).

Electron Microscopy

The single immunostaining of sections of the basal hypothalamus and pancreatic islets showed that the reactivity of AFRU, either in ME fibers or in D cells, was localized in secretory granules. These granules displayed the same ultrastructural characteristics as those of the secretory granules immunostained with the anti-SOM serum (Fig. 5).

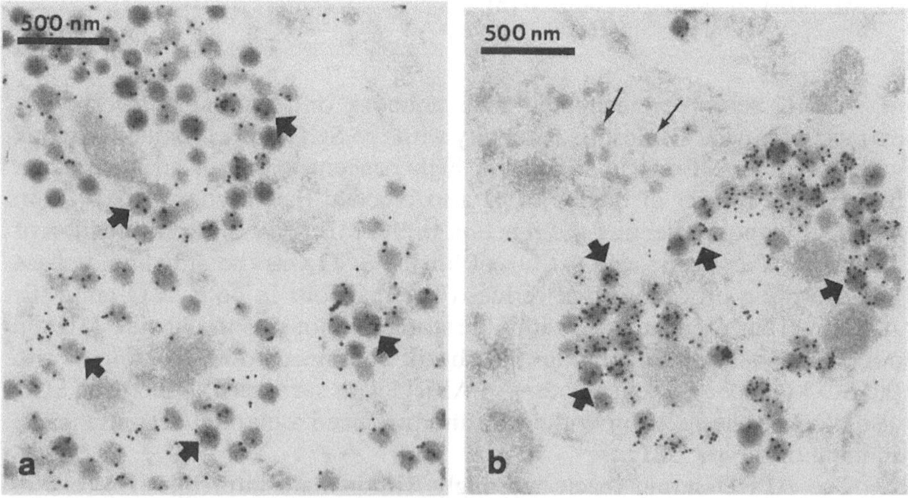

Fig. 5a,b. Ultrastructural immunostainings of ME fibers using anti-SOM (**a**) and AFRU (**b**). Both immunoreactivities are confined to secretory granules of the same characteristics (*thick arrows*). *Thin arrows* point to negative granules

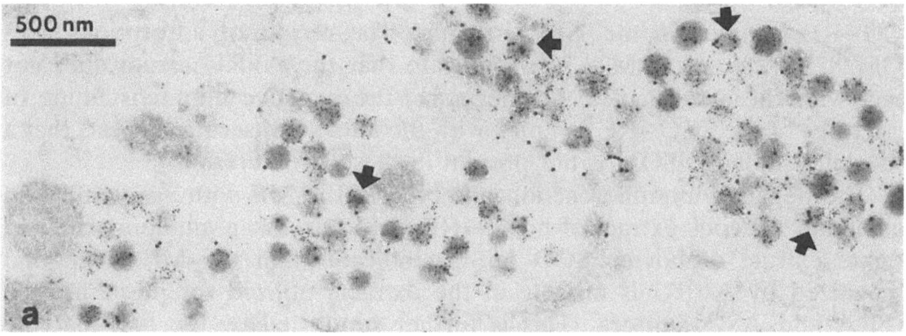

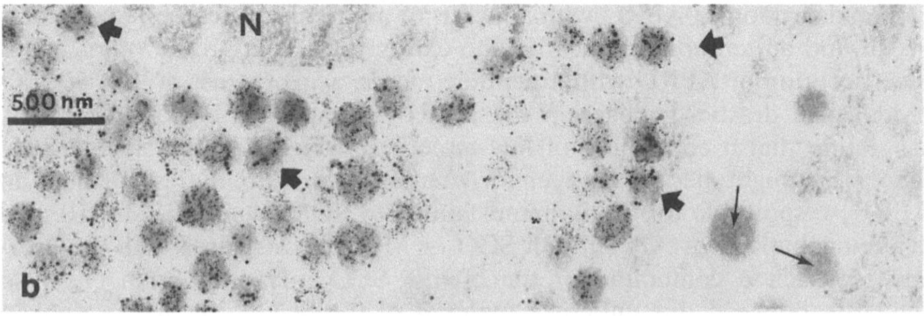

Fig. 6a,b. Double ultrastructural immunostainings of ME fibers (**a**) and a pancreatic islet (**b**) using anti-SOM (large colloidal gold particles) and AFRU (small colloidal gold particles). The same structures are immunostained by both antisera (*thick arrows*). *Thin arrows* point to negative granules of a pancreatic B cell; *N*, nucleus of a pancreatic D cell

Double immunocytochemistry using anti-SOM and AFRU sera on the same ultrathin sections confirmed that both immunoreactivities were colocalized in the same secretory granules in the ME fibers and pancreatic D cells (Fig. 6).

Discussion

These results confirm previous observations on the reactivity of antisera raised against EDTA–DTT–urea extracts of bovine RF (AFRU) with somatostatinergic structures of the basal hypothalamus (Rodríguez et al. 1987). The fact that exclusively this reaction to AFRU disappeared when the antiserum was subjected to high dilutions (at which the reaction in the SCO still persisted) suggests that the reaction with somatostatinergic structures depends only on some of the antibodies which, in this polyclonal serum, react with the SCO–RF complex. The titer of this antibody(ies) must be relatively low as compared to the bulk of antibodies recognizing the

SCO–RF complex in the AFRU serum. The results of adsorptions with SOM28 15–28 and SOM28 1–14 indicate that the AFRU serum does not react with these molecules. Furthermore, the negative immunostaining of structures of the SCO–RF complex with the anti-SOM serum suggests that a molecule similar to SOM is not present in these structures.

The negative immunoreaction observed in the ME with antisera raised against an aqueous extract of bovine RF (AFRA), or an ammonium bicarbonate extract of bovine SCO (ASO) indicates that the ME compound recognized by AFRU is missing in the extracts utilized for preparing the AFRA and ASO antisera. This is further supported by the fact that the AFRU-positive reaction in ME fibers and pancreatic D cells was not abolished by preadsorption of AFRU with aqueous or ammonium bicarbonate extracts of RF and an ammonium bicarbonate extract of SCO. On the other hand, preadsorption of AFRU with a peptidic (acetic acid) extract of bovine SCO produced a drastic decrease in the AFRU immunostaining of ME fibers, suggesting that a compound of low molecular weight, present in the SCO (and RF) might be the antigen originating the antibody(ies) present in AFRU, responsible for the immunostaining of SOM-secreting cells. Results of preincubations of AFRU with SCO or ME sections agree with the presence of such a compound in the bovine SCO and point to a chemical similarity between this unknown material of the SCO with that present in somatostatin cells of the hypothalamus and endocrine pancreas.

The findings obtained by means of single and double (anti-SOM-AFRU) ultrastructural immunocytochemistry point to the secretory nature of the peculiar material recognized by the AFRU serum. Its coexistence with SOM in the same secretory granules, both in mediobasal hypothalamic structures and in pancreatic D cells, strongly suggest that this, as yet unidentified, compound is secreted into the portal vasculature of the hypothalamus and also released from the endocrine pancreas.

Acknowledgments. This work was supported by grants from: Volkswagen-Stiftung, Germany (I/63-476), FONDECYT, Chile (88-0890 and 91-0956) and Research Office, Universidad Austral de Chile (S-89-01).

References

Bendayan M (1982) Double immunocytochemical labeling applying the protein A-gold technique. J Histochem Cytochem 30:81–85

Bradford M (1976) A rapid and sensitive method for the quantitation of microgram quantities of protein utilizing the principle of protein-dye binding. Anal Biochem 72:248–254

Jones CW, Pickering BT (1969) Comparison of the effects of water deprivation and sodium chloride imbibition on the hormone content of the neurohypophysis of the rat. J Physiol 203:449–458

Nualart F, Hein S, Rodríguez EM, Oksche A (1991) Identification and partial characterization of the secretory glycoproteins of the bovine subcommissural organ-Reissner's fiber complex. Evidence for the existence of two precursor forms. Mol Brain Res 111:227–238

Peruzzo B, Rodríguez EM (1989) Light and electron microscopical demonstration of concanavalin A and wheat-germ agglutinin binding sites by use of antibodies against the lectin or its label (peroxidase). Histochemistry 92:505–513

Rodríguez EM, Yulis CR, Peruzzo B, Alvial G, Andrade R (1984a) Standardization of various applications of methacrylate embedding and silver methenamine for light and electron microscopy immunocytochemistry. Histochemistry 81:253–263

Rodríguez EM, Oksche A, Hein S, Rodríguez S, Yulis CR (1984b) Comparative immunocytochemical study of the subcommissural organ. Cell Tissue Res 237:427–441

Rodríguez EM, Hein S, Rodríguez S, Herrera H, Peruzzo B, Nualart F, Oksche A (1987) Analysis of the secretory products of the subcommissural organ. In: Scharrer B, Korf HW, Hartwig HG (eds) Functional morphology of neuroendocrine systems. Evolutionary and environmental aspects. Springer, Berlin Heidelberg New York, pp 189–201

Rodríguez EM, Korf HW, Oksche A, Yulis CR, Hein S (1988) Pinealocytes immunoreactive with antisera against secretory glycoproteins of the subcommissural organ: a comparative study. Cell Tissue Res 254:469–480 [illegible] (1984) [illegible] of the [illegible] of the [illegible] [illegible] [illegible] [illegible] 111: [illegible]

Portmann [illegible] (1969) Light and electron microscopic [illegible] [illegible] [illegible] Histochemistry [illegible]

[illegible] (19[illegible]) [illegible] light and [illegible] Histochemistry [illegible]

[illegible] Comparative [illegible] [illegible]

[illegible]

[illegible]

VIII. Physiology of the Cerebrospinal Fluid

Cerebrospinal Fluid Secretion: The Transport of Fluid and Electrolytes by the Choroid Plexus

P.D. BROWN and C. GARNER

Introduction

The four choroid plexuses, found in the third, fourth and two lateral ventricles, secrete two-thirds or more of all cerebrospinal fluid (CSF)[1]. The source of the remaining one-third of CSF production is still unclear, but the majority is currently thought to be secreted by the arachnoid membrane (Johanson 1988). Once it is formed the CSF flows towards the fourth ventricle and then, via the foramina of Luschka and Magendie, into the cisternae on the external surface of the brain. The fluid then circulates freely around the subarachnoid spaces until it reaches the arachnoid villi, from where it returns to the vascular system via the dural sinuses.

In composition the CSF is very similar to the plasma. However, it is not simply an ultrafiltrate of the plasma; i.e. the concentrations of glucose, amino acids and proteins are substantially lower than in the blood. There are also homeostatic mechanisms which ensure that the concentrations of K^+, Ca^{2+} and HCO_3^- in the CSF remain constant, even when the plasma levels are varied. The presence of homeostatic mechanisms indicate that CSF is actively secreted by the choroid plexuses (henceforth referred to simply as the choroid plexus). The purpose of this review is to discuss this secretory process.

The Choroid Plexus: An Epithelium

One of the most compelling pieces of evidence for the role of the choroid plexus in the active secretion of CSF is that it appears morphologically to be ideally suited for the purpose. It is a highly branched structure, in which a single layer of cuboidal epithelial cells overlies a loose arrangement of connective tissue and many blood capillaries (see Fig. 1). The capillaries

---

[1] In humans the total volume of CSF is about 150 ml. This volume is replaced about four times each day, i.e. total CSF production = 600 ml day^{-1}.

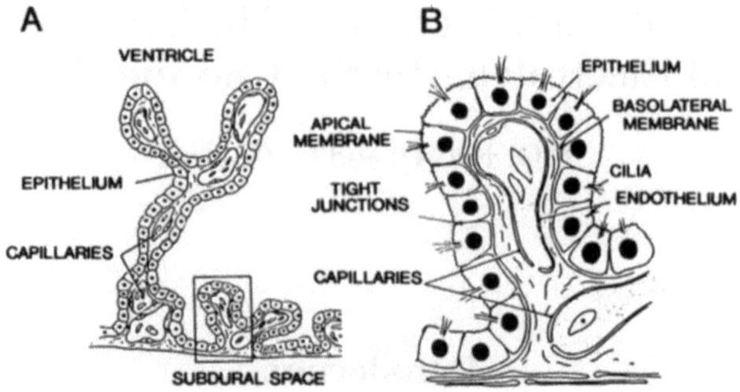

Fig. 1A,B. Diagrams showing the major anatomical features of the choroid plexus from the IVth ventricle of the frog. **A** Villi projecting into the ventricle. These consist of a core of loose connective tissue, containing numerous capillaries, which is covered by a single layer of cuboidal epithelial cells. **B** More detailed drawing of the epithelial cell layer. (From Wright 1972)

have a fenestrated endothelium and so present little resistance to the movements of ions, etc. The principal barrier between the blood and CSF is therefore the epithelium.

The fundamental feature of any epithelium which enables it actively and selectively to secrete (or absorb) fluid and electrolytes is that epithelial cells are polarised, i.e. the apical (CSF-facing) membrane has properties which are different to those of the basolateral (blood-facing) membrane (see Fig. 1). The direction of transport across the epithelium, either absorption or secretion, is determined by the differential expression of ion transport proteins in these membranes. In order to understand the cellular mechanisms which underlie secretion, the precise location (apical or basolateral) of ion transport proteins must first be determined. It is then possible to formulate models of ion movements which can be tested experimentally (see Steward and Case 1989).

One such model which is used to explain fluid secretion in many epithelia is the "neutral cotransport" model (Silva et al. 1977). In this model (Fig. 2) electroneutral Na^+-$2Cl^-$-K^+ cotransporters on the basolateral side of the cell actively accumulate Cl^-. The energy required for this process is provided by the Na^+ gradient which is in turn maintained by the activity of Na^+-K^+ ATPase. K^+ transported into the cell by Na^+-K^+ ATPase and the cotransporter leaks back out of the cell through K^+-selective channels in the basolateral membrane. Cl^- channels in the apical membrane allow Cl^- to flow out of the cell down its electrochemical gradient. Cl^- is therefore moving across the cell by what is termed a transcellular route. Na^+ is thought to follow Cl^- by diffusion via the paracellular route, i.e. through the junctional complexes between the cells. Thus, a net movement of NaCl occurs, creating an osmotic gradient which is sufficient to drive water transport. It is not clear how water moves across the epithelium, but it is probably by a combination of paracellular and transcellular routes.

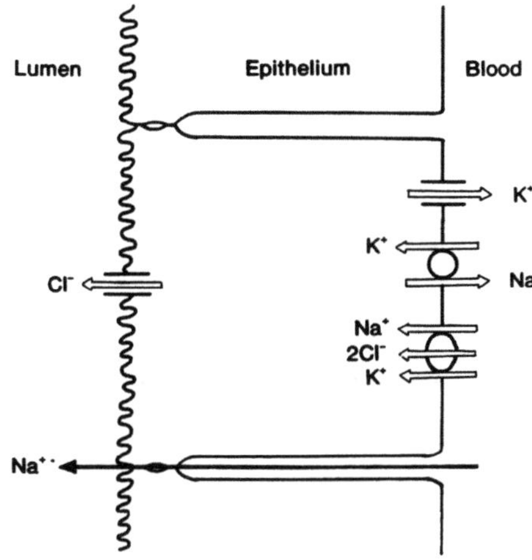

Fig. 2. The "neutral cotransport" model of fluid secretion

The "neutral cotransport" model provides a good explanation of fluid secretion in many epithelia (see Case et al. 1989). However, studies of the choroid plexus suggest that CSF secretion cannot be explained by this model. The evidence for this is as follows: (1) CSF secretion is only slightly inhibited by loop diuretics (Vogh and Langham 1981), which inhibit the Na^+-$2Cl^-$-K^+ cotransporter and markedly reduce secretion in many epithelia. (2) There is a significant role for HCO_3^-/CO_2 in CSF secretion, because inhibitors of Cl^--HCO_3^- exchange, such as the stilbene derivatives DIDS and SITS, reduce CSF secretion (Saito and Wright 1983; Deng and Johanson 1989). Carbonic anhydrase inhibitors, e.g. acetazolamide (Diamox), also reduce CSF secretion (Maren and Broder 1970). (3) Na^+-K^+ ATPase is situated on the apical rather than the basolateral membrane of the choroid plexus (Masuzawa et al. 1984; Ernst et al. 1986). Other models of ion transport must therefore be considered to explain CSF secretion by the choroid plexus.

A Model for CSF Secretion by the Amphibian Choroid Plexus

One of the most widely accepted models for CSF secretion was originally proposed by Saito and Wright (1983, 1984) on the basis of electrophysiological experiments on bullfrog choroid plexus. Their principal conclusions were that: (1) HCO_3^-, rather than Cl^- or Na^+, is the major ion which is actively transported across the epithelium; (2) active HCO_3^- secretion is increased by noradrenaline, prostaglandin E_1 and adrenocorticotrophic hormone,

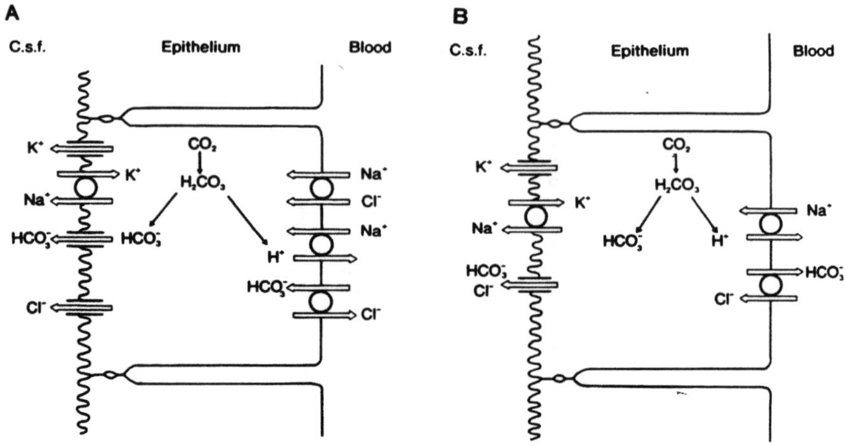

Fig. 3A,B. Models for CSF secretion by the choroid plexus. The model in **A** was proposed by Saito and Wright (1983), following experimental work on bullfrog choroid plexus. The model in **B** is a new model for CSF secretion by the mammalian choroid plexus (details in text)

which probably all act by elevating the intracellular concentration of cAMP; and (3) HCO_3^- enters the CSF through cAMP-dependent ion channels in the apical membrane of the choroid plexus. The model they proposed, which includes all of these features, is shown in Fig. 3A.

Since this model was first proposed, studies of intracellular Cl^- homeostasis with Cl^--sensitive microelectrodes have provided some indirect evidence for the presence of the transport pathways in the basolateral membrane (Saito and Wright 1987). However, none of these pathways have been characterised in detail in the amphibian choroid plexus. Perhaps the weakest part of this model, however, is the suggestion that HCO_3^- selective channels are present in the apical membrane. There is no precedent for the existence of such channels from work on other epithelial cells. Furthermore, patch-clamp studies of the apical membrane of amphibian choroid plexus have identified a number of ion channels permeable to K^+ (Christensen and Zeuthen 1987; Brown et al. 1988), Ca^{2+} (Christensen 1987) and Cl^- (Christensen et al. 1989; Birnir et al. 1988), but channels which are primarily selective to HCO_3^- have yet to be identified. While it remains possible that these channels do exist, but cannot be identified by patch-clamp methods, the more likely explanation is that Saito and Wright's model is incorrect in this detail. Another criticism directed against this model is that it is based wholly on studies of amphibian tissue and may not necessarily be applicable to mammalian choroid plexus.

Ion Transport by the Mammalian Choroid Plexus

Pumps and Carriers

The evidence for the presence of Na^+-K^+ ATPase, Cl^--HCO_3^- exchange, Na^+-H^+ exchange and Na^+-Cl^- cotransport in the mammalian choroid plexus has been reviewed quite recently by Johanson (1988, 1989). The majority of this evidence comes from in vivo studies of the effects of inhibitors on CSF production or from in vitro ion flux studies. Neither of these methods provides precise information about the subcellular location of the various ion transporters identified. However, it is now possible to localise many transporters very accurately using immunocytochemical methods. In this way Na^+-K^+ ATPase was localised to the apical membrane (Masuzawa et al. 1984; Ernst et al. 1986), and more recently Cl^--HCO_3^- exchange has been found exclusively in the basolateral membrane (Lindsey et al. 1990).

Ion Channels

There have been few electrophysiological studies of the mammalian choroid plexus, mainly because it cannot be mounted in an Ussing chamber for measurements of the transepithelial resistance and short circuit current. The small size of mammalian cells also means that they are difficult to impale with conventional microelectrodes. However, the patch-clamp technique (Sakmann and Neher 1983) has recently been used to study ion channels in the rat and mouse choroid plexus. Some of the initial data from Wright's laboratory in Los Angeles and from our own work in Manchester are discussed below.

Anion Channels

Two types of anion channel have now been identified in patch-clamp studies of the apical membrane of rat choroid plexus (Brown 1990). The predominant channel (found in about 50% of inside-out patches) has a conductance of 26 pS, and is similar in many respects to Cl^- channels which have an important role in fluid secretion in other epithelia (Gogelein 1988).

Cl^- channels with similar properties have also been observed in mouse choroid plexus by Hung et al. (1990), who reported that channel activity was increased by the neurotransmitter 5-HT. The choroid plexus epithelial cell is known to possess $5\text{-}HT_{1C}$ receptors (Yagaloff and Hartig 1985), which are coupled to inositol trisphophate production and presumably to intracellular Ca^{2+} mobilisation (Takahshi et al. 1987). The link between occupation of the $5\text{-}HT_{1C}$ receptor and Cl^- channel activation could therefore be an increase in the intracellular Ca^{2+} concentration. We have examined the Ca^{2+} sensitivity of the Cl^- channels in the rat choroid plexus (Fig. 4A). Removal of Ca^{2+} from the bath (intracellular) solution reduced the time for

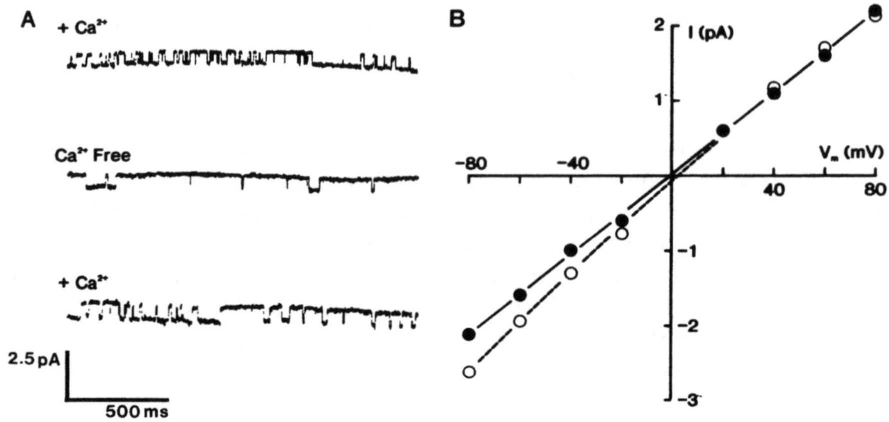

Fig. 4A,B. Anion channels in the rat choroid plexus. **A** Ca^{2+} dependence of single-channel opening in an inside-out patch. Channel activity recorded in the presence (1 m*M*) or absence (20 n*M*) of Ca^{2+} at the cytoplasmic face of the patch. At the holding potential of −40 mV, channel openings are seen as downward deflections of the current record. The open probability of the channel decreased from 0.74 to 0.08 on the removal of Ca^{2+}. This change in activity was subsequently reversed by the readdition of Ca^{2+}. **B** Current–voltage relationships for the channels in inside-out patches. The electrode contained 140 m*M* KCl and the bath contained either: 140 m*M* NaCl (•—•), or 120 m*M* NaCl + 25 mM $NaHCO_3$ (o---o). The permeability ratio of the channel to Cl^- against HCO_3^- ($P_{Cl}:P_{HCO_3}$) = 1:0.6

which the channel was open (Fig. 4A), an effect which was reversed on returning to the high Ca^{2+} concentration. Channel activity, thus, appears to be dependent on intracellular Ca^{2+} activity.

Selectivity studies have shown that these channels can discriminate between cations and Cl^- very effectively (Brown 1990). However, they may be less selective against other anions, as shown in Fig. 4B. The current–voltage relationships (Fig. 4B) were linear when the inside-out patch was bathed in solutions with equal Cl^- activity, but deviated from linearity when HCO_3^- was added to the bath solution. The simplest explanation for these results is that the channels are also permeable to HCO_3^-; the calculated permeability ratio ($P_{Cl}:P_{HCO_3}$ = 1:0.6) indicates that the channels are almost twice as permeable to Cl^- as HCO_3^-. Similar observations have been made for "Cl^- channels" in other epithelial cells (Brown et al. 1989; Tabcharani et al. 1989).

The other type of anion channel, which is observed in only 3% of inside-out patches, has a conductance of 330 pS and appears to belong to the category of "maxi"-Cl^- channels which have been observed in other epithelia (Gogelein 1988). Both types of anion channel have been observed in cell-attached patches, where they mediate Cl^- efflux in the absence of any applied potential. The extent to which they may contribute to net Cl^- secretion, however, remains to be determined.

The fact that these channels mediate Cl^- efflux suggests that Cl^- is accumulated by the epithelial cell to above the expected equilibrium concentration (to 31 m*M* rather than 21 m*M*, calculated using data from patch-clamp experiments). Furthermore, Cl^- accumulation appears to be greatly enhanced (80 m*M*) in cells that are bathed in solutions containing HCO_3^-/CO_2. The mechanism responsible for this extra accumulation of Cl^- is DIDS-sensitive (Garner and Brown 1992), suggesting that Cl^--HCO_3^- exchange plays a major role in the accumulation of Cl^- by the choroid plexus.

Cation Channels

Low conductance (13 pS) K^+ channels have been observed in the apical membrane of the rat (Garner and Brown, unpublished observations) and the mouse choroid plexus (Hung et al. 1991). They are active only in cell-attached patches and their properties have not been studied in any great detail. However, the current–voltage relationships for these channels can be used to estimate the membrane potential of the mammalian choroid plexus cell as in the order of −50 mV. This is similar to values reported for amphibian tissue (Saito and Wright 1984). These K^+ channels were observed quite frequently (in 10% of patches) and may provide a leak pathway for K^+ pumped into the cell by Na^+-K^+ ATPase.

The other cation channel which was been observed is a rectifying channel with a maximum conductance of 120 pS (Brown 1990). It is not active in inside-out patches, so detailed studies of selectivity, etc. have yet to be performed. Channels with similar properties have been observed in the amphibian choroid plexus, where Christensen (1987) reported that channel activation occurred when the membrane was "stretched" by an increase in cell volume. The opening of these channels was therefore suggested to increase Ca^{2+} influx into the cell and initiate cell volume regulatory mechanisms (Christensen 1987).

These patch-clamp studies of the mammalian choroid plexus are still far from complete. More information is needed about the properties of individual ion channels, while whole-cell studies are necessary to determine the relative contribution of the various channel types to the overall conductance of the cell. It does appear, however, that there are marked similarities, at least in terms of the types of ion channels present, between amphibian and mammalian choroid plexus. Other similarities also exist, e.g. the intracellular Cl^- activity and membrane potential estimated from the patch-clamp data. In formulating a model of CSF secretion by mammalian choroid plexus, we have therefore tried to accommodate the ideas of Saito and Wright (1984) as far as possible.

A Model of CSF Secretion by the Mammalian Choroid Plexus

Figure 3B shows a model for CSF secretion by the mammalian choroid plexus. The model is something of a hybrid between the "neutral cotransport" model (Fig. 2) and Saito and Wright's model (Fig. 3A). The principal steps are: (1) the production of HCO_3^- from CO_2 catalysed by carbonic anhydrase; (2) exchange of HCO_3^- for Cl^- at the basolateral membrane as the major pathway for Cl^- accumulation into the cell (as suggested by the patch-clamp experiments); (3) exit of Cl^- and HCO_3^- from the cell down their respective electrochemical gradients (derived mainly from the negative membrane potential difference) through the anion channels; (4) secretion of Na^+ via transcellular transport: Na^+ enters the cell in exchange for H^+ at the basolateral membrane and is extruded by Na^+-K^+ ATPase.

Na^+-Cl^- cotransport has not been included in the model because there is little direct evidence for its involvement in CSF secretion, and because the model is largely able to explain the electrophysiological data of Saito and Wright (1983, 1984) without it, i.e. in the absence of Cl^- (but not HCO_3^-) the choroid plexus can still secrete CSF. However, the situation may not be quite the same in the mammal and future experiments will need to clarify the role of Na^+-Cl^- cotransport in the mammalian choroid plexus.

Future Directions

A great deal more information is required to substantiate the existence of the individual components of the model, e.g. Na^+-H^+ exchange and the nonselective anion channels. It is also important to understand how the activity of the individual pathways is regulated and how they may interact to bring about net secretion whilst maintaining intracellular pH. Measurements of intracellular pH using fluorescent probes may be particularly useful in these investigations, as indeed they have already been in establishing the role of HCO_3^- in the secretion of saliva (Brown et al. 1989; Case et al. 1989). Experiments to investigate the effects of inhibitors on the ionic composition of CSF may also be very informative. If the model is correct it should be possible to predict changes in ion concentrations following the addition of transport inhibitors. Similar methods have been used to study ion transport in other secretory epithelia (e.g., see Case et al. 1989).

References

Birnir B, Loo DDF, Brown PD, Wright EM (1988) Whole cell currents in *Necturus* choroid plexus. FASEB J 2:8287

Brown PD (1990) Single channel patch-clamp studies of rat choroid plexus. J Physiol 423:46P

Brown PD, Loo DDF, Wright EM (1988) Calcium-activated potassium channels in the apical membrane of *Necturus* choroid plexus. J Membr Biol 105:207–219

Brown PD, Elliott AC, Lau KR (1989) Indirect evidence for the presence of non-specific anion channels in rabbit mandibular salivary gland acinar cells. J Physiol 414:415–431

Case RM, Lau KR, Steward MC, Brown PD, Elliott AC (1989) Epithelial electrolyte transport: inter-relationship between H^+, HCO_3^- and Cl^-. Biochem Soc Trans 17:803–805

Christensen O (1987) Mediation of cell volume regulation by Ca^{2+} influx through stretch-activated channels. Nature 330:66–68

Christensen O, Zeuthen T (1987) Maxi K^+ channels in leaky epithelia are regulated by intracellular Ca^{2+}, pH and membrane potential. Pflugers Arch 408:249–260

Christensen O, Simon M, Randlev T (1989) Anion channels in a leaky epithelium: a patch-clamp study of choroid plexus. Pflugers Arch 415:37–46

Deng QS, Johanson CE (1989) Stilbenes inhibit exchange of chloride between blood, choroid plexus and cerebrospinal fluid. Brain Res 501:183–187

Ernst SA, Palacios JR, Siegel GJ (1986) Immunocytochemical localization of Na^+, K^+ – ATPase catalytic polypeptide in mouse choroid plexus. J Histochem Cytochem 34:189–195

Garner C, Brown PD (1992) The effect of CO_2/HCO_3^- buffered saline on Cl^- channel activity in isolated rat choroid plexus. J Physiol 446, 350P

Gogelein H (1988) Chloride channels in epithelia. Biochim Biophys Acta 947:521–547

Hung BCP, Loo DDF, Wright EM (1990) Serotonin regulates Cl channels in mouse choroid plexus. Biophys J 57:83a

Hung BCP, Loo DDF, Wright EM (1991) Serotonin (5HT) inhibits K^+ channels in the apical membrane of mouse choroid plexus. Biophys J 59:649a

Johanson CE (1988) The choroid plexus-arachnoid membrane-cerebrospinal fluid system. In: Boulton AA, Baker GB, Waltz W (eds) Neuromethods; the neuronal microenvironment. Humana, New Jersey, pp 33–104

Johanson CE (1989) Potential for pharmacologic manipulation of the blood-cerebrospinal fluid barrier. In: Neuwelt EA (ed) Implications of the blood-brain barrier and its manipulation, vol 1. Plenum, New York, pp 223–260

Lindsey AE, Schneider K, Simmons DM, Baron R, Lee BS, Kopito RR (1990) Functional expression and subcellular localization of an anion exchanger cloned from choroid plexus. Proc Natl Acad Sci USA 87:5278–5282

Maren TH, Broder LE (1970) The role of carbonic anhydrase in anion secretion into cerebrospinal fluid. J Pharmacol Exp Ther 172:197–202

Masuzawa T, Ohta T, Kawamura M, Nakahara N, Sata F (1984) Immunohistochemical localization of Na^+,K^+,-ATPase in the choroid plexus. Brain Res 302:357–362

Sakmann B, Neher E (1983) Single channel recording. Plenum, New York

Saito Y, Wright EM (1983) Bicarbonate transport across the frog choroid plexus and its control by cyclic nucleotides. J Physiol 336:635–648

Saito Y, Wright EM (1984) Regulation of bicarbonate transport across the brush border membrane of the bullfrog choroid plexus. J Physiol 350:327–342

Saito Y, Wright EM (1987) Regulation of intracellular chloride in bullfrog choroid plexus. Brain Res 417:267–272

Silva P, Stoff J, Field M, Fine L, Forrest JN, Epstein FH (1977) Mechanism of active chloride secretion by shark rectal gland: the role of Na-K-ATPase in chloride transport. Am J Physiol 233:F298–F306

Steward MC, Case RM (1989) Principles of ion and water transport across epithelia. In: Davison JS (ed) Gastrointestinal Secretion. Wright, London, pp 1–31

Tabcharni JA, Jensen TJ, Riordan JR, Hanrahen JW (1989) Bicarbonate permeability of the outwardly rectifying anion channel. J Membr Biol 112:109–122

Takahashi T, Neher E, Sakmann B (1987) Rat brain serotonin receptors in *Xenopus* oocytes are coupled by intracellular calcium to endogenous channels. Proc Natl Acad Sci USA 84:5063–5067

Vogh BP, Langham MR (1981) The effect of furosemide and bumetanide on cerebrospinal fluid formation. Brain Res 221:171–183

Wright EM (1972) Mechanisms of ion transport across the choroid plexus. J Physiol 226: 545–571

Yagaloff KA, Hartig PR (1985) ^{125}I-lysergic acid diethyl-amide binds to a novel serotonergic site on rat choroid plexus epithelial cells. J Neurosci 5:3178–3183

Physiology of Cerebrospinal Fluid Circulation: Amphibians, Mammals, and Hydrocephalus

H.C. JONES

General Aspects

The chemical environment of the brain is buffered from external influences by the blood–brain barrier at the brain capillary endothelium, the blood–cerebrospinal fluid (CSF) barrier at the choroid plexus epithelium and by a barrier membrane, the arachnoid, which envelops the brain and CSF. The CSF is secreted by the choroid plexuses which are sited inside the cerebral ventricles (two lateral ventricles, third and fourth ventricles). The composition of the CSF is closely regulated, having a lower K^+ and Ca^{2+} concentration and a higher Na^+ and Mg^{2+} concentration than would be expected from a simple plasma ultrafiltrate. Proteins and nonelectrolytes such as urea, glucose and amino acids are also low in CSF. One function of the CSF, together with the blood–brain barrier, therefore, is to maintain brain ion homeostasis and provide the correct chemical environment for neuronal function. A second function is to provide buoyancy for the brain by maintaining a hydrostatic pressure on both its internal and external aspects. Thus the human brain weighs approximately 1300 g in air but only 50 g in water and the CSF is assumed to provide protection from injury. A third, less convincing, function of the CSF may be to provide a flow pathway for the transportation of waste products and pharmacologically active substances. In support of this, many secretory/sensory circumventricular organs are situated in specialised ependyma adjacent to the CSF. However, because CSF turnover is slow and measured in hours rather than minutes, only very local transportation would have a rapid response time.

The mammalian CSF system (reviewed in Davson et al. 1987) consists of both an internal and an external fluid with CSF flow in the direction of lateral to third to fourth ventricle, out into the subarachnoid space through the foramina of Luschka and Magendie, sited in the lateral recesses and the roof of the fourth ventricle, followed by circulation in the subarachnoid spaces around the spinal cord and brain. Ultimate absorption of CSF takes place via two main routes, into dural venous sinuses and into the lymphatic system. The CSF system in other vertebrates differs anatomically from the mammalian one to a variable extent. Although all vertebrates, with the possible exception of the hagfish, have choroid plexuses and ventricular CSF, the external CSF is variable. The terrestrial group, birds, reptiles and amphibians, has an external CSF similar to mammals but flow continuity between the two has been demonstrated only in amphibians. Birds and reptiles have a large ependymal tela separating the fourth ventricle and the external CSF, which although complete, is also thin and probably allows exchange of CSF (e.g. Jones and Dolman 1979). Aquatic submammalian vertebrates, teleosts, elasmobranchs, cyclostomes and others, also have an external barrier layer, but it is variable in nature (e.g. Bundgaard and Cserr 1991, Momose et al. 1988, Rovainen et al. 1971) and the ion composition of the external fluid, where studied, has similarities to CSF (Bundgaard and Cserr 1981).

Amphibian CSF Physiology

The circulation of CSF has been studied in an amphibian, *Rana pipiens*, using an infusion technique whereby a small volume of high-molecular weight fluorescent Dextran solution is infused into the lateral ventricle at a rate similar to the CSF secretion rate, and the subsequent distribution studied by examination of the frozen heads or by fluorescence microscopy (Jones 1978, 1980a). After 15 min the fluorescence is seen throughout the ventricles and in the subarachnoid space overlying the fourth ventricle and rostral spinal cord (Fig. 1). No fluorescence was found within the central canal of the spinal cord. With longer times the fluorescence extends throughout the whole length of the spinal subarachnoid space and subsequently throughout the brain subarachnoid space. The marker is totally cleared from the ventricles in 5 h and from the whole system within 8 h. Similar circulation has been observed in the bullfrog, *Rana catesbeiana* (Tornheim and Foltz 1979).

The internal and external CSF in amphibians are connected by a unique system of microscopical pores in the rhombencephalic posterior tela which forms the roof of the fourth ventricle caudal to the choroid plexus (Jones 1979; Tornheim and Michaels 1979). The ependymal cells of the tela form an incomplete layer with round or oval fenestrations up to 100 μm in diameter (Fig. 2). The fenestrations are bordered by rounded or pleomorphic ependymal cells with coiled basal lamina and with variable amounts of loosely arranged and fenestrated pial cells on the extraventricular side (Jones 1979, 1982; Tornheim and Michaels 1979). During development in the frog, pores form from an undifferentiated fourth ventricle roof, coinciding with the development of the choroid plexus, and their number increases with age (Jones and Jopling 1983). It seems likely that a chemical or mechanical signal is responsible for pore formation, and this is supported by the observation that in adult frogs the shape of the ependymal pores can be influenced by exposure to chemical agents (Jones and Taylor 1983).

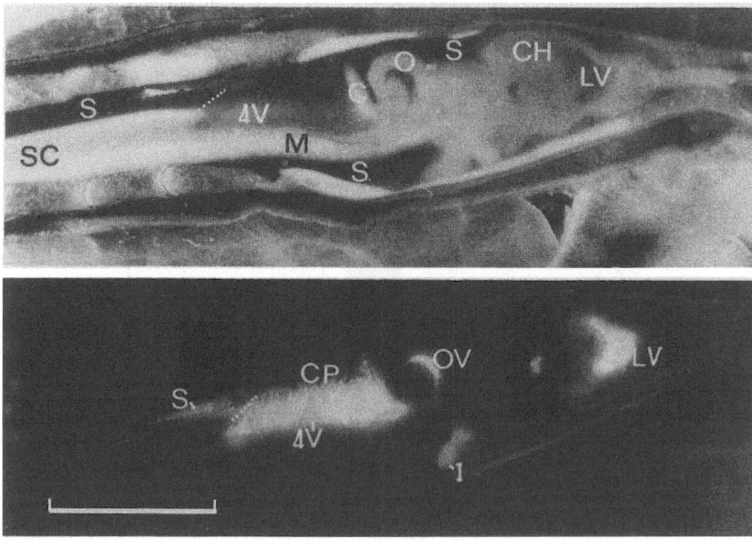

Fig. 1. The head of *Rana pipiens* 15 min after infusion of 1 µl fluorescein-labelled Dextran (MW 150000) into the lateral ventricle (*LV*). The head was frozen and cut sagittally and photographed both in ordinary light (*top*) and in ultraviolet light for distribution of fluorescence (*bottom*). The posterior tela is marked as a *dotted line* and fluorescence occurs throughout the ventricular system and also in the dorsal subarachnoid space (*S*) adjacent to the tela. *SC*, spinal cord; *4V*, fourth ventricle; *M*, medulla; *C*, cerebellum; *O*, optic lobe; *CH*, cerebral hemisphere; *LV*, lateral ventricle; *CP*, choroid plexus; *OV*, optic ventricle; *I*, infundibular recess. *Scale bar* = 5 mm

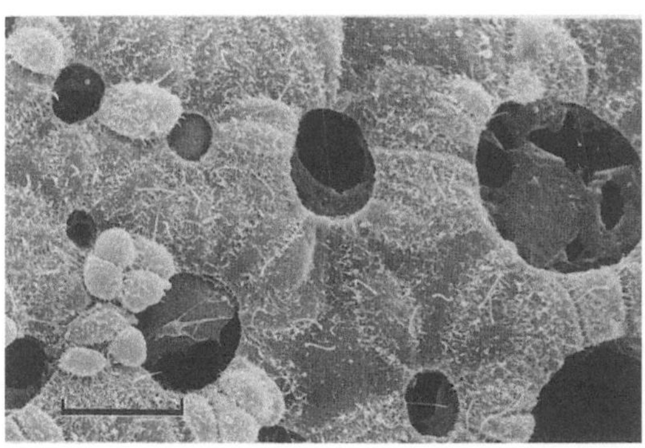

Fig. 2. Scanning electron micrograph of part of the ventricular surface of the posterior tela in *Rana pipiens*. The pores are variable in diameter and are bordered by ependymal cells which are flattened or, in some cases, rounded. *Scale bar* = 20 µm

The physiology of CSF absorption has been studied with the constant-rate infusion technique (Davson et al. 1970). This is based on Poiseuille's law which states that at a steady state pressure:

$$\text{flow rate} = \text{pressure/resistance to flow}$$

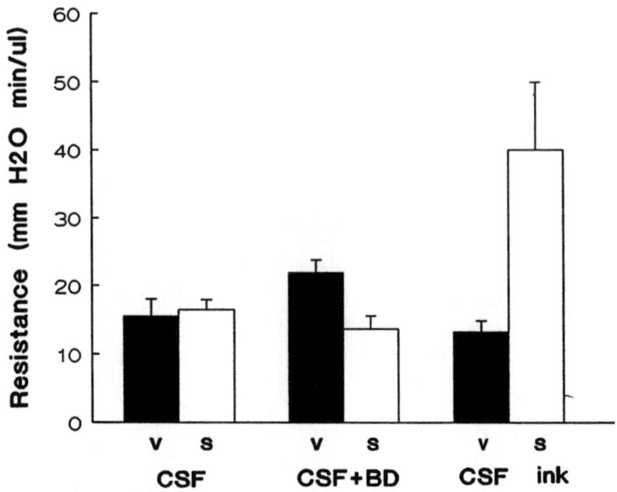

Fig. 3. Resistance to cerebrospinal fluid absorption in *Rana pipiens* measured by the constant-rate infusion technique. With artifical cerebrospinal fluid (*CSF*) as the infusion fluid, there was no significant difference between the measurement made from the lateral ventricle (*v*, *closed bars*) and from the subarachnoid space (*s*, *open bars*). With 0.5% blue Dextran (*BD*) added to the CSF, resistance was higher from the lateral ventricles, probably to an increase in viscoscity of the infusion fluid. When Indian ink was added to the infusion CSF, the resistance increased by more than twofold, probably because ink particles were blocking the absorption sites. Values are means +/− SEM, $n = 6–10$

For a series of infusion rates into the CSF, resistance to flow (or absorption) equals the gradient of a plot of steady-state pressure against infusion rate. The resistance to absorption in *Rana pipiens* was measured from infusions into the lateral ventricles and found to give a linear plot with a gradient of 15.5 mm H_2O min μl^{-1} (Fig. 3; Jones and Taylor 1984). This value is similar to other small vertebrates such as turtle (Heisey and Michael 1971), mouse (Jones 1985) and rat (Mann et al. 1980; Jones et al. 1987b). The same result was obtained when the infusion was made into the subarachnoid space, which indicates that the pores of the posterior tela do not provide any resistance to flow and that both methods measure CSF absorption from the subarachnoid space. The resistance to absorption was increased by the addition of blue Dextran or Indian ink to the infusion solution, probably through an increase in the viscosity in the case of blue Dextran and through particulate clogging of the absorption sites with Indian ink. Although CSF absorption sites have not been positively identified in amphibians, there is a rapid appearance in blood of radiolabelled substances infused into the CSF and the concentration was higher in venous blood (Jones and Taylor 1984), suggesting the existence of a direct route into dural veins as occurs in mammals.

CSF Physiology During Mammalian Development

In mammals the CSF flows out of the ventricular system through the foramina in the roof and lateral recesses of the fourth ventricle. These foramina become patent at a relatively late stage in development, which in the rat, for example, is on the last day of gestation and the day of birth (Strong and Alban 1932; Jones and Sellars 1982). The choroid plexus, however, develops by day 14 of pregnancy (Chamberlain 1973) and although not structurally or functionally mature until after birth (Keep et al. 1986; Jones and Keep 1987, 1988; Keep and Jones 1990) probably secretes an immature CSF during the late fetal stage. Early studies on pig fetuses by Weed (1917) suggested that fluid escapes through the roof of the fourth ventricle, and Weed thought that this escape contributed to the formation of the subarachnoid space. This prompted us to examine the structure of the roof of the fourth ventricle in rat fetuses by scanning electron microscopy (Jones 1980b). The study showed that at 16 days gestation the ventricle roof consisted of developing choroidal folds anteriorly and a complete layer of smooth ependyma posteriorly, but at 17 days gestation the posterior roof becomes pouch shaped with small pores up to 10 µm in diameter and resembling those seen in amphibians. The pores enlarge as gestation proceeds but are only transient because by the time of birth most of the membranous roof is obliterated by the overlying cerebellum and at this time the lateral foramina become patent.

Infusion experiments were carried out with fluorescent Dextran using a similar technique to that for amphibians but with rat fetuses exteriorised from the uterus of anaesthetized females with the placental circulation intact (Jones and Sellars 1982). In the majority of 16 day gestation fetuses the fluorescence was confined to the ventricles within the times tested which, from the start of infusion, varied between 15 and 45 min, but in fetuses 17 days gestation and over, the fluorescence was also distributed outside the brain in the subarachnoid space (Fig. 4), consistent with the development of pores at this time. With short times (15 min) the spread was confined to the dorsal space over the rhombencephalon, but with longer times there was preferential spread along the subarachnoid space of the spinal cord and subsequently into the space around the brain. In postnatal rats there was a gradual transition to exiting via a lateral route instead of the medial rhombencephalic roof, consistent with the opening of the foramina of Luschka.

The resistance to absorption of the CSF has been measured in developing rats using the constant rate infusion technique described above (Jones et al. 1987b). There was a significant change with age, with low resistance in fetuses, a peak high resistance at 1 day postnatal, followed by a gradual decrease to adult values of 6–8 mm H_2O min μl^{-1} at 30 days after birth (Fig. 5). The resistance to absorption was the same whether the measurements were made from the lateral ventricle or from the subarachnoid space showing

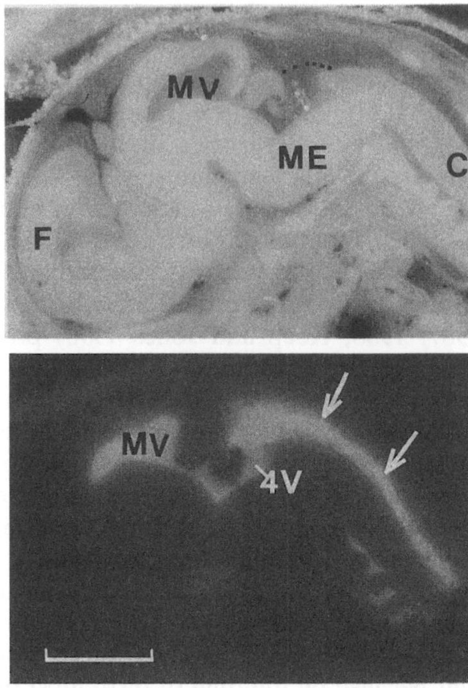

Fig. 4. The head of an 18-day rat fetus 11 min after the start of an infusion of 1 µl fluorescein-labelled Dextran into the mesencephalic ventricle (*MV*). The head was frozen and cut sagittally and photographed in normal light (*top*) and in ultraviolet light (*bottom*). The fluorescence can be seen in the mesencephalic and fourth ventricle (*4V*) and in the subarachnoid space (*arrows*) dorsal to the hindbrain and spinal cord (*C*). *F*, forebrain; *ME*, medulla; *dotted line* marks the position of the posterior tela. *Scale bar* = 2 mm

that as for frogs, there was no resistance to flow out of the ventricular system at any stage. It was subsequently shown that the high resistance in the early postnatal period could be explained by a simultaneous increase in dural venous sinus pressure which occurs in response to increases in CSF pressure. This would have the effect of reducing the pressure gradient for absorption across the sinus wall (Jones and Gratton 1989). These findings suggest that CSF absorption is restricted in young rats, and possibly in other immature animals as well (Jones 1985), and that increases in intracranial pressure will be compensated less readily in young animals because of the reduced capacity for CSF absorption.

Congenital Hydrocephalus

Inherited hydrocephalus has arisen spontaneously a number of times in both mice and rats (Bruni et al. 1988); a substantial proportion of these cases

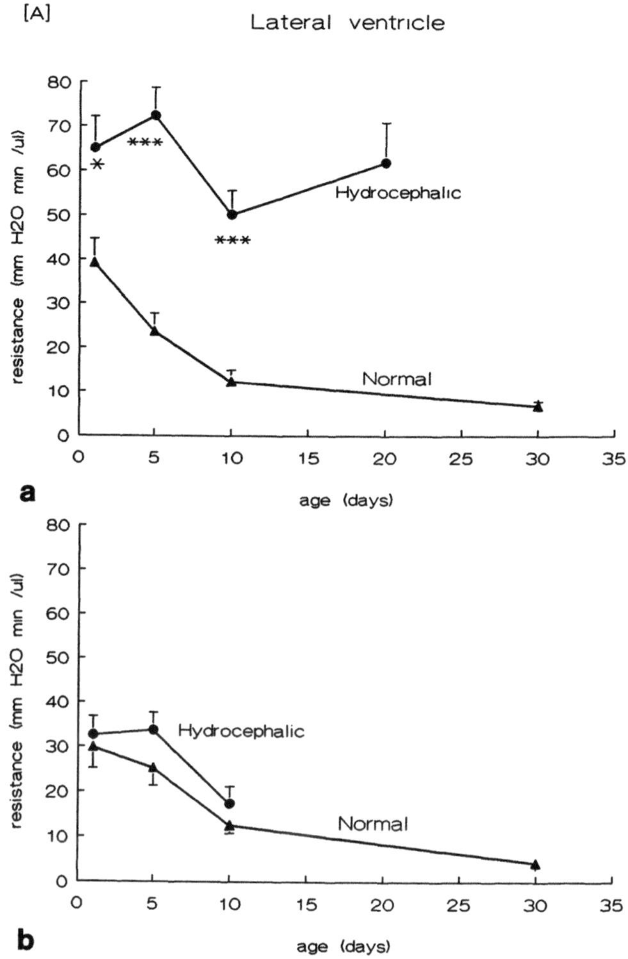

Fig. 5a,b. Resistance to CSF absorption at different postnatal ages in normal rats (*filled triangles*) and hydrocephalic, H-Tx rats (*filled circles*). Resistance from the lateral ventricles (**a**) is very similar to that from the cisterna magna (**b**) in normal rats and both decline with age. Hydrocephalic rats have a significantly higher resistance from the lateral ventricles but are similar to normal from the cisterna magna. This is consistent with an obstruction to flow in the cerebral aqueduct. Values are means +/− SEM, n = 6–10

have been attributed to a narrowing or stenosis of the cerebral aqueduct during gestation. Since the aqueduct is a narrow channel between the third and fourth ventricles, it is a particularly vulnerable site where even a minor developmental abnormality could obstruct CSF flow and result in hydrocephalus. We have studied the hydrocephalic SUMS/NP mouse and also the H-Tx rat, both of which have aqueduct abnormalities in late gestation (Fig. 6; Jones et al. 1987a; Jones and Bucknall 1988). Infusion of fluorescent Dextran into the lateral ventricles of hydrocephalic SUMS/NP mice showed that in the early postnatal stage the reduced aqueduct allowed the passage of fluorescence in some individuals, but in mice older than 10 days there was no spread out of the lateral and third ventricles. Similar experiments in H-Tx rats showed that the reduced aqueduct allowed some flow in fetuses and early neonates but by 5 days after birth there was a complete obstruction to markers placed in the lateral ventricles. The resistance to CSF absorption from the lateral ventricles in hydrocephalic SUMS/NP mice is significantly higher than for controls, particularly in mice 4 days old and over. Resistance to absorption from outside the ventricular system in the cisterna magna, on the other hand, is not affected by the hydrocephalus (Jones 1985). A similar situation exists in the lateral ventricles of the H-Tx rat where a significantly higher resistance to absorption was measured as early as the day before birth and it continued to be at least twofold higher than normal in the postnatal period (Fig. 5; Jones and Bucknall 1987). As for the SUMS/NP mouse, there was no defect in absorption from the cisterna magna in H-Tx rats. These results, therefore, are entirely consistent with an obstruction in the ventricular system, namely the cerebral aqueduct as described above.

The relationship of the cerebral aqueduct to the subcommissural organ (SCO) is a close one, such that CSF as it flows along the aqueduct inevitably passes close to the secretory cells and to Reissner's fibre. Thus although the function of the SCO is not known, one possibility is that it relates to the circulation of CSF. In support of this, a large proportion of animal models for hydrocephalus have an abnormal SCO (Table 1). This includes the two models (SUMS/NP mouse and H-Tx rat) studied by us (e.g. Fig. 6) and other models of inherited hydrocephalus (D'Amato et al. 1986; Takeuchi et al. 1987, 1988), all of which have a narrow or stenosed aqueduct. Animals with induced hydrocephalus also have a stenosed aqueduct and reduced

Table 1. Status of subcommissural organ in hydrocephalus

I. Inherited hydrocephalus with narrowed or stenosed aqueduct		
1. Rat (Wistar)	SCO variable	(D'Amato et al. 1986)
2. Rat (CWS/Idr)	SCO small	(Takeuchi et al. 1988)
3. Rat (H-Tx)	SCO small	(Jones and Bucknall 1988)
4. Rat (LEW/jms)	SCO not reported	(Sasaki et al. 1983)
5. Mouse (MT/HOKIdr)	SCO absent	(Takeuchi et al. 1987)
6. Mouse (SUMS/np)	SCO small	(Jones et al. 1987a)
II. Inherited hydrocephalus with normal aqueduct		
1. Mouse (Hy-3)	SCO normal	(Jones, unpublished)
III. Induced hydrocephalus with narrowed or stenosed aqueduct		
1. Rat (X-ray)	SCO reduced	(Takeuchi and Takeuchi 1986)
2. Rat (folic acid deficiency)	SCO reduced	(Overholzer et al. 1954)
3. Rat (MNU)	SCO reduced	(Takeuchi and Takeuchi 1987)
IV. Induced hydrocephalus with open aqueduct		
1. Rat (kaolin)	SCO reduced	(Irigoin et al. 1990)
2. Rat (viral)	SCO reduced	(Irigoin et al. 1990)

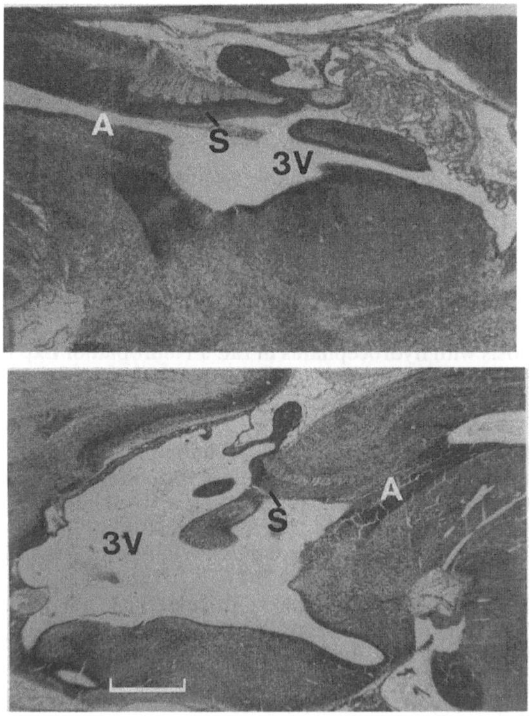

Fig. 6. Midsagittal sections through the brains of two 20-day gestation rat fetuses. The normal littermate (*top*) has a patent aqueduct (*A*) and a subcommissural organ (*S*) which extends caudally into the aqueduct for at least 1 mm. The hydrocephalic littermate (*bottom*) had enlarged lateral ventricles (not shown), a stenosed aqueduct and an abnormally small subcommissural organ. *3V*, third ventricle. Brains were embedded in wax, sectioned at 6 μm and stained with Mallory's triple stain. *Scale bar* = 0.5 mm

SCO (Overholzer et al. 1954; Takeuchi and Takeuchi 1986, 1987) whereas two others have an open aqueduct with a reduced SCO (Irigoin et al. 1990). One exception appears to be the Hy-3 mouse which has no obvious change in aqueduct or SCO (Jones, unpublished observations). These observations, although suggestive of a role relating to CSF circulation, do not allow us to determine whether there is a primary defect of the SCO which leads to hydrocephalus or whether a disturbance of CSF circulation in hydrocephalus leads to a functionally reduced SCO.

References

Bruni JE, Del Bigio MR, Cardoso ER, Persaud TVN (1988) Hereditary hydrocephalus in laboratory animals and humans. Exp Pathol 35:239–246
Bundgaard M, Cserr HF (1981) A glial blood-brain barrier in elasmobranchs. Brain Res 226:61–73
Bundgaard M, Cserr HF (1991) Barrier membranes at the outer surface of the brain of an elasmobranch, *Raja erinacea*. Cell Tissue Res 265:113–120
Chamberlain JG (1973) Analysis of developing ependymal and choroidal surfaces in rat brains using scanning electron microscopy. Dev Biol 31:22–30
D'Amato CJ, O'Shea KS, Hicks SP, Glover RA, Annesley TM (1986) Genetic prenatal aqueductal stenosis with hydrocephalus in rat. J Neuropathol Exp Neurol 45:665–682
Davson H, Hollingsworth G, Segal MB (1970) The mechanism of drainage of the cerebrospinal fluid. Brain 93:665–678
Davson H, Welch K, Segal MB (1987) The physiology and pathophysiology of the cerebrospinal fluid. Churchill Livingstone, Edinburgh
Heisey SR, Michael DK (1971) Cerebrospinal fluid formation and bulk absorption in the freshwater turtle. Exp Neurol 31:258–262
Irigoin C, Rodríguez EM, Heinrichs M, Frese K, Herzog S, Oksche A, Rott R (1990) Immunocytochemical study of the subcommissural organ of rats with induced postnatal hydrocephalus. Exp Brain Res 82:384–392
Jones HC (1978) Continuity between the ventricular and subarachnoid cerebrospinal fluids in an amphibian, *Rana pipiens*. Cell Tissue Res 195:153–167
Jones HC (1979) Fenestration of the epithelium lining the roof of the fourth cerebral ventricle in Amphibia. Cell Tissue Res 198:129–136
Jones HC (1980a) Circulation of marker substances in the cerebrospinal fluid of an amphibian, *Rana pipiens*. Cell Tissue Res 211:317–330
Jones HC (1980b) Intercellular pores between the ependymal cells lining the roof of the fourth cerebral ventricle in mammalian fetuses. Z Kinderchir 31:309–316
Jones HC (1982) The ultrastructure of the rhombencephalic posterior tela and adjacent tissues in an amphibian, *Rana pipiens*. J Anat 134:91–102
Jones HC (1985) The cerebrospinal fluid pressure and resistance to absorption during development in normal and hydrocephalic mutant mice. Exp Neurol 90:162–172
Jones HC, Bucknall RM (1987) Changes in cerebrospinal fluid pressure and outflow from the lateral ventricles during development of congenital hydrocephalus in the H-Tx rat. Exp Neurol 98:573–583
Jones HC, Bucknall RM (1988) Inherited prenatal hydrocephalus in the H-Tx rat: a morphological study. Neuropathol Appl Neurobiol 14:263–274
Jones HC, Dolman GS (1979) The structure of the roof of the fourth ventricle in pigeon and chick brains by light and electron microscopy. J Anat 128:13–29
Jones HC, Gratton JA (1989) The effect of cerebrospinal fluid pressure on dural venous pressure in young rats. J Neurosurg 71:119–123
Jones HC, Jopling CAC (1983) The development of interependymal pores in the rhombencephalic posterior tela in late embryonic, larval and metamorphosing stages of *Rana pipiens*. Dev Brain Res 7:121–130
Jones HC, Keep RF (1987) The control of potassium concentration in the cerebrospinal fluid and brain interstitial fluid of developing rats. J Physiol 383:441–453
Jones HC, Keep RF (1988) Brain fluid calcium concentration and response to acute hypercalcaemia during development in the rat. J Physiol 402:579–593
Jones HC, Sellars RA (1982) The movement of fluid out of the cerebral ventricles in fetal and neonatal rats. Z Kinderchir 37:130–133
Jones HC, Taylor CM (1983) Morphological changes in amphibian ventricular ependyma after infusion with Evans blue dye. J Physiol 339:48–49P
Jones HC, Taylor CM (1984) Absorption of the cerebrospinal fluid and intracranial compliance in an amphibian *Rana pipiens*. J Physiol 353:405–417

Jones HC, Dack S, Ellis C (1987a) Morphological aspects of the development of hydrocephalus in a mouse mutant (SUMS/NP). Acta Neuropathol (Berl) 72:268–276
Jones HC, Deane R, Bucknall RM (1987b) Developmental changes in cerebrospinal fluid pressure and resistance to absorption in rats. Dev Brain Res 33:23–30
Keep RF, Jones HC (1990) A morphometric study on the development of the lateral ventricle choroid plexus, choroid plexus capillaries and ventricular ependyma in the rat. Dev Brain Res 56:47–53
Keep RF, Jones HC, Cawkwell RD (1986) A morphometric analysis of the development of the fourth ventricle choroid plexus in the rat. Dev Brain Res 27:77–85
Mann JD, Mann ES, Cookson SL (1980) Differential effects of pentobarbital, ketamine hydrochloride and enflurane anaesthesia on CSF formation rate and outflow resistance in the rat. In: Miller JD, Becker DP, Hochwald G, Marmarou A, Shulman K (eds) Intracranial pressure IV. Springer, Berlin Heidelberg New York, pp 466–471
Momose Y, Kohno K, Ryuzo I (1988) Ultrastructural study on the meninx of the goldfish brain. J Comp Neurol 270:327–336
Overholzer MD, Whitley JR, O'Dell BL, Hogan AG (1954) The ventricular system in hydrocephalic rat brains produced by a deficiency of vitamin B12 or of folic acid in the maternal diet. Anat Rec 120:917–933
Rovainen CM, Lemcoe GE, Peterson A (1971) Structure and chemistry of glucose-producing cells in meningeal tissue of the lamprey. Brain Res 30:99–118
Sasaki S, Goto H, Nagano H, Furuya K, Omata Y, Kanazawa K, Suzuki K, Sudo K, Collmann H (1983) Congenital hydrocephalus revealed in the inbred rat LEW/Jms. Neurosurgery 13:548–554
Strong RM, Alban H (1932) The development of the lateral apertures of the fourth ventricle in the albino rat brain. Anat Rec 52:39
Takeuchi IK, Takeuchi YK (1986) Congenital hydrocephalus following X-irradiation of pregnant rats on an early gestational day. Neurobehav Toxicol Teratol 8:143–150
Takeuchi IK, Takeuchi Y (1987) Dysgenesis of subcommissural organ in the congenital hydrocephalic rats induced by prenatal methylnitrosourea (MNU) treatment. Teratology 36:444
Takeuchi IK, Kimura R, Matsuda M, Shoji R (1987) Absence of subcommissural organ in the cerebral aqueduct of congenital hydrocephalus spontaneously occurring in MT/HOK1dr mice. Acta Neuropath (Berl) 73:320–322
Takeuchi IK, Kimura R, Shoji R (1988) Dysplasia of subcommissural organ in congenital hydrocephalus spontaneously occurring in CWS/Idr rats. Experientia 44:338–340
Tornheim PA, Foltz FM (1979) Circulation of the CSF in the bullfrog, *Rana catesbeiana*. Anat Rec 194:389–404
Tornheim PA, Michaels JE (1979) Fine structure of the rhombencephalic tela of the bullfrog, *Rana catesbeiana*. Cell Tissue Res 202:479–491
Weed LH (1917) The development of the CSF spaces in pig and in man. Contrib Embryol 14:1–116

IX. Experimental Aspects

Effects of Hibernation and Hypothermia on the Secretory Activity of the Subcommissural Organ

F. Nürnberger and P.A. Peña Salazar

Introduction

During the past decades, the subcommissural organ (SCO) has been the subject of numerous investigations focussed mainly on structural aspects. To date, we have a fairly precise morphological idea of this circumventricular organ (see Oksche 1969; Rodríguez et al. 1984a,b, 1987). Several aspects of the synthesis and processing of the secretory product of the SCO have been studied rather extensively. With respect to a compartmentation of the processing pathway of the subcommissural secretion product, Rodríguez et al. (1986) observed a differential cleavage pattern of the sugar residues of the subcommissural glycoprotein during the intracellular passage through various organelles (pre- and post-Golgi) and also during the ventricular transport phase along the Reissner's fiber (RF). The processing of the secreted glycoprotein appears to go on until RF disappears via specialized structures at the end of the central canal (see Hofer et al. 1984).

In marked contrast to the striking secretory activity of the SCO, there is so far no clear-cut evidence for a functional involvement of this organ in the control of systemic physiological processes. Several hypotheses comprising, for example, an influence on osmoregulation, thermoregulation, or a preventive role concerning the obliteration of the central canal (for review, see Leonhardt 1980) were only partly substantiated by experimental investigations. To date, no experimentally induced change in the functional state of the organism has been fully correlated with specific alterations in the secretory activity of the SCO.

However, season-related changes are apparent in the synthesis and release of the subcommissural secretion product. Oksche (1962) described a conspicuous increase in the content of selectively stainable secretory material in the SCO of *Rana temporaria* during the torpid state of hibernation. Similar results were also presented by d'Uva et al. (1976) in the reptilian species *Lacerta s. sicula*.

The English term "hibernation" does not provide a precise description of a particular physiological phenomenon. In invertebrates and lower vertebrates, especially amphibians and reptiles (i.e., poikilotherms), hibernation represents a passive decline in the metabolic activity related to the ambient temperature. Often, this drop occurs only after a preparatory phase, characterized by a retreat into an adequate hibernation den and removal of crystallization nuclei from the body liquids. In mammals, the term "hibernation" describes two different phenomena, i.e., (1) a kind of resting period during the winter season without development of torpor as found, for example, in bears inhabiting higher latitudes and (2) a precisely regulated physiological condition at a low metabolic level. In the latter case, the hibernators can arouse from hibernation utilizing an internal heat production without exogenously delivered energy.

With this information in mind, it seemed to be worthwhile to investigate the secretory capacity of the SCO during different functional states of regulated mammalian hibernation. For analysis of possible causal relationships, it was mandatory to compare the secretory activity of the SCO with the reactivity pattern of other secretory elements of the brain, especially with the apparatus of peptidergic secretory neurons.

The Reactivity Pattern of Selected Systems of Peptidergic Neurons in Relation to Hibernation

The different representatives of the peptidergic system display a rather striking correlation to the individual functional states of hibernation. Repeatedly, within one particular neuropeptide system a differential reactivity pattern could be observed. The enkephalin-immunoreactive elements in the caudate nucleus and putamen did not show a major response when changing from the euthermic, nonhibernating state into the phase of hibernation, whereas the enkephalin-immunoreactive perikarya and nerve fibers of hypothalamus and lateral septum significantly increased their enkephalin synthesis and content as demonstrated by in situ hybridization and immunocytochemistry (Nürnberger et al. 1991). A comparable differential reactivity pattern was also observed for the vasopressin-immunoreactive system. The majority of the magnocellular vasopressin-immunoreactive perikarya decreased their activity during hibernation, whereas the suprachiasmatic component and a small population of paraventricular perikarya increased their immunoreactivity compared to euthermic controls (Nürnberger 1983; Nürnberger et al. 1985). To make this picture even more complex, the increase in immunoreactivity for vasopressin could be observed only in a particular group of hibernators–the long-term hibernators, which exhibit hibernation phases of several months and hibernation periods of more than 1 week. In short-term hibernators, e.g., the golden hamster, the immunoreactivity of the magnocellular neurons increases during the hibernation state, in comparison to euthermic controls, but the immunoreactivity in the other regions (septum, suprachiasmatic nucleus) remains rather constant (Nürnberger et al. 1982).

The differential immunoreactivity pattern, reflecting the synthesis and secretion of neuropeptides, obviously depends on the location of this component within the neuroendocrine system and mirrors its functional role at this particular site. Accordingly, the immunoreactivity of the neuropeptide-Y projections to the suprachiasmatic nucleus significantly decreased during hibernation, indicating a lack of input from the lateral geniculate body, whereas the neuropeptide-Y immunoreactivity remained virtually unchanged within the effector nuclei of the vagal nerve, which is known to be constantly active during both euthermia and hibernation.

As for the intrinsic central peptidergic neurons, a differential pattern of secretory activity was evident also for the neuroendocrine hypothalamo-adenohypophysial system. Several releasing factors were drastically decreased in their pathways to the portal vessels, whereas somatostatin did not exhibit such change (Nürnberger et al. 1986).

These alterations in the secretory activity of peptidergic neurons are in good accord with the bioelectric activity of the respective neuronal populations. This holds true also for different transmitter systems, especially for the serotonin system, which showed a general increase in its activity, and the noradrenergic system, which also appeared to display signs of increased activity. In summary, the secretory performance of peptidergic neurons depends very much on the demands of the particular physiological processes during hibernation.

The Secretory Capacity of the SCO During Euthermia, Hibernation, and Hypothermia (Figs. 1–3)

The present investigation was performed with four different species of mammals. Hibernating hedgehogs (*Erinaceus europaeus*), ground squirrels (*Spermophilus richardsonii*, *S. columbianus*), and European hamsters (*Cricetus cricetus*) were compared with nonhibernating, euthermic controls of the respective species. In addition, these two groups of Columbian and Richardson's ground squirrels were compared with hypothermic animals. The ground squirrels were made hypothermic by exposing them to cooled air at approx. 5°C containing also approx. 2% halothane (for methodological details, see Jourdan and Wang 1987). After the animals had reached a body temperature equivalent to that during hibernation, at which the cold was no longer defeated by shivering, the anesthetic was removed and the animals were sacrificed after at least 12 h in deep hypothermia.

All animals were sacrificed by perfusion with Bouin's fixative during deep pentobarbital anesthesia. The brain tissue was embedded in paraffin and cut in serial sections (7 μm). The secretory activity of the SCO was studied by the application of an antibody against urea-extracted RF material (FRU, Rodríguez et al. 1984b). The respective A(anti)FRU antisera were described and characterized by Rodríguez et al. (1984b). To obtain further evidence concerning the activity state of the SCO, the density of RNA staining (Einarson 1951) and the size of cell nuclei were quantified.

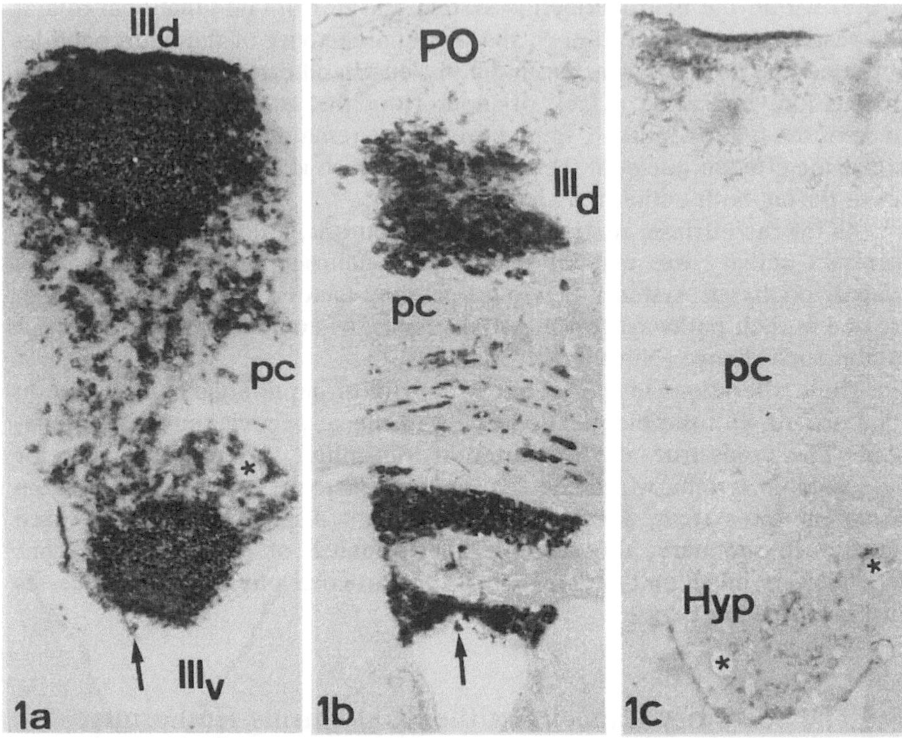
IIId
pc
IIIv
1a
PO
IIId
pc
1b
pc
Hyp
1c

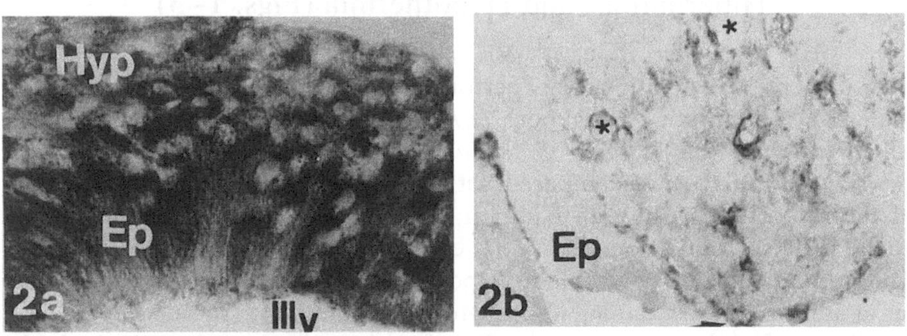
Hyp
Ep
2a
IIIv
Ep
2b

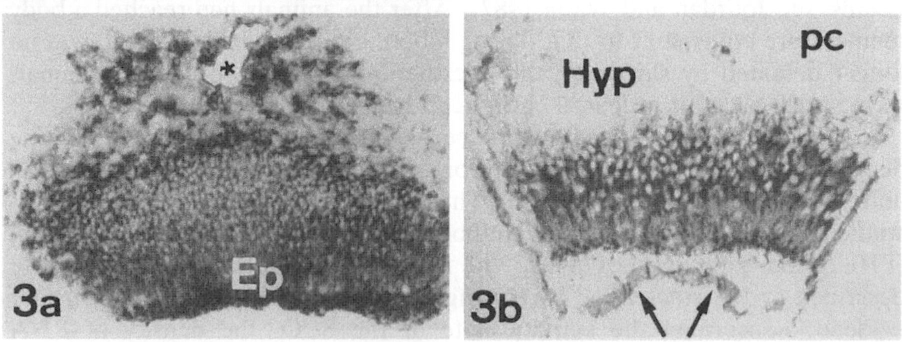
3a
Ep
Hyp
pc
3b

In euthermic ground squirrels and hamsters, the immunostainings revealed a reactivity pattern of the SCO known from other rodent species. In addition to a conspicuously thickened ependyma, adjacent hypendymal cells spread along the basis of the SCO and penetrated the posterior commissure. In hedgehogs, the immunoreactivity is much weaker than in the other mammalian species. In accord with Rodríguez and colleagues (1984a), in the present hedgehog material, the hypendymal layers stained more extensively than the principal ependymal component of the SCO. In this species, most of the immunostaining could be observed in the suprapineal epithelia of the choroid plexus (see Rodriguez et al. 1984a).

In all rodent species examined, in contrast to the hedgehog, a distinct, clearly stained RF could be observed. The origin of RF from the secretory product of the ependymal layer of the SCO was clearly evident from the visualization of filamentous structures connecting the secretory ependyma with the condensed material in RF.

A hypothermic state lasting for several hours did not significantly alter the pattern of immunoreactivity of the SCO. All parts of the organ including RF were immunostained to an extent as characteristic of the euthermic state.

In marked contrast to the euthermic individuals of all species investigated, however, the immunoreactivity for the secretion product of the SCO was, as a rule, drastically decreased during hibernation. Most prominent were these changes in both ground squirrel species; the intense staining of the SCO cytoplasm known from euthermic animals was almost completely missing during hibernation. Only the apical portion of the SCO ependyma and some cell processes directed toward the hypendymal blood vessels still contained a slight amount of AFRU-immunoreactive material.

The hibernation-correlated changes in hedgehogs, which showed AFRU-immunoreactive elements exclusively in the hypendymal formation of their SCO, displayed a similar pattern. During hibernation, these hypendymal cells were even less immunoreactive. The slightest differences were observed in European hamsters. In this latter species, the immunoreactivity for the

←

Fig. 1a–c. AFRU-immunoreactive cells in the subcommissural organ (SCO) of euthermic (**a**), hypothermic (**b**), and hibernating (**c**) ground squirrels, *Spermophilus columbianus*. Note the immunoreactive cell clusters intercalated with the fiber bundles of the posterior commissure in **a** and **b**, ×80. *Ep*, ependymal component of the SCO; *Hyp*, hypendymal component of the SCO; *pc*, posterior commissure; *PO*, pineal organ; *IIId*, *IIIv*, dorsal and ventral portion of the third ventricle; *asterisk*, hypendymal blood vessel, *arrow*, Reissner's fiber

Fig. 2a,b. AFRU-immunoreactive cells in the SCO of euthermic (**a**) and hibernating (**b**) *Spermophilus richardsonii*, ×350. See Fig. 1 for abbreviations

Fig. 3a,b. Cells immunoreactive for AFRU in the SCO of euthermic (**a**) and hibernating (**b**) European hamsters (*Cricetus cricetus*). Note the fibrous secretion product connecting the apical surface of the SCO and Reissner's fiber in **b**. **a**, ×130; **b**, ×180. See Fig. 1 for abbreviations secretory material was rather similar in both hibernating and euthermic specimens.

Reissner's fiber was still detectable during hibernation; however, its stainability was considerably decreased in contrast to euthermic animals. The intensely stained fibrous material connecting RF with the apical pole of the epithelial cells of the SCO was missing in the lethargic state.

Conclusions

The findings obtained from the present investigation clearly indicate a significant decrease in immunoreactive material in the SCO of hibernating animals. This result, together with the smaller size of cell nuclei and less intense RNA-staining in the cells of the SCO, speaks in favor of a generally decreased activity of the organ during hibernation. To date, such striking change in the secretory activity of the SCO as the difference existing between hibernating and euthermic mammals could not be observed in any other contrasting physiological situation. It is interesting that the hibernation-related drop in the activity of the SCO cannot be interpreted as an effect of the deep body temperature during hibernation; in hypothermic animals with an almost identical low body temperature, the immunoreactivity of the secretion product of the SCO shows no principal difference to that in euthermic controls.

The restriction of AFRU-stained material exclusively to the apical pole of the ependymal cells of the SCO and to the hypendymal endfeet adjacent to the local blood vessels is suggestive of a complete hibernation-related depletion of most of the intracellular pools of the secretory product. Only at sites displaying exocytotic release of the secretory material, the specific immunoreaction was still detectable. As mirrored by this distribution pattern, the principal pathway for processing the secretion product of the SCO might be functional also during hibernation (see Rodríguez et al. 1987). Since RF is still visible during hibernation, the apical accumulation of secretory material appears to indicate the need for a continuous replacement of this material. The immunoreactive endfeet of ependymal cells establishing contact to blood vessels speak in favor of a functional role of this connection.

The secretion pattern of the SCO during mammalian hibernation differs quite remarkably from that in amphibian and reptilian hibernators, which store high amounts of secretory material in their SCO. It is still a matter of speculation whether these augmented deposits of secretion are mobilized in the course of processes accompanying the termination of the state of temperature-dependent lethargy.

Major differences in the secretory activity of the SCO, however, were observed only in the long-term hibernators, such as Richardson's and

Columbian ground squirrel. In European hamsters, which generally express shorter hibernation periods, the difference in immunoreactivity of the SCO is only slight. This might be explained by the shorter time intervals between the onset of hibernation and the time of sacrifice. However, according to our experimental protocol with perfusion of the animals on the third day of hibernation, the immunoreactivity observed most probably reflects a physiological situation. Both the much shorter overall phase of hibernation allowing the regeneration of the secretory product during the intermittent arousal periods and the generally persisting activity level of autonomic regulation within the brain of short-term hibernators might also help to explain the differences in the immunoreactivity of the SCO between ground squirrels and hamsters.

The inhibition of the secretory activity of the SCO during hibernation is most likely induced by various input systems to this circumventricular organ. As shown for several other species, the SCO receives vasopressin projections from the paraventricular nucleus of the hypothalamus and serotonin projections from the raphe nuclei. With respect to hibernation, the serotonin input might be of particular interest for the inhibition of the SCO, since the serotonin terminals display a higher immunoreactivity in hibernating animals. Moreover, serotonin expresses a series of other inhibitory effects during hibernation, such as the suppression of the body temperature and the inhibition of the suprachiasmatic circadian output (Schindler and Nürnberger 1990).

Although the changes in secretory activity of the SCO are, in principle, uniform in all species of hibernators so far examined, their value for defining a general function of the SCO is rather restricted. To date, even for the physiology of hibernation, the role of the SCO is unknown and a possible involvement in the control of body temperature, osmoregulation or other autonomic processes is still open to discussion.

Acknowledgements. The authors are grateful to Dr. E.M. Rodríguez, who provided us with AFRU serum, and to the Deutsche Forschungsgemeinschaft (grant Nu 36/2–3) and the Deutscher Akademischer Austauschdienst (grant 322/504/020/9 to P.A.P.S.) for their support.

References

D'Uva V, Garcia G, Ciarletta A (1976) The subcommissural organ of the lizard *Lacerta s. sicula* Raf. Ultrastructure and secretory cycle. J Submicrosc Cytol 8:175–191

Einarson L (1951) On the theory of gallocyanin-chromalum staining and its application for quantitative estimation of basophilia. Acta Pathol Microbiol Scand 28:82–102

Hofer H, Meinel W, Erhardt H, Wolter A (1984) Preliminary electron-microscopical observations on the ampulla caudalis and the discharge of the material of Reissner's fibre into the capillary system of the terminal part of the tail of ammocoetes (Agnathi). Gegenbaurs Morphol Jahrb 130:77–110

Jourdan ML, Wang LCH (1987) An improved technique for the study of hypothermia physiology. J Therm Biol 12:175–178
Leonhardt H (1980) Ependym und Circumventriculäre Organe. In: Oksche A, Vollrath L (eds) Neuroglia I. Handbuch der mikroskopischen Anatomie des Menschen, vol IV, part 10. Springer, Berlin Heidelberg New York, pp 177–665
Nürnberger F (1983) Der Hypothalamus des Igels (*Erinaceus europaeus* L.) unter besonderer Berücksichtigung des Winterschlafes. Cytoarchitektonische und immuncytochemische Studien. Thesis, University of Marburg
Nürnberger F, Blähser S, Merker G (1982) Reactivity pattern of vasopressin neurons in long-term hibernators. Pflügers Arch [Suppl] 392:R31
Nürnberger F, Rorstad OP, Lederis K (1985) The hypothalamo-neurohypophysial system and hibernation in the ground squirrel, *Spermophilus richardsonii*. In: Kobayashi H, Bern HA, Urano A (eds) Neurosecretion and the biology of neuropeptides. Japan Sci Soc Press, Tokyo/Springer, Berlin Heidelberg New York, pp 518–520
Nürnberger F, Lederis K, Rorstad OP (1986) Effects of hibernation on somatostatin-like immunoreactivity in the brain of the ground squirrel (*Spermophilus richardsonii*) and European hedgehog (*Erinaceus europaeus*). Cell Tissue Res 243:263–271
Nürnberger F, Lee TF, Jourdan ML, Wang LCH (1991) Seasonal changes in methionine-enkephalin immunoreactivity in the brain of a hibernator, *Spermophilus columbianus*. Brain Res 547:115–121
Oksche A (1962) Histologische, histochemische und experimentelle Studien am Subkommissuralorgan von Anuren (mit Hinweisen auf den Epiphysenkomplex). Z Zellforsch 57:240–326
Oksche A (1969) The subcommissural organ. J Neuro Visc Rel [Suppl] IX: 111–139
Rodríguez EM, Oksche A, Hein S, Rodríguez S, Yulis R (1984a) Spatial and structural interrelationship between secretory cells of the subcommissural organ and blood vessels. An immunocytochemical study. Cell Tissue Res 237:443–449
Rodríguez EM, Oksche A, Hein S, Rodríguez S, Yulis (1984b) Comparative immunocytochemical study of the subcommissural organ. Cell Tissue Res 237:427–441
Rodríguez EM, Herrera H, Peruzzo B, Rodríguez S, Hein S, Oksche A (1986) Light- and electron microscopic immunocytochemistry and lectin histochemistry of the subcommissural organ: evidence for processing of secretory material. Cell Tissue Res 243:545–559
Rodríguez EM, Oksche A, Rodríguez S, Hein S, Peruzzo B, Schoebitz K, Herrera H (1987) The subcommissural organ and Reissner's fiber: Fine structure and cytochemistry. In: Gross PM (ed) Circumventricular organs and body fluids, vol II. CRC Press, Boca Raton, pp 3–41
Schindler CU, Nürnberger F (1990) Hibernation-related changes in the immunoreactivity of neuropeptide systems in the suprachiasmatic nucleus of the ground squirrel, *Spermophilus richardsonii*. With reference to observations in the hedgehog, *Erinaceus europaeus*. Cell Tissue Res 262:293–300

The Subcommissural Organ: Immunohistochemistry and Potential Relations to Salt/Water Balance

W.B. Severs, C.D. Balaban, B.A. Morrow, C.L. Snyder, and L.C. Keil

Introduction

General Comments

The subcommissural organ (SCO) seems to be an "orphan" among the circumventricular organs (CVO). It attracts relatively little research activity and there is no clear-cut indication of what normal physiological function(s) it contributes to the day-to-day life of an animal. By contrast, other CVOs with more clearly defined roles attract considerable research effort. For example, the area postrema and the nearby nucleus tractus solitarius are well known for their involvement in cardiovascular reflex mechanisms (Miselis et al. 1987). Similarly, there is considerable information, including anatomical connectivity, about the subfornical organ and its linkage to water ingestion and vasopressin secretion (Lind 1987).

The proposed functions of the SCO and its associated Reissner's fiber (RF) include a "volume" or "stretch" receptor within the cerebroventricular system, a "detoxification" function for chemicals entering the ventricular cavities, a role in salt and water homeostasis (reviews, Dundore 1985; Palkovits 1987; Severs et al. 1987), and potential roles in the physiology of reproduction (Limonta et al. 1982) and sleep (Sallamon et al. 1984). More recent research has linked the SCO to certain forms of both congenital and experimental hydrocephalus (see below). In a sense, hydrocephalus is an extension of the concept that the SCO affects salt/water balance.

Common Technical Problems

Major impediments to studying the SCO, especially in relation to salt/water balance, include the small size, irregular shape, and difficulty in specifically accessing the structure as it narrows along the dorsal portion of the third ventricle into the cerebral aqueduct. Also, there are heterogenous cell types and the SCO is partitioned into distinct ependymal and hypendymal layers. Relatively little is known about neuroanatomical input and output circuitry. The most distinct innervation is the serotonergic input from the raphe nuclei (Palkovits et al. 1987). Also, although the list of neurochemicals available as potential information exchangers is expanding, evaluation of the SCO is not included in many studies.

General problems related to analysis of salt/water balance include the fact that most individual parameters related to fluid homeostasis are multifactorally controlled and highly redundant. This can make difficult the analysis of the role of the SCO, and evaluation under both basal and stimulated conditions is necessary for functional studies. Also, salt/water metabolism is highly reactive to "stresses" that may be imposed by experimental protocols, and this may be a confounding variable in both anesthetized and conscious animals.

Overview

This chapter touches on several of the subjects raised. Included are (a) recent immunohistochemical data about the presence of tyrosine hydroxylase (TH), substance P (SP), somatostatin (SRIF), and vasopressin in the SCO, (b) an assessment of the effect of aldosterone in the SCO area on salt/water homeostasis, and (c) an evaluation of whether the SCO–ependymal barrier is altered in a new model of intracranial hypertension.

Recent Immunohistochemical Studies on SCO Neurochemicals

General Comments

The SCO is often not included in neurochemical studies designed to identify potential neurotransmitters and neuromodulators in brain structures. The SCO has been described as showing "... no indication that the organ receives any substantial catecholaminergic input...", whereas the serotonergic input is described as "massive" (Bouchaud and Bosler 1986). We report here that TH-positive axons and varicosities are consistently present in the SCO, along with SRIF and vasopressin – neuropeptides reported to be elevated in patients with high intracranial pressure (Sorensen 1986; Stepien et al. 1986) – and SP, a neuropeptide associated with serotonin pathways (Green 1989).

Immunohistochemical Observations

The immunochemical studies were conducted on brains from adult male Sprague-Dawley and Long-Evans rats. After pentobarbital injection (75–100 mg/kg i.p.) the rats were perfused transcardially with a 50 m*M* phosphate-buffered (pH 7.2–7.4) 0.9% saline solution (PBS), followed by either (a) paraformaldehyde-lysine-sodium metaperiodate (PLP) fixative (McLean and Nakane 1974) or (b) sequential infusions of 4% 1-ethyl-3-(3-dimethylaminopropyl)-carbodiimide HCl in 50 m*M* PBS (pH 7.2–7.4) and

4% paraformaldehyde in 50 m*M* PBS (pH 7.2–7.4). The brains were postfixed for 2–4 h at 4°C and cryoprotected for 24–72 h in a 30% sucrose-PBS solution at 4°C. The brains were sectioned at 40–50 µm on a freezing microtome in either coronal or horizontal planes, and free-floating sections were processed for immunohistochemistry using standard ABC reaction methods (Balaban et al. 1989; Billingsley et al. 1986).

Briefly, the primary antibodies were polyclonal antibodies against PC12 cell TH (kindly provided by Dr. Donald Kuhn); see Billingsley et al. (1986) for specificity of this antibody for a single protein. Antibodies against SRIF, SP, and vasopressin were obtained from Chemicon Co (USA). A second anti-vasopressin antibody was previously described (Severs et al. 1987). The dilutions used ranged from 1:1000–1:2500 (in PBS containing 0.1% Triton X-100). The second antibody (biotinylated anti-rabbit IgG, 1:500 dilution) and Vectastain Elite ABC avidin–horseradish peroxidase reagents were obtained from Vector Laboratories; a diaminobenzidine chromogen was used.

Figure 1 illustrates the regionally distinct distributions of TH-positive axons and SP and SRIF-positive neurons in the SCO. TH-positive fibers were most apparent in horizontal sections of the SCO. The fine-caliber TH-positive axons intruded into the hypendymal layer of the SCO from the overlying posterior commissure. These axons typically entered the basal portion of the ependymal layer and divided in the transverse plane to form both varicose and clublike endings. The plexus of TH-positive axons was dense near the rostral pole of the SCO; it was sparse caudally. SP-immunoreactive (IR) cells were observed in the lateral wings of the hypendymal layer. SRIF-like IR cells were present immediately subjacent to the lateral wings of the hypendymal layer and along the lateral margin of the ependymal layer of the SCO. Finally, sparse varicose vasopressin-IR axons entered the regions containing SRIF and SP-IR cells (not illustrated).

These data expand the list of potential neurochemicals available for information exchange. It should be emphasized that although the density of IR catecholaminergic (TH) and peptidergic (SRIF, SP, and vasopressin) inputs are of a lesser magnitude than that of serotonin (Palkovits 1987), this does not exclude a high level of amplification and an important role in information processing. Future studies are needed to identify the sources of catecholaminergic and peptide input and to evaluate their activity under conditions where the SCO is considered to have functional significance.

The SCO and Aldosterone Actions

General Aspects

Among the potential effects of the SCO that relate to salt/water metabolism (review, Severs et al. 1987), there are numerous experiments suggesting that

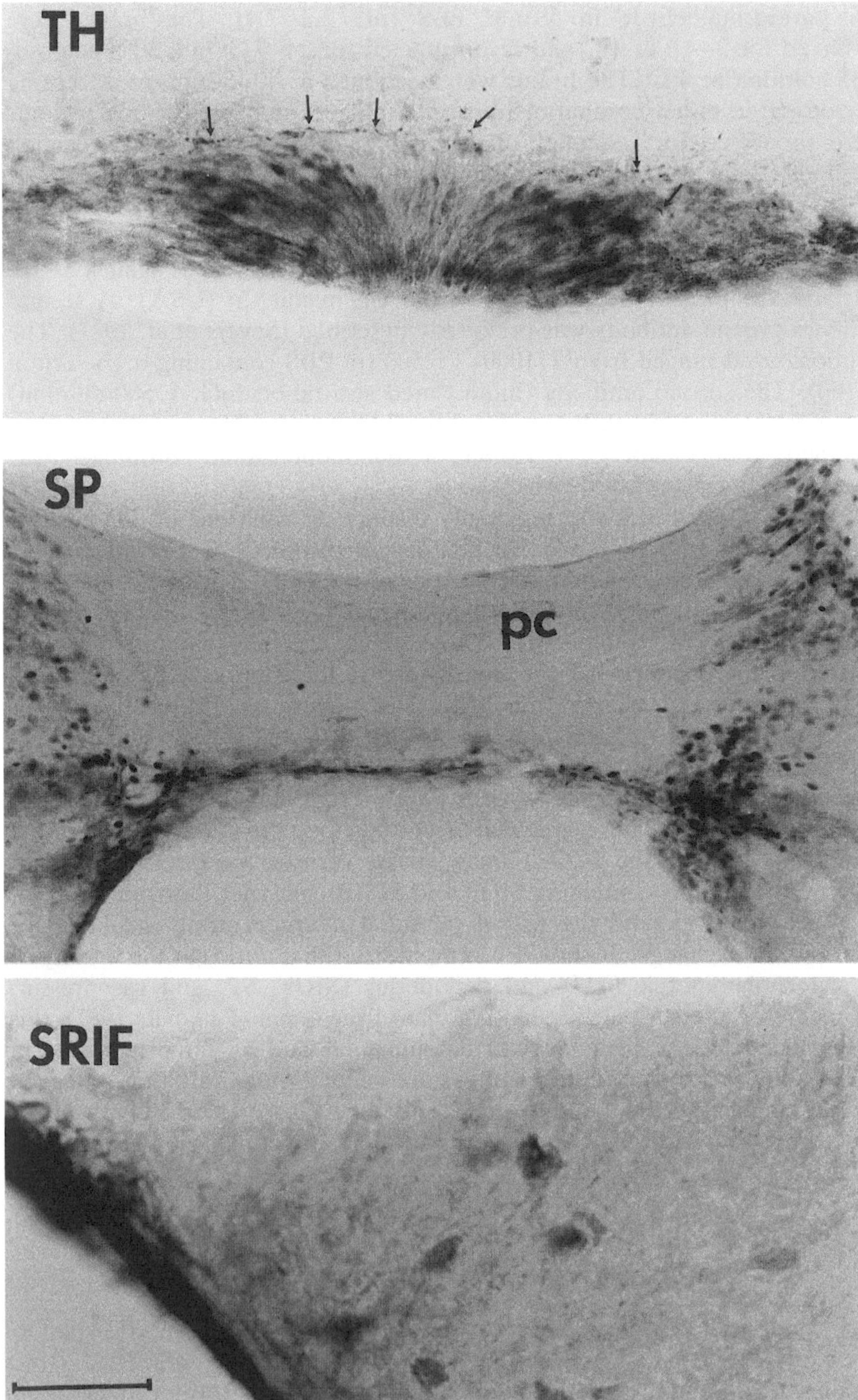
TH
SP
pc
SRIF

the SCO stimulates aldosterone secretion from the adrenal cortex and alters zona glomerulosa morphology. Further, SCO secretory material is depleted by epinephrine, the main adrenal medullary hormone (Gilbert and Armstrong 1966). In order to evaluate a potential endocrine feedback loop, we infused aldosterone into the SCO area or a lateral cerebral ventricle [intracerebral ventricular (ICV) route] of conscious rats (5 ng/μl/h) for 6 days and monitored multiple parameters associated with salt/water metabolism, as well as cross-sectional area of adrenal gland structures (Dundore et al. 1984; Dundore et al. 1987).

Central Aldosterone Effects

Figure 2 summarizes the effects of local aldosterone infusions into the SCO area, or ICV, on adrenal cross-sectional areas. There were no changes in the adrenal gland after the ICV route. Our cumulative experience with local SCO infusions shows that only the adrenal medulla changes: there is a clear reduction ($p < 0.05$) in the cross-sectional area that does not occur with solvent infusions, or when the infusion cannula is misplaced up to 1 mm in either rostral or caudal directions. It is noteworthy that if the adrenal cortex cross-sectional area, total or zonal, is "normalized" to the adrenal medullary cross-sectional area, a statistically significant increase occurs. However, the "normalizing" procedure yields an incorrect conclusion. The apparent increase in cortical/medullary area was caused by decreasing the denominator (medulla), and not by stimulating the cortex.

Figure 3 compares variables associated with consummatory behavior and urine output after SCO or ICV aldosterone infusions. The important points that emerge from these studies include the following. Aldosterone reduced both food and water intake compared to vehicle controls after the ICV, but not the SCO route of administration. The ICV route did not affect the urinary output of water, sodium, potassium, or the Na^+/K^+ ratio. By contrast, SCO infusions increased absolute Na^+ excretion and raised the Na^+/K^+ ratio ($p < 0.05$).

Data from experiments where the SCO cannula was misplaced showed that sodium excretion increased when the cannula tip was confluent with CSF at the roof of the caudal third cerebral ventricle or the pineal recess.

Fig. 1. Photomicrographs illustrating tyrosine hydroxylase (*TH*)-, substance P (*SP*)- and somatostatin (*SRIF*)-like immunoreactivity in the SCO. The *upper panel* shows a low magnification view of TH-immunoreactive axons (*arrows*) entering the SCO from the posterior commissure in a horizontal section. The micrograph is oriented such that the rostral edge of the section is directed downward; the SCO is sectioned at approximately the middle of its rostrocaudal extent. The *lower two panels* illustrate SP- and SRIF-like immunoreactive cells in coronal sections. The *calibration bar* =50 μm in the SP and SRIF panels, and 200 μm in the TH panel

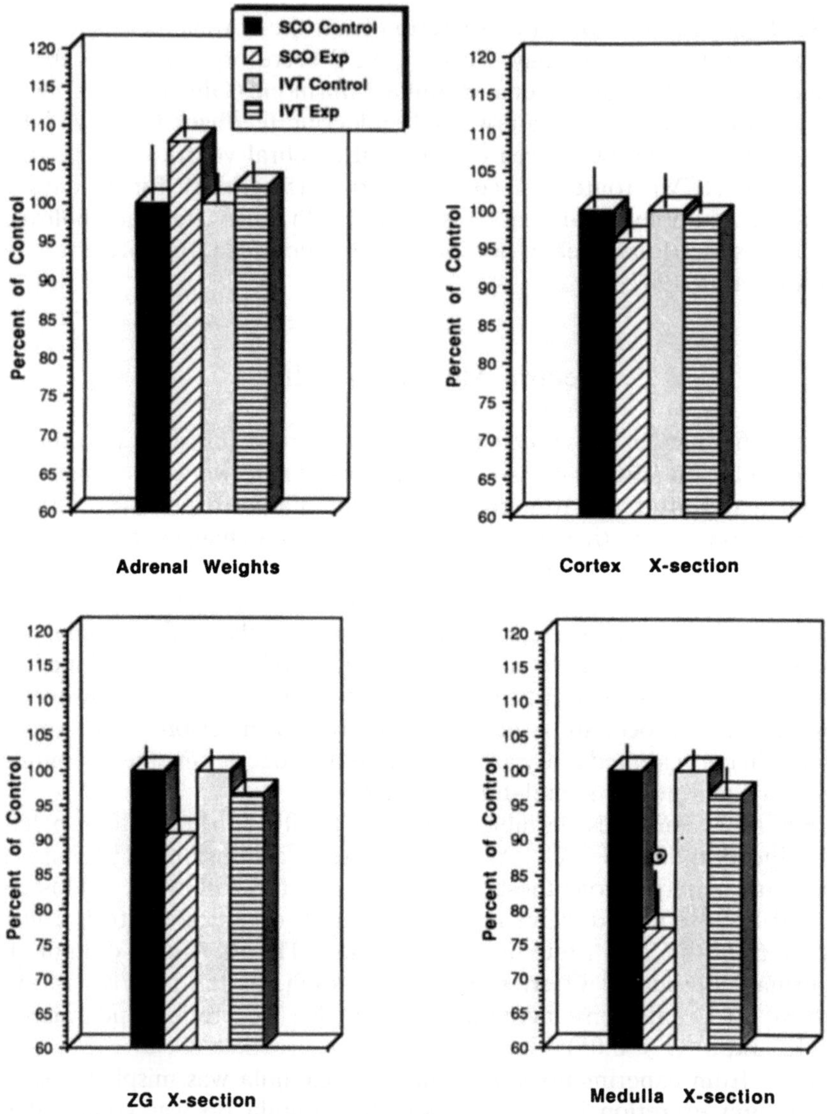

Fig. 2. Adrenal weights and cross-sectional areas after intracerebroventricular (*IVT*) or subcommissural organ area (*SCO*) infusions of D-aldosterone (5 ng/h × 6 days). Data are shown as % control + SEM. The only change ($p < 0.05$) was a decrease in the medulla of rats where the infusion cannula was properly placed in the SCO. Adapted from Dundore et al. (1984, 1987)

Tritiated aldosterone, infused under these conditions, showed minimal movement from the targetted SCO site and no isotope was in the pineal gland. Thus, tissue associated with increased sodium excretion may include the SCO, habenula, and periventricular regions near the rostral margin of

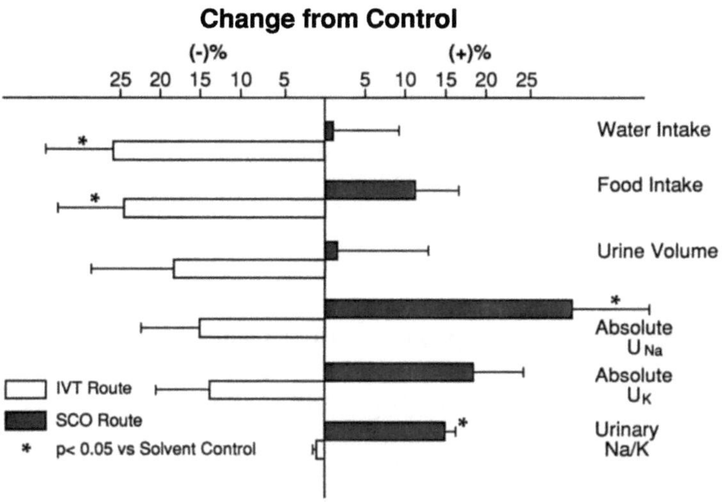

Fig. 3. Comparison of consummatory behavior, urine volume, and electrolytes after intracerebroventricular (*IVT*) or subcommissural organ area (*SCO*) infusions of D-aldosterone (5 ng/h × 6 days). Data are shown as % change over the 6 days from the solvent-infused controls + SEM. The prominent effect of the IVT route was a decrease in consummatory behavior ($p < 0.05$), whereas the SCO route was clearly associated with natriuresis and an increase in the urinary Na/K ratio ($p < 0.05$). Adapted from Dundore et al. (1984, 1987)

the cerebral aqueduct. It is important to note that the natriuresis evoked by aldosterone in this brain region is *opposite* to the peripheral effect of the mineralocorticoid on sodium homeostasis.

Other biochemical changes with this approach of evaluating the responsiveness of the SCO area to local aldosterone infusions include a highly specific increase in plasma epinephrine concentration ($p < 0.05$), which did not occur in rats receiving solvent or aldosterone via cannulae that were misplaced rostrally. Plasma norepinephrine and dopamine concentrations were unchanged. Adrenal corticosterone content increased ($p < 0.05$) after SCO infusions, and rats receiving aldosterone via misplaced cannulae had intermediate values. However, this measure was more variable. No changes in the above parameters occurred after ICV infusions of the same dose of aldosterone (Dundore et al. 1987).

Conclusion

These studies clearly establish that aldosterone exerts an effect in the SCO and/or adjacent brain areas (excluding the pineal gland), resulting in an overall increase in sodium excretion without changes in plasma sodium or potassium concentrations. This may indicate a gradual reduction in the extracellular fluid compartment since increases in Na^+ excretion were not discernable on a daily basis. The specific effects of aldosterone on the SCO that reduced adrenal medullary cross-sectional area and raised plasma epinephrine concentration are somewhat paradoxical. It is possible that the increase in plasma epinephrine was a compensatory mechanism to offset inhibition of the SCO by the local aldosterone infusion, because epinephrine is stimulatory to SCO secretory activity (Gilbert and Armstrong 1966).

The SCO, Hydrocephalus, and Intracranial Pressure

Background

Recently, a new issue related to the SCO and salt/water balance has emerged, namely, does the relationship between the SCO and hydrocephalus indicate a regulatory role in intracranial salt/water balance and/or pressure? The SCO has been described as absent in congenital hydrocephalus in MT/HoIdr mice (Takeuchi et al. 1987), and maldeveloped in rats made hydrocephalic by prenatal exposure to X-irradiation (Takeuchi et al. 1990). A dysplasia of SCO has been noted in spontaneously hydrocephalic CWS/Idr rats (Takeuchi et al. 1988).

Rodríguez et al. (1990), using ICV injections of anti-RF antibody in rats, reported that the RF detached from the SCO after about 4 h and underwent fragmentation. A newly growing RF appeared to extend into the aqueduct about 8 days later. The use of this antibody to study the SCO of rats made hydrocephalic by ICV kaolin or infection with Borna disease virus revealed a partial depletion of SCO secretory material and an impaired ability of secreted material to assemble into an RF, irrespective of whether kaolin-induced occlusion occurred (Irigoin et al. 1990). Whereas these data support an association of the SCO with hydrocephalus, they provide no mechanistic information.

Specifically, none of the experiments cited above provided data on the status of intracranial pressure when the SCO was taken for evaluation. We were interested in whether a defect exists in ependymal barrier properties of the SCO in the presence of abnormally elevated intracranial pressure. Such information would clarify whether abnormalities in the SCO of hydrocephalic animals were associated with high pressure, per se.

SCO Ependyma and CSF Pressure

Here, we report preliminary results of an experiment in which fluorescent latex microspheres (Molecular Probes Inc. No. L-5203, 30 nm, 1:6 dilution with artificial CSF) were injected ICV into conscious, freely moving rats

Table 1. Protocol for microsphere study of CSF pressure

1. Prepare male SD rats (300–400 g) with lateral ventricle cannula 2–4 days earlier under pentobarbital (50 mg/kg i.p.) anesthesia.
2. Connect conscious rats to CSFp recording system in metabolism cages (Morrow et al. 1990). After ~10 min adaptation, record 5 min CSFp (group 1), for groups 2 and 3, continue recording until microspheres given.
3. After baseline recording in groups 2 and 3, infuse artifical CSF ICV @ 8 µl/min × 10 min. Take rats for microspheres after 15 min (2) or 4 h (3).
4. Inject microspheres (10 µl/10 sec) ICV; wait 10 min, give pentobarbital 50 mg/kg i.p.; at 15 min give artificial CSF 10 µl/10 s and begin transcardiac perfusion (80 mmHg) with neutral buffered saline followed by 10% neutral buffered formalin. Postfix including 30% sucrose for cryoprotection; cut and mount frozen sections.

with or without an earlier 10 min ICV infusion of 8 µl/min artificial CSF. Such infusions raise CSFp during the infusion itself and, between hours 2 and 4, CSFp rises to a peak about 3× baseline (Morrow et al. 1992; Morrow et al. 1990). The microspheres, similar in size to viral particles, have been successfully employed to monitor phagocytotic processes (Hook and Odeyale 1989). The protocol is summarized in Table 1. Briefly, three groups were used: (1) uninfused controls, or infused and used (2) 15 min or (3) 4 h after the ICV infusion. After microsphere administration, all rats were treated identically (Table 1).

Figure 4 shows that the physical effects of the infusion per se were similar in groups 2 and 3, which are, in turn, similar to values reported

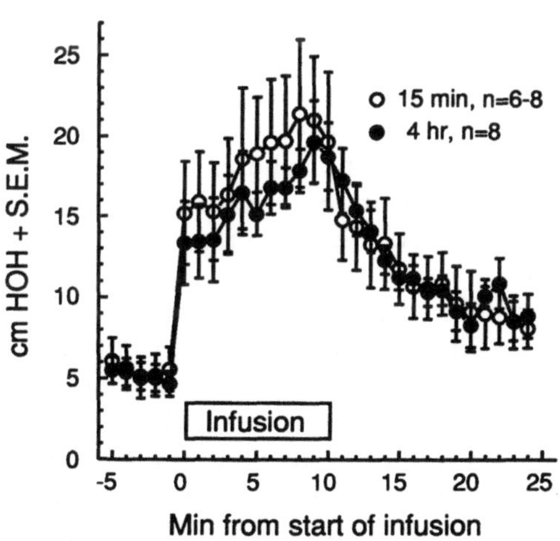

Fig. 4. Cerebrospinal fluid pressure (CSFp, Y-axis) is shown before, during, and after the intracerebroventricular infusion of 8 µl/min × 10 min artificial CSF to groups 2 (*open circles*) and 3 (*closed circles*). This pattern is common in rats (Morrow et al. 1990). The rise in both groups was similar ($p < 0.05$) during the infusion, and the CSFp of both groups were slightly but significantly ($p < 0.05$) elevated 15 min after the infusion (ANOVA & Newman-Keul's range test)

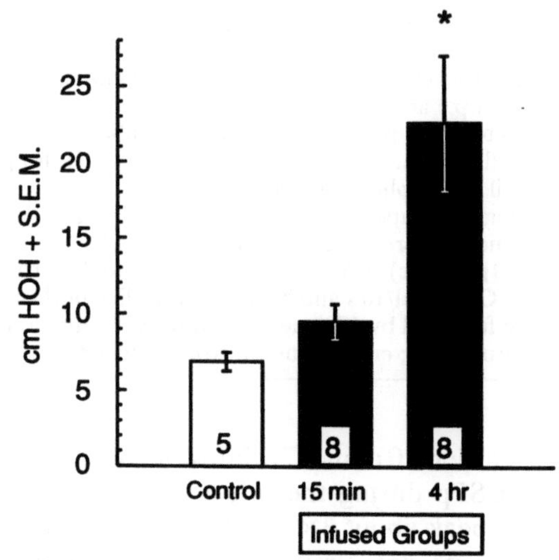

Fig. 5. Cerebrospinal fluid pressure (CSFp) of rats at the time of microsphere administration. Note that only rats given fluorescent microspheres at 4 h had elevated CSFp ($p < 0.05$, ANOVA)

under similar conditions (Morrow et al. 1992). Figure 5 shows the CSFp of the three groups at the time of microsphere administration. Note that only group 3 has elevated CSFp. Pressure does not rise until 2–4 h (Morrow et al. 1992). Also, under conditions similar to groups 1 and 2, resistance to CSF outflow and intracranial compliance would be expected to be normal and similar; however, group 3 would be expected to have elevated resistance to outflow and reduced intracranial compliance (Morrow et al. 1992). Microspheres from the brains of these rats crossed ependymal tissue poorly, with no group specificity. However, the spheres accumulated on the ependymal surface of the SCO. By contrast, cells in the choroid plexus of the fourth cerebral ventricle internalized microspheres, indicating active uptake from the CSF (see Fig. 6).

Under the conditions described, the latex spheres were not freely diffusable from the CSF to subependymal fluids although group 3 had CSF pressures about 2.5-fold higher than baseline values. Dye particles did, however, appear to move into subarachnoid CSF as they were visually observed on the surfaces of the cerebellum, pineal gland, and pituitary stalk when the brains were dissected. Thus, elevated CSFp, per se, does not evoke pressure-dependent changes in microsphere movement between CSF and subependymal interstitial fluid. Hence, no unspecific defect in transependymal fluid exchange occurs in this model of intracranial hypertension.

Future work on the functional relationship between the SCO and hydrocephalus should include measurements of both CSFp and the extracellular volume of the brain. This would clarify the role of the SCO in both high and

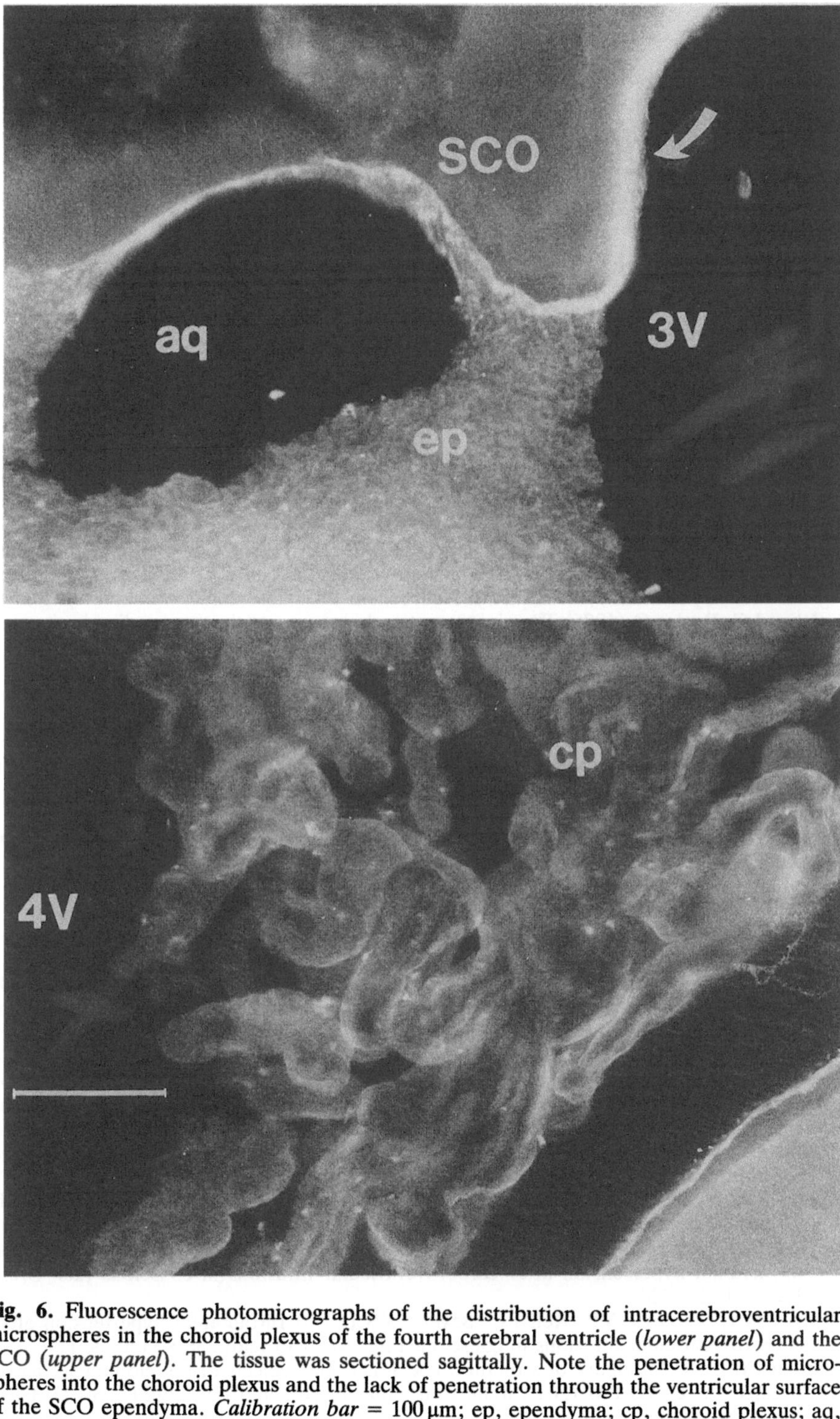

Fig. 6. Fluorescence photomicrographs of the distribution of intracerebroventricular microspheres in the choroid plexus of the fourth cerebral ventricle (*lower panel*) and the SCO (*upper panel*). The tissue was sectioned sagittally. Note the penetration of microspheres into the choroid plexus and the lack of penetration through the ventricular surface of the SCO ependyma. *Calibration bar* = 100 µm; ep, ependyma; cp, choroid plexus; aq, aqueduct; *3V*, third ventricle; *4V*, fourth ventricle; *SCO*, subcommissural organ

normal pressure hydrocephalus. Even though CSFp was not measured, the kaolin studies with anti-RF antibody (Irigoin et al. 1990) revealed no morphological differences based on occlusiveness. This again suggests that elevated pressure is not necessary for the linkage of the SCO to hydrocephalus. Perhaps, because SCO morphology is affected by low sodium diets (Palkovits 1987), a specific ion(s) is regulated by the SCO that controls *intracranial* fluid – electrolyte distribution or pressure mechanisms.

References

Balaban CD, Billingsley ML, Kincaid RL (1989) Evidence for trans-synaptic regulation of calmodulin-dependent cyclic nucleotide phosphodiesterase in cerebellar Purkinje cells. J Neurosci 9:2374–2381

Billingsley ML, Balaban CD, Berresheim U, Kuhn DM (1986) Comparative studies on the distribution of protein O-carboxymethyltransferase and tyrosine hydroxylase in rat brain by immunocytochemistry. Neurochem Int 8:255–265

Bouchaud C, Bosler O (1986) The circumventricular organs of the mammalian brain with special reference to monoaminergic innervation. Int Rev Cytol 105:283–327

Dundore RL (1985) Studies on the subcommissural organ area in the rat: the effects of aldosterone infused into the central nervous system, Ph D Thesis, Pennsylvania State University, Hershey

Dundore RL, Wurpel JND, Balaban CD, Keil LC, Severs WB (1984) Central effects of aldosterone infused into the rat subcommissural organ region. Neurosci Res 1:341–351

Dundore RL, Wurpel JND, Balaban CD, Harrison TS, Keil LC, Seaton JF, Severs WB (1987) Site-dependent central effects of aldosterone in rats. Brain Res 401:122–131

Gilbert GJ, Armstrong EP (1966) The effect of epinephrine upon subcommissural organ secretion in the rat. Neurology 16:236–241

Green JP (1989) Histamine and serotonin. In: Siegel G, Agranoff B, Albers RW, Molinoff P (eds) Basic neurochemistry, 4th edn. Raven, New York, p 253

Hook GR, Odeyale CO (1989) Confocal scanning fluorescence microscopy: a new method for phagocytosis research. J Leukocyte Biol 45:277–282

Irigoin C, Rodríguez EM, Heinrichs M, Frese K, Herzog S, Oksche A, Rott R (1990) Immunocytochemical study of the subcommissural organ of rats with induced postnatal hydrocephalus. Exp Brain Res 82:384–392

Limonta P, Maggi I, Martini L, Piva F (1982) Role of the subcommissural organ in the control of gonadotrophin secretion in the female rat. J Endocrinol 95:207–214

Lind RW (1987) The subfornical organ: neural connections. In: Gross PM (ed) Circumventricular organs and body fluids, vol I. CRC Press, New York, p 27

McLean IW, Nakane PK (1974) Periodate-lysine paraformaldehyde fixative: a new fixative for immunoelectron microscopy. J Histochem Cytochem 22:1077–1083

Miselis RR, Shapiro RE, Hyde TM (1987) The area postrema. In: Gross PM (ed) Circumventricular organs and body fluids, vol II. CRC Press, New York, pp 185–207

Morrow BA, Holt MR, Starcevic VP, Keil LC, Severs WB (1992) Mechanism of delayed intracranial hypertension after cerebroventricular infusions in conscious rats. Brain Res 570:218–224

Morrow BA, Starcevic VP, Keil LC, Severs WB (1990) Intracranial hypertension after cerebroventricular infusions in conscious rats. Am J Physiol 258:R1170–R-1176

Palkovits M (1987 Summary of structural and functional aspects of the circumventricular organs. In: Gross PM (ed) Circumventricular organs and body fluids, vol II. CRC Press, New York, pp 209–218

Rodríguez S, Rodríguez EM, Jara P, Peruzzo B, Oksche A (1990) Single injection into the cerebrospinal fluid of antibodies against the secretory material of the subcommissural organ reversibly blocks formation of Reissner's fiber: immunocytochemical investigations in the rat. Exp Brain Res 81:113–124

Sallamon M, Buda C, Janin M, Jouvet M (1984) Effects of lesion of subcommissural organ on sleep in the cat. Neurosci Lett 49:12–126

Severs WB, Dundore RL, Balaban CD (1987) The subcommissural organ and Reissner's fiber: Physiological regulation. In: Gross PM (ed) Circumventricular organs and body fluids, vol II. CRC Press, New York, pp 43–58

Sorensen PS (1986) Studies of vasopressin in the human cerebrospinal fluid. Acta Neurol Scand 74:81–102

Stepien H, Stawawy A, Cieslak D, Brzezinski J, Pawlikowski M (1986) Increased CSF levels of somatostatin in patients with brain tumors and intracranial hypertension. Horm Metab Res 18:555–557

Takeuchi YK, Takeuchi IK (1990) Critical period for induction of congenital hydrocephalus and dysplasia of subcommissural organ by prenatal X-irradiation in rats. Experientia 46:446–449

Takeuchi IK, Kimura R, Matsuda M, Shoji R (1987) Absence of subcommissural organ in the cerebral aqueduct of congenital hydrocephalus spontaneously occurring in MT/HokIdr mice. Acta Neuropathol (Berl) 73:320–322

Takeuchi IK, Kimura R, Shoji R (1988) Dysplasia of subcommissural organ in congenital hydrocephalus spontaneously occurring in CWS/Idr rats. Experientia 44:338–340

Immunological Blockade of the Subcommissural Organ–Reissner's Fiber Complex

S. Rodríguez, L. Quiñones, and E.M. Rodríguez

Introduction

After the first publications describing the existence of the subcommissural organ (SCO), numerous hypotheses with respect to the function of this circumventricular organ have been proposed. Most of these hypotheses were based on indirect evidence mainly obtained by use of morphological procedures. This resulted in a number of contradictory and unrelated statements (see Leonhardt 1980; Severs et al. 1987).

Since the SCO is a structure formed by a single or a few layers of secretory cells covering an uneven region of the ventricular wall and these secretory cells are located at the borderline between the CSF and neural tissue and intercalated with the latter in a very complex manner, investigators face serious difficulties when attempting to manipulate the SCO experimentally. Ablation of the SCO by means of electrolytic lesions necessarily implies damage to the adjacent structures, especially the posterior commissure and the pineal organ (Brown and Afifi 1965; Bougnon et al. 1965; Palkovits 1965; Sallanon et al. 1984). Also, an opening of the ventricular cavity, followed by escape of CSF may lead to uncontrolled changes in CSF pressure and flow. This led us to use a completely different approach to interfere with the function of the SCO, i.e., the application of antibodies to immunologically block its secretion into the cerebrospinal fluid (CSF).

The material secreted by the SCO into the CSF first condenses on the SCO surface as a distinct layer (pre-RF material; Rodríguez et al. 1986) and then becomes assembled to form Reissner's fiber (RF). This extracellular location of the SCO secretion allows the anti-RF antibodies to interact with the discharged secretion. Since pre-RF and RF can be distinguished in tissue sections, the antigen–antibody complexes formed "in vivo" should also be visible in tissue sections. Based on these two facts, an antiserum against the glycoproteins of the bovine RF was injected into the lateral ventricle of rats and rabbits with the aim to interfere with RF formation. The sites of binding were detected in tissue sections by the use of the immunoperoxidase method employing anti-rabbit immunoglobulin G (IgG) as primary antibody.

Materials and Methods

One hundred and twenty adult male rats (Sprague Dawley, Holtzman strain) and 20 rabbits (2–2.5 kg body weight) were used. In ether-anesthetized rats, a cannula was stereotaxically positioned in the lateral ventricle and, by use of a perfusion pump, 20 µl of an antibody-containing solution infused at a rate of 1 µl/min. In cases with a postinjection interval longer than 1 h, the animals were put back into their cages and kept under constant temperature and photoregime (L:D = 12:12). These animals had free access to food and water.

The following antibody solutions were used: (1) Antiserum against bovine RF developed in rabbits (AFRU: A = antiserum; FR = fiber of Reissner; U = urea; see Rodríguez et al. 1984). This antiserum was partially purified by precipitation with 2.0 *M* ammonium sulphate (pH 6.8, 22°C, 16 h) and centrifugation at 10000 *g* (Clausen 1981). The precipitate was dissolved in phosphate-buffered saline (PBS), pH 7.4. This resulted in a protein concentration of 10 µg/µl, as determined by the Lowry method. (2) Antiserum against bovine RF developed in rats (RAFRU; R = rat). (3) Antiserum against a partially purified extract of bovine SCO developed in rabbits. For this antiserum the acronym ASO is used (see Rodríguez et al. 1988). ASO was partially purified in the same way as indicated above for AFRU. (4) Antiserum against arginine vasopressin (anti-AVP) developed in rabbits. (5) Rabbit IgG, whole molecule (Sigma St. Louis, MO, USA), dissolved in PBS at a protein concentration of approximately 8 µg/µl. Antisera (1)–(4) were raised in our laboratory.

The 120 rats were divided into the following experimental groups:

Group I: Rats injected with AFRU and killed shortly after the injection: 0 min, 20 min, 1, 4, 8, 12 and 24 h.

Group II: Rats injected with AFRU and killed at long postinjection intervals: 2, 3, 5, 8, 15, 21 and 30 days.

Group III: Rats perfused with AFRU and killed at very long postinjection intervals: 2, 3, 4, 5, 6, 7 and 8 months.

Group IV: Rats injected with ASO and killed at 0 min, 20 min, 2 h and 4 h after the injection.

Group V: Rats injected with anti-rabbit IgG and killed 20 min, 2 h, 8 h, and 1 week after the injection.

Group VI: Rats injected with anti-AVP and killed 20 min and 2 h after the perfusion.

The 20 rabbits were divided into two experimental groups:

Group I: Rabbits perfused via the lateral ventricle with 100 µl of AFRU or RAFRU at a rate of 2 µl/min. These animals were killed at 0 min, 1, 12, and

24 h and 1 week after the injection. Rabbits not subjected to any treatment were used as controls.

Group II: Rabbits used to raise antibodies against the bovine RF. These animals were immunized with an extract of RF dissolved in a medium containing ethylene diamine tetraacetic acid, DL-dithiothreitol and urea (FRU). For details concerning isolation and extraction of RF, and production and characterization of the antibodies, see Rodríguez et al. 1984.

Under anesthesia, the rats and rabbits were intravascularly perfused with a washing solution followed by Bouin's fixative. The brain and the spinal cord were dissected out and immersed in Bouin's fluid for 2 days. After dehydration in increasing concentrations of ethanol, the material was embedded in Paraplast. The area of the brain containing the SCO, sylvian aqueduct and the fourth ventricle was serially sectioned.

The cervical, thoracic and lumbar segments of the spinal cord were embedded separately. Adjacent, 8-μm-thick sections were processed for the following immunostaining procedures: (1) Complete immunostaining procedure: The immunoperoxidase method was applied according to Sternberger et al. (1970). Anti-RF sera developed in rabbits (AFRU, dilution 1:1000) and rats (RAFRU, dilution 1:1000) were used as primary antisera. When AFRU was used, the secondary antibody was an anti-rabbit IgG raised in

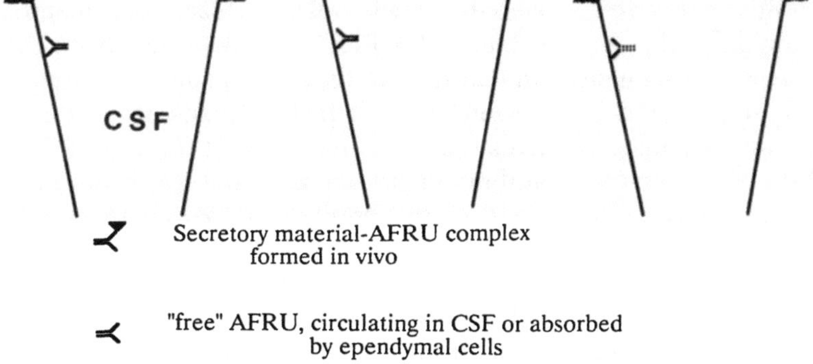

Fig. 1A–C. Schematic representation of the immunostaining (*IMS*) procedures used. *Solid triangles* and *solid IgG molecules* represent SCO secretory material and IgG molecules, which are immunoreactive. *Triangles* and *IgG molecules* indicated by *broken lines* represent secretory material and IgG molecules, which are not immunoreactive. *Triangles pointing upwards* represent intracellular secretory material. *SO*, subcommissural organ; *CSF*, cerebrospinal fluid sheep (dilution 1:30), and the PAP complex (dilution 1:75) was prepared with an antiperoxidase raised in rabbit. When RAFRU was employed as primary antiserum, an anti-rat IgG developed in rabbits (dilution 1:50) and a rat PAP (dilution 1:50) were used. (2) Incomplete immunostaining procedure: The sections were immunostained as indicated above but the incubation in AFRU or RAFRU was omitted (Fig. 1).

The complete immunostaining technique using AFRU and rabbit-PAP was applied for the demonstration of the secretory material of the SCO + the injected antibody. The procedure using RAFRU and rat-PAP allowed to visualize the secretory material, but it did not reveal the injected antibodies. The incomplete immunostaining procedure demonstrated exclusively the injected antibody (Fig. 1).

By applying the complete immunostaining with AFRU and RAFRU or the incomplete method to adjacent serial sections it was possible to visualize the distribution of the injected antibody and its relationship to the secretory material released by the SCO.

Results

After a single injection into the rat CSF of antibodies against the secretory material of the subcommissural organ the following events were observed: (1) The injected antibody bound selectively to the pre-RF and RF (Fig. 2a,b). (2) The antibody was present in the brain cavities for at least 8 h. (3) Pre-RF showed antibody binding during 24 h following the injection of AFRU. During the second and the third postinjection day, pre-RF became progressively free of antibody, thus indicating that during this period the pre-RF material was renewed by newly released secretory material. (4) Approximately 4 h after injection of AFRU, RF detached from the SCO and underwent fragmentation. Clusters of these fragments were found in the sylvian aqueduct and fourth ventricle. (5) In the fragments of the original RF the antibody injected against RF remained bound for a week; after this period the RF fragments could no longer be visualized. (6) From the second postperfusion day until the end of the observation period (8 months after perfusion), RF was missing in the central canal of the spinal cord. (7) One day after the injection, a new RF began to grow from the rostral end of the SCO; this new RF displayed a normal appearance and its lack of bound AFRU made it distinguishable from the "old" RF. Eight days after the antibody injection the new RF reached the fourth ventricle (see Rodríguez et al. 1990).

One month after the antibody injection RF had reached the entrance of the central canal of the cervical spinal cord. A few μm beyond the entrance to the central canal the RF material formed a structure resembling

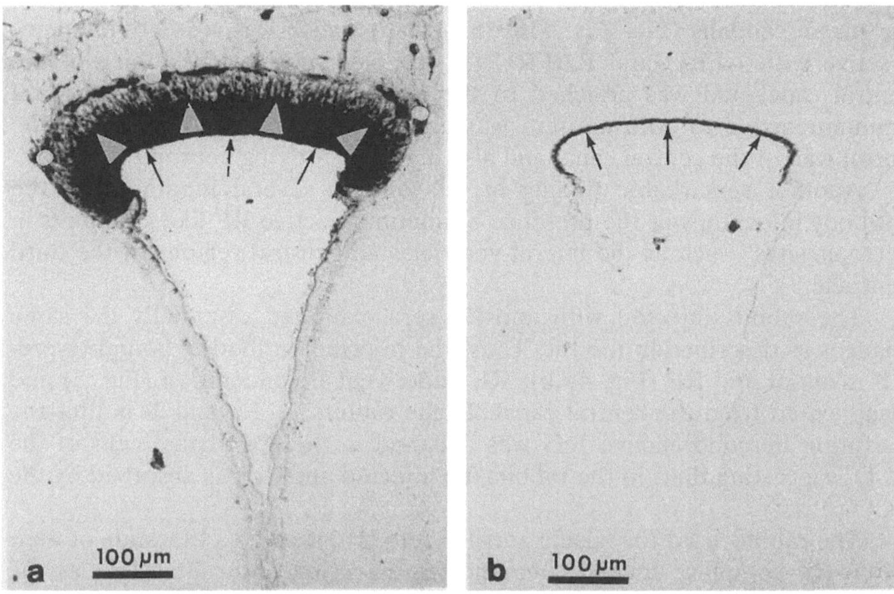

Fig. 2a,b. Rat SCO 1 h after injection of AFRU into the CSF. Complete immunostaining using AFRU (**a**) shows immunoreactive pre-RF material (*arrows*), and heavily stained ependymal (*white triangles*) and hypendymal cells (*white dots*). In the adjacent section, immunostained with the incomplete procedure (**b**), the injected antibody is bound exclusively to the pre-RF material (*arrows*)

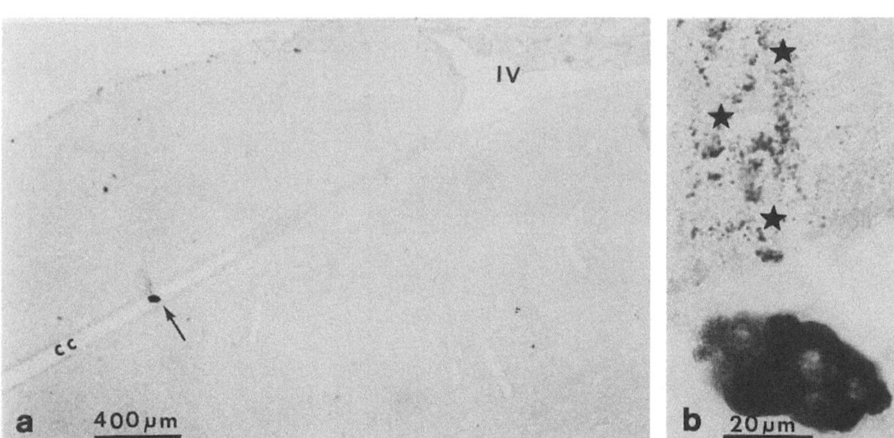

Fig. 3a,b. Complete immunostaining using AFRU of a sagittal section through the fourth ventricle (*IV*) and the central canal (*cc*) of a rat sacrificed 1 month after injection of AFRU. The RF terminates as a strongly immunoreactive mass at the entrance to the central canal (**a**, *arrow*). AFRU-immunoreactive material is also seen in the ependyma lining the dorsal wall of the central canal and in the underlying neuropil (**b**, *stars*)

the massa caudalis (Fig. 3a). This "massa rostralis" was strongly immunoreactive with AFRU and RAFRU. It occupied most of the lumen of the central canal and was attached to the dorsal wall of the canal. Material immunoreactive to both antisera was observed in the ependyma lining the dorsal wall of the central canal and also in the underlying neuropil (Fig. 3b).

Another remarkable finding in rats one or several months after the antibody injection was the presence of immunoreactive RF-like structures in "ectopic sites" such as the lateral ventricles and rostral regions of the third ventricle.

The rabbits injected with anti-RF sera exhibited essentially the same changes as described in the rat. Thus, the injected antibodies bound to pre-RF material and RF (Fig. 4a,b); RF underwent fragmentation (Fig. 5) and disappeared from the central canal. In the rabbits, 1, 12, and 24 h after the perfusion immunoreactive IgG was detected in the ependymal cells of the SCO, suggesting that, in the rabbit, the injected antibody is absorbed by the SCO cells.

The rabbits used for raising anti-RF sera also showed a blockade of their SCO–RF complex. Indeed, immunostaining using anti-rabbit IgG as the

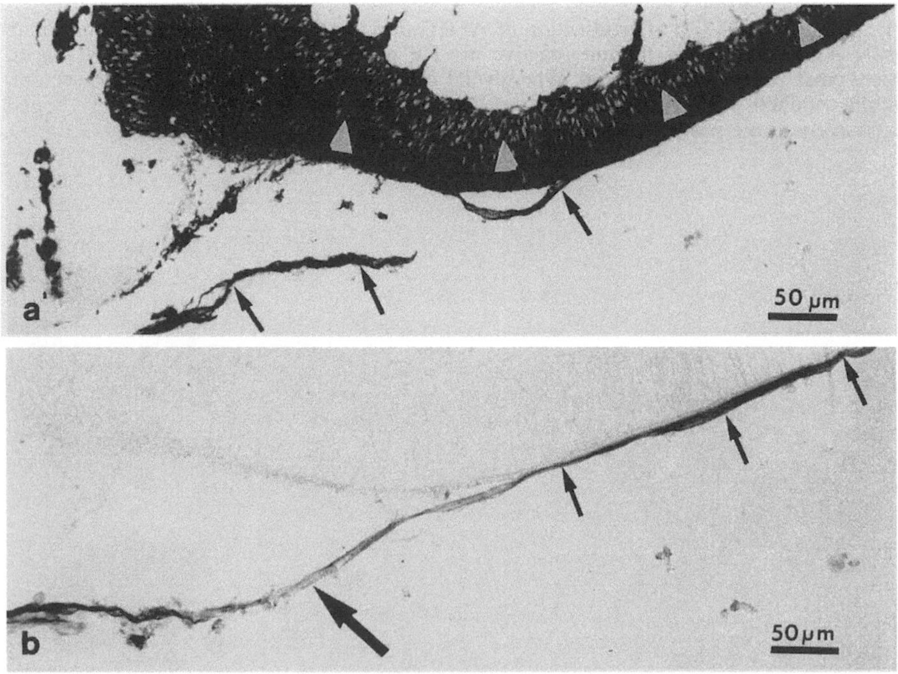

Fig. 4a,b. Rabbit SCO 12 h after injection of RAFRU. Complete immunostaining using RAFRU (**a**) shows immunoreactive material in intracellular (*white triangles*) and extracellular (*arrows*) locations. Immunostaining with the incomplete procedure (**b**) reveals the sites of binding of the injected antibody in the pre-RF material (*small arrows*) and RF (*large arrow*)

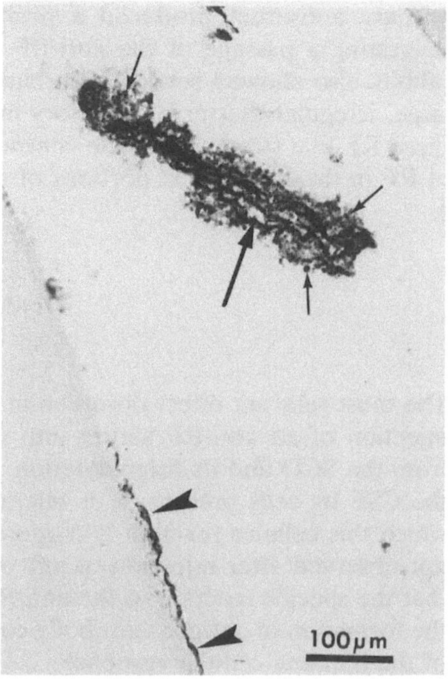

Fig. 5. Section through the sylvian aqueduct of a rabbit sacrificed 24 h after injection of AFRU. Complete immunostaining using AFRU shows a strongly immunoreactive fragment of RF (*arrow*), surrounded by infiltrating cells (*small arrows*). AFRU-immunoreactive material is also found on the surface of the ependyma (*arrowheads*)

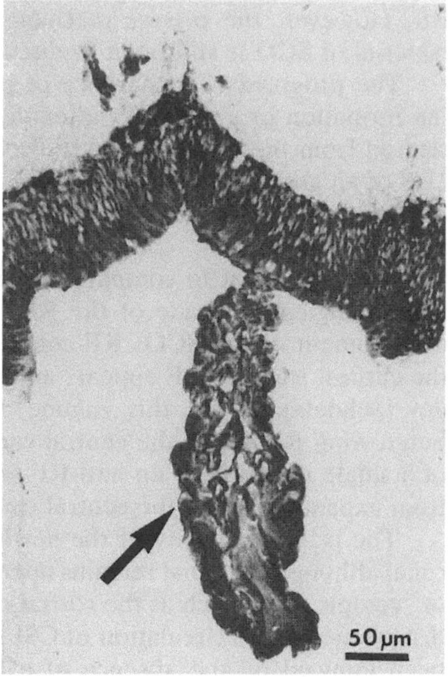

Fig. 6. Section of the SCO of a rabbit used to raise antibodies against the bovine RF, sacrificed 8 months after the beginning of immunization. Complete immunostaining shows an atypical mass of RF-material (*arrow*)

primary antiserum produced a positive reaction in the pre-RF and RF, suggesting a passage of the anti-RF antibodies from blood to CSF. These rabbits also showed severe disturbance in the SCO–RF complex (Fig. 6): large, irregular masses of secretory material in the sylvian aqueduct; two or three RF-like structures in the central canal of the spinal cord; and absence of RF in the caudalmost portions of the spinal cord.

Discussion

The most relevant effect observed in this experimental study is that a single injection of an anti-RF serum into the CSF leads to a detachment of RF from the SCO and its fragmentation. The fragments of RF are cleared from the CSF by cells present in or migrating into the CSF. The mechanism by which this cellular reaction is triggered is unknown. Since this reaction was not observed after injection of any of the other IgGs, it may be concluded that the specific reaction of the anti-RF antibodies with pre-RF and RF, and the formation of antigen–antibody complexes are characteristic prerequisites of the immune cellular response.

A likely explanation for the detachment of RF from the SCO is that the antibodies bound to the pre-RF material may interfere with the aggregation or packaging of the secreted glycoproteins, thus preventing the formation of RF. However, the precise mechanism of polymerization of the secretory material of SCO is still open to discussion.

The progressive appearance of pre-RF material free of antibodies and the formation of a new RF, after the IgG molecules have been completely cleared from the CSF, lead to the conclusion that a single injection into the CSF of an anti-RF antibody produces a reversible effect on the RF assembly but does not interfere with the capacity of the SCO to synthetize and release RF glycoproteins.

It is of interest to compare the situation in adult rats suffering from an immunological blockade of the RF formation with the regular ontogenetic development of the SCO–RF complex in newborn rats. In the latter case, the earliest stage of RF appears and starts to grow during the first postnatal day (Schoebitz et al., this volume). During the second postnatal day, the outgrowing RF enters the central canal. One may wonder about the effects of a single injection of an anti-RF serum, preventing the newly formed RF from expanding along the central canal of the spinal cord.

The lack of capacity of the newly formed RF to grow along the central canal although the canal remains open, and the presence of RF-like structures in "ectopic sites" such as the rostral portion of the third ventricle may reflect disturbance in the circulation of CSF. In turn, such a disturbance might have been induced by the absence of a RF during the first days following the injection of the anti-RF antibodies. The formation of a "rostral mass" of secretory material at the entrance to the central canal indicates: (1) that RF material continues to arrive at the level of the central canal, and (2) that this material, instead of proceeding along the canal in the shape of a regular RF, forms a local mass of twisted fibrous structures, probably because of altered hydrodynamic properties of the CSF at the entrance to the central canal. Evidence in favor of this possibility is presented by Fernández-Llebrez et al. (this volume). The lack of formation of RF in hydrocephalic rats also supports the view that the CSF may play an important role in RF formation (Irigoin et al. 1990).

The present investigation has shown that a single injection of an antiserum against RF into the ventricular CSF provides a useful experimental model displaying the following consecutive phases: (1) complete lack of a regular RF (first 24 h); (2) a short initial segment of the RF restricted to the aqueduct (first week); (3) a longer RF extending from the SCO to the entrance of the central canal of the spinal cord (first month on); (4) permanent lack of RF in the central canal in spite of the features listed under (3).

The present experimental model may provide a useful tool for investigating functional aspects of the SCO–RF complex. Thus, in rats, over 5 days following a single injection of an anti-RF serum into the CSF, a decrease in urine volume and water intake was observed in comparison to control animals injected with IgG (Rodríguez 1991). Furthermore, these rats showed a slowing down of the CSF circulation along the central canal (Cifuentes 1991; Fernández-Llebrez et al., this volume).

The fact that rabbits used for raising antibodies against RF glycoproteins exhibit distinct alterations of their SCO–RF complex indicates that the circulating antibodies have gained access to the ventricular CSF. This finding has to be taken into account when designing future experiments. A crucial experiment would be to block RF formation during the ontogenetic development; this group of animals would then never have a chance to develop a RF. Taking into account all these considerations, we proceeded to induce pregnancy in female rabbits, which had previously been immunized by application of RF glycoproteins and displayed high titers of anti-RF antibodies. Four baby rabbits were delivered by such females. These newborn rabbits showed a high titer of circulating anti-RF antibodies. Further studies of these newborn rabbits are in progress, with the particular aim of investigating the SCO under such experimental conditions.

Acknowledgments. This work was supported by grant I/63 476 from the Volkswagen-Stiftung, Federal Republic of Germany; grant 91/0956 from FONDECYT, Chile, and grant S-89-01 from the Dirección de Investigaciones, Universidad Austral de Chile.

References

Bougnon C, Lenys R, Lenys D (1965) Recherches sur d'éventuelle corréllations entrel'organe sous-commissural et la zone glomérulaire surrénalienne, productice d'aldosterone. Ann Sci Univ Besançon 2, Série Med 1:43–59
Brown DD, Afifi AK (1965) Histological and ablation studies on the relation of the SCO and rostral midbrain to sodium and water metabolism. Anat Rec 153:255
Cifuentes M (1991) Circulación del LCR a través del canal central medular de la rata. Influencia del sistema OSC-RF. Ph.D. Thesis, University of Malaga
Clausen J (1981) Immunochemical techniques for the identification and estimation of macromolecules. In: Work TS, Work E (eds) Laboratory techniques in biochemistry and molecular biology. Elsevier/North-Holland Biomedical, New York
Irigoin C, Rodríguez EM, Heinrichs M, Frese K, Herzog S, Oksche A, Rott R (1990) Immunocytochemical study of the subcommissural organ of rats with induced postnatal hydrocephalus. Exp Brain Res 82:384–392
Leonhardt H (1980) Ependym und circumventriculäre Organe: In: Oksche A, Vollrath L (eds) Neuroglia I. Handbuch der mikroskopischen Anatomie des Menschen, vol 4, part 10. Springer, Berlin Heidelberg New York, pp 117–665
Palkovits M (1965) Morphology and function of the subcommissural organ. Stud Biol Hung 4:1–105
Rodríguez EM, Oksche A, Hein S, Rodríguez S, Yulis R (1984) Comparative immunocytochemical study of the subcommissural organ. Cell Tissue Res 237:427–441
Rodríguez EM, Herrera H, Peruzzo B, Rodríguez S, Hein S, Oksche A (1986) Light-and electron-microscopic lectin histochemistry of the SCO: evidence for processing of the secretory material. Cell Tissue Res 243:545–559
Rodríguez EM, Korf HW, Oksche A, Yulis CK, Hein S (1988) Pinealocytes immunoreactive with antisera against secretory glycoproteins of the subcommissural organ: a comparative study. Cell Tissue Res 254:469–480
Rodríguez S (1991) Immunological blockage of the SCO-RF complex. Ph.D. Thesis, University of Málaga
Rodríguez S, Rodríguez EM, Jara P, Peruzzo B, Oksche A (1990) Single injection into the cerebrospinal fluid of antibodies against the secretory material of the subcommissural organ reversibly blocks formation of Reissner's fiber: immunocytochemical investigations in the rat. Exp Brain Res 81:113–124
Sallanon M, Buda C, Janin M, Jouvet M (1984) Effect of lesion of SCO on sleep in cat. Neurosci Lett 49:123–126
Schoebitz K, Garrido O, Heinrichs M, Speer L, Rodríguez EM (1986) Ontogenetic development of the chick and duck subcommissural organ. An immunocytochemical study. Histochemistry 84:31–41
Severs WB, Dundore RL, Baadan CD (1987) The SCO and Reissner's fiber: Physiological regulation. In: Gross PM (ed) Circumventricular organs and body fluids, vol II. CRC Press, Boca Raton, pp 43–58
Sternberger LA, Hardy PH, Cuculis Jr, Meyer HG (1970) The unlabeled antibody enzyme method of immunohistochemistry. Preparation and properties of soluble antigen-antibody complex (horseradish-peroxidase-antiperoxidase) and its use in identification of spirochetes. J Histochem Cytochem 18:315–333

The Effect of Immunological Blockade of Reissner's Fiber Formation on the Circulation of Cerebrospinal Fluid Along the Central Canal of the Rat Spinal Cord

P. Fernández-Llebrez, M. Cifuentes, J.M. Grondona, J. Pérez, and E.M. Rodríguez

Introduction

Although the subcommissural organ–Reissner's fiber (SCO–RF) complex was first described almost a century ago, the function of this conspicuous glandular structure of the brain remains enigmatic. Several functional hypotheses have been proposed based on indirect evidence. Briefly, the SCO–RF has been suspected to be involved in: (1) osmoregulation (Palkovits et al. 1965; Leatherland and Dodd 1968), (2) detoxification of the cerebrospinal fluid (CSF) (Olsson 1958; Hess and Sterba 1973; Diederen et al. 1983), (3) mechanoreception (Kolmer 1921), and (4) morphogenesis of the vertebral column and the spinal cord (Hauser 1969; Rühle 1971). However, none of these hypotheses has been properly substantiated.

It has been suggested that polymerization of the SCO glycoproteins into a RF may depend on an adequate flow of CSF through the ventricles (Oksche 1961). The possible influence of CSF on the formation of a RF has been reinforced by two observations: (1) the SCO grafted under the kidney capsule secretes into interstitial cavities glycoproteins that do not polymerize into a fiber (Rodríguez et al. 1989), and (2) animals suffering from congenital or experimental hydrocephalus do not possess a RF (Irigoin et al. 1990; Takeuchi et al. 1987). Recently, Rodríguez et al. (1990) developed an experimental model in which rats are permanently deprived of RF in the central canal of the spinal cord after a single injection of an anti-RF serum into the ventricular CSF. Using this model, the hypothesis that RF may influence the circulation of CSF in the central canal of the spinal cord was tested in our laboratory.

Materials and Methods

The brain of male rats ($n = 47$) (Sprague Dawley, Holtzman strain) was perfused via the right lateral ventricle with 30 µl of 3% horseradish peroxidase (HRP) solution. The perfusion lasted for 10 min (3 µl/min). Twenty of these rats had been previously (11 days, $n = 17$; or 9 months, $n = 3$) intracerebroventricularly perfused with 20 µl of the immunoglobulin G (IgG) fraction of an antiserum against bovine RF (AFRU, Rodríguez et al. 1984) for 10 min. As described by Rodríguez et al. (1990), this single injection leads to fragmentation and disappearance of RF, and to the outgrowth of a new RF that never enters the central canal (S. Rodríguez et al., this volume). The central nervous system of the normal and RF-deprived rats was fixed 13 min, 20 min, 1 h, 2 h, and 4 h after the HRP injection by vascular perfusion with Karnovsky's fixative. After perfusion, the brain and the entire spinal cord were dissected out. The following segments of the spinal cord were processed separately: (1) C = cervical (C_1 to C_6); (2) AT = anterior thoracic ($T_1 - T_4$); (3) MT = middle thoracic (T_5–T_8); (4) PT = posterior thoracic (T_9–T_{12}); and (5) LS = lumbo-sacral region (L_1 to filum terminale).

The brain and spinal cord material was sectioned (30–50 µm thick) with a Vibratome. The brain was cut in frontal, the spinal cord in transverse and horizontal planes. Peroxidase was reacted according to Graham and Karnovsky (1966), and the sections were mounted in glycerol or in Aquatex (Merck) for light-microscopic examination. Some sections were further

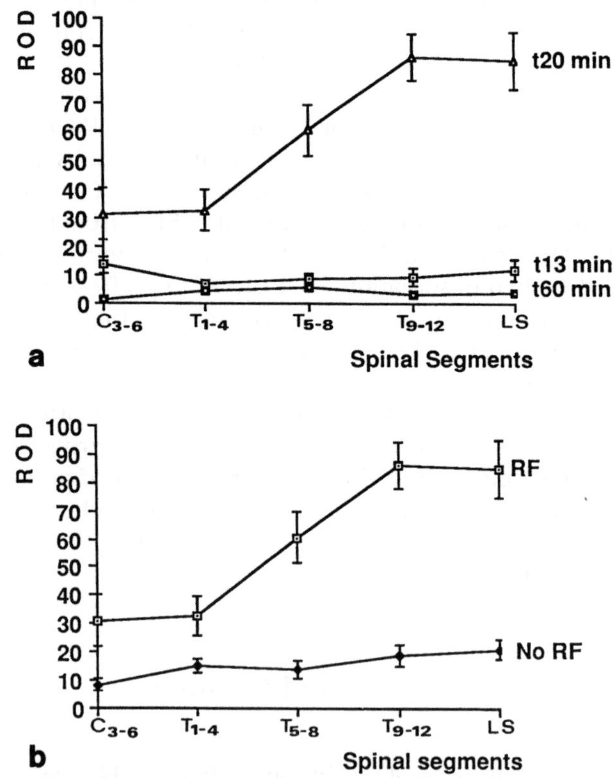

Fig. 1. a Relative optic density (ROD) (*ordinate*) of the central canal region in different spinal segments (*abscissa*) of control rats killed 13, 20 and 60 min after peroxidase perfusion. **b** Relative optic density (ROD) (*ordinate*) of the central canal region in different spinal segments (*abscissa*) of control (*RF*) and Reissner's fiber-deprived (*No RF*) rats killed 20 min after peroxidase perfusion

processed for electron microscopy. For this purpose, the Vibratome sections were fixed in 1% O_sO_4, dehydrated and embedded in Araddite.

Semiquantitative Studies

The amount of HRP present in the region of the central canal was measured by densitometry using an image-analyzing system IBAS-2000 (Kontron). This system digitalizes the images, and a value of gray, from 0 = black to 255 = white, is assigned to each point. The images were obtained by means of a video camera connected to a light microscope. The analysis was performed in an area of 100 × 100 pixels that included the central canal. In each section, a region of gray matter lacking any HRP staining was measured as background (Xb); this value was subtracted from the values obtained at the different points of the studied region (Xp) that comprise the central canal. Thus, the relative optic density (ROD) was measured as Xp-Xb.

From all specimens in all experimental groups three transverse sections, randomly chosen of each of the spinal segments, were analyzed. The total number of analyzed regions (N) in each spinal segment was N = 3n, "n" indicating the number of animals in each group. For each group and spinal segment the mean and standard deviation were calculated.

The values of ROD obtained were submitted to the test of normality according to Kolmogorov-Smirnov ($p < 0.01$) (Sokal and Rohlf 1981), and then an analysis of variance was applied. For comparison of the values in the different regions a Student *t*-test ($p < 0.01$) was employed.

Results and Discussion

Flow of CSF Along the Central Canal

A principal observation of the present study is that, in the rat, the CSF flows along the central canal of the spinal cord. Both in normal and RF-deprived rats the tracer was found in all portions of the spinal cord (Fig. 1). Since in some of the specimens sacrificed 13 min after the HRP injection the tracer was present in the filum terminale and the spinal cord had a length of ~12 cm, the CSF appears to circulate at an approximate velocity of 1 cm/min. There are only few studies on the circulation of CSF along the central canal of the spinal cord. Bradbury and Lathem (1965) investigated this problem in monkeys, rabbits, guinea pigs, rats and lampreys. Since only in the rabbit a flow of CSF could be demonstrated, these authors considered this "exceptional fact as a relic of vertebrate evolutionary history". Bradbury and Lathem used various tracers and sacrificed their rabbits at different time intervals after the injection of the tracer. However, in the other species studied, including the rat, only Evans blue at long survival times (2 h) was used. According to our results in the rat, 2 h after the injection of the tracer almost all HRP has been cleared from the central canal; the amount of HRP was highest at the 20-min postinjection interval. Bradbury and Lathem (1965) found a weak mark of Evans blue in the central canal of rats 2 h after the injection of the tracer.

Distribution of the HRP in the Central Canal Region

The injected HRP extending via the central canal diffused toward the subependymal neuropil. The tracer was detected in the following locations (Figs. 2a, 3): (1) in the lumen of the central canal where it was bound to RF and the luminal surface of the ependymal cells; (2) in the intercellular spaces of the ependymal layer and the subependymal neuropil; (3) in the perivascular basement membrane of the subependymal vessels, including their numerous labyrinthine extensions reaching the basal region of the ependymal cells; (4) in pinocytotic vesicles present in the ependymal cells and endothelial cells of subependymal capillaries. Some ependymal cells were completely loaded with HRP.

When comparing the distribution and dynamics of the HRP in the brain and the spinal cord two facts deserve particular attention: (1) Whereas HRP penetrated only a few micrometers into the subependymal neuropil of the central canal, it diffused several millimeters deep into the brain tissue. (2) The HRP was cleared from the central canal region very soon (1–2 h) (Fig. 1), while in the brain parenchyma it remained detectable over a period of 24 h (Wagner et al. 1974). According to our observations one may assume that around the central canal a very efficient clearance system is established. The tracer occurring in the CSF of the central canal apparently diffused among the ependymal cells and filled the numerous labyrinths of basement membrane; via this pathway HRP gained access to the perivascular spaces and the meninges or was transported to the bloodstream in a transcellular fashion across the endothelium (Wagner et al. 1974; Matakas et al. 1978). Similar routes have been proposed previously for the brain (Wagner et al. 1974). Moreover, the ependyma proper may contribute to the disappearance of a part of the HRP by means of pinocytosis and lysosomal degradation.

The HRP injected into the lateral ventricle also invaded the spaces of the outer CSF-compartment. As has been repeatedly shown in the brain (Rennels et al. 1985; Cserr 1988), the tracer intruded into the nervous parenchyma from the outer CSF via the perivascular spaces of the penetrating vessels. However, the feasibility of this route has been recently questioned (Krisch et al. 1984). The present study clearly shows that in the rat spinal cord two type of vessels can be distinguished. Type-A vessels penetrate the spinal cord from the dorsolateral aspect. At any of the postinjection intervals studied, HRP was visualized around these vessels only up to the level of the gray matter. It appears that an effective barrier preventing further passage

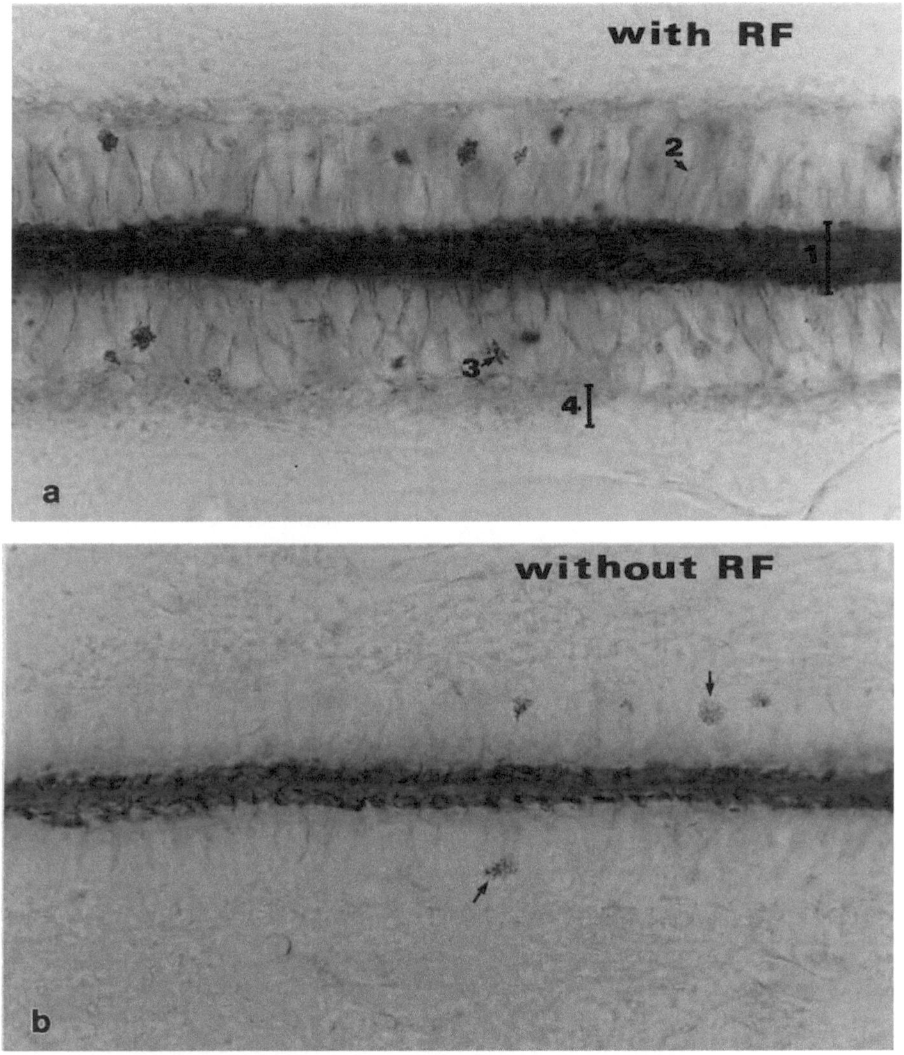

Fig. 2a,b. Horizontal Vibratome section through the thoracic spinal cord of a control rat (*with RF*, **a**) and one deprived of RF (*without RF*, **b**) killed 20 min after peroxidase perfusion. **a** The central canal is filled with the tracer (*1*) which penetrates the intercellular spaces of the ependymal cells (*2*) to reach the labyrinths of the basal lamina (*3*) and the subependymal neuropil (*4*). ×900. **b** Note tracer within the central canal. Only some labyrinths appear weakly labeled (*arrows*). ×900

of HRP along the perivascular spaces is located in the gray matter. Type-B vessels penetrate the spinal cord from the ventral medial sulcus and run toward the central canal where, in close proximity to the ependyma, they turn and extend into other regions. Along these vessels HRP expanded at least up to the level of this curvature and then diffused from the wall of the vessel into the surounding neuropil (Fig. 4). Since the central canal was close to

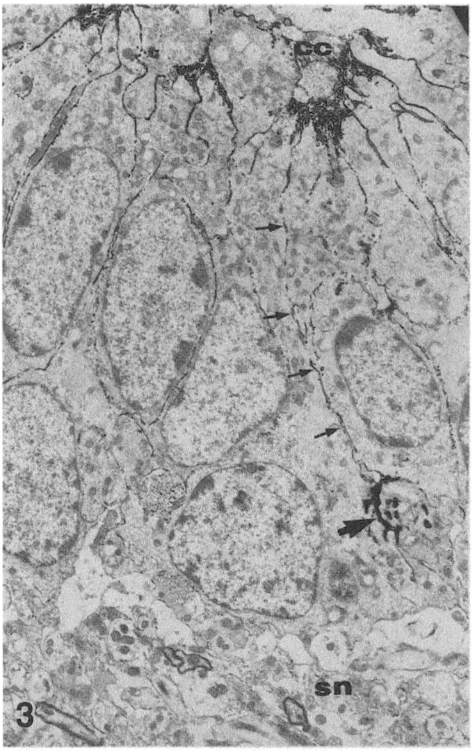

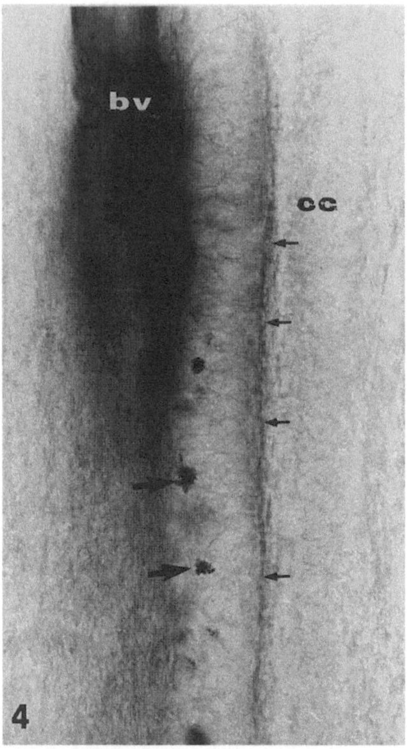

Fig. 3. Electron-microscopic survey of the ependyma of the central canal, at the thoracic level, of a control rat sacrificed 20 min after peroxidase perfusion. Note the positive reaction at the ependymal surface, in the intercellular spaces (*small arrows*) and in the labyrinth (*large arrow*); *cc*, central canal; *sn*, subependymal neuropil. ×3000

Fig. 4. Horizontal Vibratome section through the thoracic spinal cord of a control rat killed 2 h after peroxidase perfusion. The wall of a blood vessel (*bv*) lying close to the ependyma of the central canal (*cc*) appears filled with peroxidase. The tracer diffuses into the subependymal neuropil, marks the neighboring basal membrane labyrinths (*large arrows*), penetrates the intercellular spaces of the ependyma, and reaches the luminal surface of the ependyma (*small arrows*). ×600

these vessels, the tracer penetrated the labyrinths of the basement membrane and then the intercellular spaces of the ependymal layer, reaching the lumen of the central canal. This can be clearly seen in animals sacrificed after 2 h when almost all HRP has been cleared from the central canal. In conclusion, in the rat spinal cord an open communication between the inner (central canal) and the outer (meningeal) CSF spaces exists via the intercellular spaces, the basement membrane and its extensions, and along the vessels penetrating from the ventral medial sulcus. This particular property may be of importance for the fluid homeostasis in the spinal cord. In species lacking an open central canal, a system similar to that described above for the rat, established by vessels penetrating from the ventral medial sulcus, could operate and be sufficient for the fluid homeostasis in the spinal cord.

In our study, in a population of ependymal cells of the conus medullaris the entire cytoplasm was filled with HRP. At the electron-microscopic level, HRP was visualized in the cytoplasm but not found within organelles. Since, in this location, images of cell death were not observed at longer survival times, it is thought that this massive uptake of HRP was not noxious for these cells. These ependymocytes displayed long basal processes in contact with the wall of subependymal and other distant vessels. Since they were present in all animals killed 13 min, 20 min and 1 h after the HRP injection and always located in the same region, it is tempting to speculate that they constitute a specialized population of spinal tanycytes. Tanycytes have been described before in the spinal cord of the rat (Rafols and Goshgarian 1985; Bruni and Reddy 1987).

CSF Flow Alterations in RF-Deprived Rats

The main conclusion arising from our study is that the lack of RF in the central canal of the rat spinal cord results in changes in the flow of the CSF. In both control and RF-deprived animals, HRP was found at all spinal levels, even at the shortest postinjection interval of 13 min (Fig. 1b). However, at any of the experimental time periods investigated and in all segments of the spinal cord the amount of HRP was significantly lower in RF-deprived rats (compare Figs. 2a and b). Our observations suggest that: (1) the velocity of a small (laminar) fraction of the CSF along the central canal was unaffected by the absence of RF; (2) however, the bulk flow of the CSF was drastically decreased in the rats lacking a RF in the central canal.

Thus in the rat central canal more than one driving force for the circulation of the CSF may exist. One force, operating along the whole length of the central canal, may be responsible for the rapid spreading of a portion of the CSF; the other force may account for the bulk flow of the CSF along the central canal. This suggestion is in accord with the fact that in both control and RF-deprived animals HRP was present in the caudalmost portions of the central canal after 13 min and in both cases, although in RF-deprived rats in much lower amount, the bulk of the tracer took an additional 7 min to reach the caudal end of the spinal cord. As also calculated for the rabbit (Bradbury and Lathem 1965), the velocity estimated for the CSF along the rat central canal (1 cm/min) is in the range of that described for movement of particles along the ciliated surfaces of frog esophagus and rabbit trachea (Hill 1957). Therefore, it can be postulated that the rapid spreading of the CSF along the central canal depends on the beat of the ependymal cilia that may not be affected by the absence of RF. At variance, the bulk flow of the CSF may depend on other factors such as production of

CSF, arterial pulsations in the brain, reabsorption, etc. The maintenance of this flow appears to depend on the presence of an RF in the central canal, since this type of flow decreased considerably in RF-deprived animals.

It seems likely that the absence of the RF leads to a CSF turbulence at the entrance of the central canal that, in turn, results in alterations in the flow of CSF within the central canal. This agrees well with the observation that in rats chronically deprived of RF (5–9 months) the newly formed RF reached the entrance of the central canal, but at this point twisted and did not proceed further along the central canal (S. Rodríguez et al. 1990). The contortion of the RF may be a consequence of the turbulences arising at this point. In these animals (9 months after the injection of the AFRU) the flow of the CSF along the central canal remained retarded and the results resembled those in animals 11 days after the AFRU injection. An interesting finding in these animals is the increase in the extent of basement membrane labyrinths. This morphological change may somehow be related to the altered flow of CSF along the central canal.

In brief, in the rat central canal the CSF circulates at a rate of 1 cm/min. This flow rate may be affected by the absence of RF. The ependyma, the labyrinths of the basement membrane, the blood vessels penetrating from the ventral medial sulcus and, probably, also certain subependymal vessels are structures that seem to be involved in the fluid homeostasis in the rat spinal cord.

Acknowledgments. This work was supported by the following grants: DGICYT, PB90/0804 and CICYT, MAR 88-0751, Madrid, Spain to P.F.LL; Instituto de Cooperación Iberoamericana (1989–90) to J.M.P.F., Madrid; Volkswagen-Stiftung I/63476, FRG, FONDECYT 089188, Chile, and Dirección de Investigaciones de la Universidad Austral de Chile, S-89-01 to E.M.R

References

Bradbury MWB, Lathem W (1965) A flow of cerebrospinal fluid along the central canal of the spinal cord of the rabbit and communications between this canal and the sacral subarachnoid space. J Physiol 181:785–800

Bruni JE, Reddy K (1987) Ependyma of the central canal of the rat spinal cord: a light and transmission electron microscopic study. J Anat 152:55–70

Cserr HF (1988) Role of secretion and bulk flow of brain interstitial fluid in brain volume regulation. Ann N Y Acad Sci 529:9–20

Diederen JH, Vullings HGB, Rombout JHWM, de Gunst-Schoonderwoerd ATM (1983) The subcommissural organ-liquor fibre complex: the binding of catecholamines to the liquor fibre in frogs of the *Rana esculenta* complex. Acta Zool (Stockh) 64:47–53

Graham RC, Karnovsky M (1966) The early stages of absorption of injected horseradish peroxidase in the proximal tubules of mouse kidney: ultrastructural cytochemistry by a new technique. J Histochem Cytochem 14:291–302

Hauser R (1969) Abhängigkeit der normalen Schwanzregeneration bei *Xenopus*-Larven von einem diencephalen Faktor im Zentralkanal. Wilhelm Roux Arch Entwicklungsmech Org 163:221–247
Hauser R (1972) Morphogenetic action of the subcommissural organ on tail regeneration in *Xenopus* larvae. Wihelm Roux Arch Entwickl Mech Org 169:170–184
Hess J, Sterba G (1973) Studies concerning the function of the complex subcommissural organ-liquor fibre: the binding ability of the liquor fibre to pyrocatechin derivates and its functional aspects. Brain Res 58:303–312
Hill JR (1957) The influence of drugs on ciliary activity. J Physiol 139:157–166
Irigoin C, Rodríguez EM, Heinrichs M, Frese K, Herzog S, Oksche A, Rott R (1990) Immunocytochemical study of the subcommissural organ of rats with postnatal induced hydrocephalus. Exp Brain Res 82:384–392
Kolmer W (1921) Das "Sagittalorgan" der Wirbeltiere. Z Anat 60:652–717
Krisch B, Leonhardt H, Oksche A (1984) Compartments and perivascular arrangement of the meninges covering the cerebral cortex of the rat. Cell Tissue Res 238:459–474
Leatherland JF, Dodd JM (1968) Studies on the structure, ultrastructure and function of the subcommissural organ Reissner's fibre complex of the European eel *Anguilla anguilla* L.Z Zellforsch 89:533–549
Matakas F, Stechele S, Keller F (1978) Microcirculation within the cerebral extracellular space. In: Cervós-Navarro J, Betz E, Ebhardt G, Ferszt R, Wüllenweber R (eds) Advances in neurology vol 20: Pathology of cerebrospinal microcirculation. Raven, New York, pp 125–131
Newberne PM (1962) The subcommissural organ of the vitamin B_{12} deficient rat. J Nutr 76:393–414
Oksche A (1961) Vergleichende Untersuchungen über die sekretorische Aktivität des Subkommissuralorgans und den Gliacharakter seiner Zellen. Z Zellforsch 54:549–612
Olsson R (1958) Studies on the subcommissural organ. Acta Zool (Stockholm) 39:71–102
Olsson R (1961) Subcommissural ependyma and pineal organ development in human fetuses. Gen Comp Endocrinol 1:117–123
Palkovits M, Monos E, Fachet J (1965) The effect of the subcommissural organ lesions on aldosterone production in the rat. Acta Endocrinol (Copenh) 48:169–176
Rafols JA, Goshgarian G (1985) Spinal tanycytes in the adult rat: a correlative Golgi gold-toning study. Anat Rec 211:75–86
Rennels ML, Gregory TF, Blaumanis OR, Fujimoto K, Grady PA (1985) Evidence for a "paravascular" fluid circulation in the mammalian central nervous system, provided by the rapid distribution of tracer protein throughout the brain from the subarachnoid space. Brain Res 326:47–63
Rodríguez EM, Oksche A, Hein S, Rodríguez S, Yulis R (1984) Comparative immunocytochemical study of the subcommissural organ. Cell Tissue Res 237:427–441
Rodríguez EM, Oksche A, Rodríguez S, Hein S, Peruzzo B, Schoebitz K, Herrera H (1987) The subcommissural organ and Reissner's fiber: Fine structure and cytochemistry. In: Gross PM (ed) Circumventricular organs and body fluids, vol 2. CRC Press, Boca Raton, pp 3–41
Rodríguez EM, Rodríguez S, Schoebitz K, Yulis CR, Hoffmann P, Manns V, Oksche A (1989) Light- and electron-microscopic investigation of the rat subcommissural organ grafted under the kidney capsule, with particular reference to immunocytochemistry and lectin histochemistry. Cell Tissue Res 258:499–514
Rodríguez S, Rodríguez EM, Jara P, Peruzzo B, Oksche A (1990) Single injection into the cerebrospinal fluid of antibodies against the secretory material of the subcommissural organ reversibly blocks formation of the Reissner's fiber: immunocytochemical investigation in the rat. Exp Brain Res 81:113–124
Rühle HJ (1971) Anomalien im Wachstum der Achsenorgane nach experimenteller Ausschaltung des Komplexes SCO-Reissnerscher Faden. Untersuchungen am Rippenmolch (*Pleurodeles waltli* Michah 1830). Acta Zool (Stockh) 52:23–68
Sokal RR, Rohlf FJ (1981) Biometry: the principles and practice of statistics in biological research. Freeman, San Francisco, pp 440–445

Takeuchi IK, Takeuchi YK (1986) Congenital hydrocephalus following X-irradiation of pregnant rats on an early gestational day. Neurobehavioral Toxicol Teratol 8:143–150
Takeuchi IK, Kimura R, Matsuda M, Shoji R (1987) Absence of subcommissural organ in the cerebral aqueduct of congenital hydrocephalus spontaneously occurring in MT/Hokldr mice. Acta Neuropathol (Berl) 73:320–322
Wagner HJ, Pilgrim Ch, Brandl J (1974) Penetration and removal of horseradish peroxidase injected into the cerebrospinal fluid: role of cerebral perivascular spaces, endothelium and microglia. Acta Neuropathol (Berl) 27:299–315

X. Discussion Sessions

Discussion Session I[1]

Signal Pathways

Chaired by A. Oksche

A. Oksche: This discussion requires structural organization to find a common denominator for a number of contributions that represent a rather wide spectrum of structural and functional aspects of the subcommissural organ (SCO). Since cell biology and biochemistry are a central topic of the following discussion section, in the present context emphasis should be placed on secretory polarity and the neural and vascular pathways of the SCO. Thus, this discussion could be entitled "Signal pathways to and from the SCO."

The effector pathways of the SCO are involved in the discharge of the secretory material. The most spectacular secretory pathway of the organ is Reissner's fiber (RF) with its termination in the ampulla caudalis; another is the less well understood projection of secretion-containing cell processes and cell chains to the contact area with the leptomeninges. Both polar-oriented pathways have access to the cerebrospinal fluid (CSF) and the capillary network. RF has been rather well explored but do we really understand its significance and function since a part of the SCO secretion is released into the CSF in a dissolved form? Is the formation of RF primarily a physicochemical problem of aggregation of secreted molecules, or is it a matter of CSF hydrodynamics, or both? Is the leptomemingeal connection of secretory structures to the outer, subarachnoid compartment of CSF a functional alternative to RF? With respect to the neural afferents, it is difficult to understand why the SCO of the rat receives a distinct, functionally significant serotoninergic input, but the SCO of the mouse displaying rather similar morphologic features is free of serotonin-containing nerves. All these problems are somehow linked to the phylogeny and ontogeny of the SCO.

R. Olsson: The enigmatic aspect about RF is that it represents a very ancient, extremely conservative structure of the central nervous system of chordates. It is apparently important to shift the mass of the secretory material condensed in RF as far away as possible from the site of synthesis. In young avian embryos, the fiber may extend even into the amniotic fluid via the open canalis neurentericus. However, we can assume that it exerts its specific function within the neural tube. There are other peculiar things,

[1] For ultrastructural details of the subcommissural organ, see Figs. 1–3.

e.g., the very close topographical relationship between the pineal organ and the SCO, as observed in the cyclostome *Lampetra* that has well-developed pineal and parapineal organs. But in another representative of cyclostomes, the hagfish *Myxine*, the SCO is very spectacular and the pineal organ completely missing.

A. Oksche: In my opinion, there is no evidence for a general, clear-cut morphologic pattern establishing a structural or functional unity of the pineal with the SCO. With respect to RF, it makes sense to me that after release into the ventricular cavity the secretion of the SCO is maintained for a certain period of time, even for days, in the form of a depot, and not dissolved immediately in the CSF. This would allow a slow, gradual release or uptake of chemical compounds. However, then one has to ask, how this process is controlled. Is the vascular system the final target of the secretory material and its agent(s)? And what is the meaning of the conspicuous species-dependent differences.

F. Nürnberger: If RF indeed can be regarded as a kind of depot, the SCO would have to receive a signal about the actual need of the agent in order to adapt its synthetic activity to the functional events in the terminal area. It is somehow difficult to imagine RF as a depot of agents for fluctuating, fast changing physiological demands; it is easier to accept a permanent physiological need, which might arise or be conveyed at the level of the terminal portion of the spinal cord.

A. Oksche: Does anybody know the time course of the loss of RF material by continuous escape of molecules into the CSF? Probably, an in vitro experiment would help to solve this problem.

G. Sterba: Are all structural lipoproteins in the membranes of nerve cells produced by these neurons themselves or is there a possibility that some specialized lipoproteins are manufactured at a different site, then transported to the nerve cells and finally incorporated into their membranes. The substances produced by the SCO occur in the CSF in a soluble or nonsoluble form. The way from the ventricles into the brain tissue is very easy because it is open for all kinds of molecules. It may be that the nervous system needs only a limited amount of a substance that is produced in the SCO in abundance. How to deal with the surplus material? It may be that the material which is not actually utilized becomes aggregated and, in this way, produces RF. Subsequently, this substance is transported via the central canal, and if there is need of the material in the nearby spinal cord tissue, certain components are detached from RF and can penetrate into the nervous system. And the remainder is finally released into the bloodstream.

A. Oksche: There are also other aspects to be considered. In the ampulla caudalis, the secretory material is aparently not released into a part of the nervous system containing neurons, but discharged via a very thin ependymal tube. Since according to Hofer and Rodríguez, in the filum terminale of lampreys the secretion gains access to sinuslike blood vessels, one may assume a remote neuronal target area. With all these problems in mind, I feel that it would be worthwhile to find out in the central nervous system the binding site(s) of the secretion product of the SCO. After the possible detection of these binding sites, one would have to ask whether these targets are readily accessible from the CSF via the ependymal cells or through the intercellular gaps. Dr. Rodríguez, in your investigations ependymal cells lining the central canal have been shown to contain material immunoreactive to antisera against RF material. This may speak in favor of an ependymal uptake of this material. But where is the final target, the binding site?

B. Krisch: Could it be that RF functions in the sense of a guiding structure directed toward the ultimate end of the central canal that harbors its massa caudalis? Could it play a role in the circulation and resorption of the CSF? Furthermore, are degrading enzymes involved in RF mechanisms? Some of the incongruent results discussed could be explained by modified degrading processes.

A. Oksche: Thank you, Dr. Krisch for emphasizing two important points. One of these points is concerned with CSF circulation and resorption; the other focuses our attention on degrading enzymes. One would expect a particular activity at the level of the terminal cisterna and also downstream toward the terminal site.

B. Krisch: The degrading process may also be essential in connection with the soluble component of the secreted product.

A. Oksche: This approach is of great immediate interest since neurobiologists have become increasingly aware that in the case of neuropeptides the degrading enzymes play a crucial role. What happens with the RF material that is not used and not stored in the ampulla caudalis?

E. Rodríguez: In fact, at least in a grant proposal, I have advanced the hypothesis that the secretory material released by the SCO into the CSF either in the form of a structured entity such as RF or in a soluble state might be involved in the resorption of CSF. This idea occurred to me when looking at the massa caudalis in the lamprey. When one examines the dorsal wall of the ampulla with the large intercellular spaces in the ependyma and the lack of a basal lamina at this site, the open communication between the lumen of the ampulla and the local capillaries resembles somehow the fine structure of the arachnoid villi. With respect to possible targets for RF material circulating in the CSF, including its subarachnoid fraction, I considered that this compound might have some effect at sites of reabsorption of the CSF into the bloodstream. We have been able to block the appearance of RF in the central canal by injection of a specific antibody; in my opinion this is a good experimental model that could provide some causal answer. Concerning the second functional aspect mentioned by Dr. Krisch, a number of enzymes are found under normal conditions in the CSF, proteases and glycosidases. Curiously, these glycosidases apparently do not affect the sugar moiety of the glycoproteins which are present in RF. Only at the end of RF do the glycoproteins lose their sialic acid residues and become degradable; galactose is then exposed as the terminal residue. Thus, at least with respect to the cleavage of sugar residues, the glycosidases present in the CSF can only partly be effective. But I also think of the reverse possibility that the complex compound representing RF could have an enzymatic activity of its own.

W. Severs: A question to the structural investigators: what is the status of the hypendymal secretion in the SCO, is it real or not?

E. Rodríguez: I think that the secretory capacity of the hypendymal cells is as real as the secretion of ependymal cells. Now I feel more easy about the secretory activity of the hypendymal elements because they are clearly immunoreactive with the same type of antisera as the SCO ependyma. Furthermore, they possess virtually the same binding properties for different lectins; however, their location is remote from the ventricular cavities. The fine structure of the hypendyma resembles that of the ependymal cells. There might be a difference with respect to the innervation of the hypendymal elements. But they emit processes, and by using Vibratome sections one can follow very fine branches of these processes to their endings either on a blood vessel or at the brain surface. The latter may play the same role as the ependymal terminals in releasing a soluble material into the subarchnoid CSF.

A. Oksche: In my opinion, there is a gradual transition from ependymal to hypendymal cells. All sorts of intermediate cell forms and mixed arragements can be observed. The hypendymal cells can send their processes to the leptomeninges, but they can also terminate on intrinsic blood vessels in analogy to ependymal cells. In conclusion, it is one family of cellular elements. From comparative studies I have gained the impression that if the posterior commissure becomes very thick, an individual process of the ependyma is unable to ascend to the leptomeningeal border; in such cases there is an increase in hypendymal cells, and the hypendyma is very highly vascularized, especially in mammals.

W. Severs: What is the possibility of a SCO secretion being destined to the bloodstream or being destined to be reabsorbed by processes from the SCO out of the bloodstream?

E. Rodríguez: Is there a compound of SCO secretion that may reach the bloodstream and does this compound have any relation with the compound released into the ventricle and later assembled into RF? I can describe a crucial observation in the canine SCO – to me the most glandular type of SCO I know. In the dog, in large accumulations of secretory cells, one cannot distinguish exactly between hypendmal and ependymal elements; it is a glandular structure, highly vascularized, just like an endocrine gland. When we performed immunocytochemistry using anti-RF serum at the electron-microscopic level, we observed only very few immunoreactive processes containing secretory vesicles and contacting blood vessels. Most of the processes containing granules were free of immunoreaction. We don't know if anybody has succeeded in producing an antibody reactive with that granular material, which has a very close spatial relationship to the vascular wall. I feel more comfortable with the idea that something is released into the blood that might be different from the compound released into the ventricle.

A. Oksche: I confirm that the SCO of the dog is a highly glandular structure. It contains within some of its characteristic cell columns also elements, which – according to ultrastructural criteria – have an epithelial, glandular appearance and no longer display the regular features of hypendymal glialike cells. All sorts of transitional forms may occur.

A. Meiniel: I return to the problem of vascular secretion: it has been stated that the basal secretory protein might be different from the apical. To this problem some additional information can be provided. With the use of a monoclonal antibody we were able to label the basal portion of the SCO as well as the apical portion. This suggests that the same protein may be released at the basal and the apical aspect of the SCO.

A. Castañeyra Perdomo: Concerning the ependymal and hypendymal cells, I have seen ependymal processes in contact with hypendymal processes. In the mouse, I have found three types of SCO-cells.

P. Fernández-Llebrez: We have observed, in reptiles, an unusual population of SCO cells which do not react with the anti-RF serum. The granules of these cells are released toward the bloodstream, but they are apparently also discharged into the CSF. We suggest that these granules may be responsible for the modification of the secretory material when RF is formed. Another interesting aspect is concerned with species that lack RF. In some species, like the hedgehog, we even miss an open central canal in the spinal cord. Other species, like some squirrels, are apparently devoid of an SCO and thus also lack RF. Although the SCO–RF complex is a conservative structure, there are vertebrates free of these differentiations. Such animals could provide models for functional studies, stimulating a search for other brain structures that may take over the SCO–RF functions.

A. Oksche: A very small cavity with some circulating CSF or CSF-like fluid is apparently sufficient to enable the formation of a RF. Since we are prepared to accept that the secretory material is also shifted toward the blood vessels or, more precisely, to the perivascular space, and subsequently gains access to the brain, we may deal with a number of different mechanisms. For this reason, I prefer to speak of "effector signal pathways," irrespective of the commuting medium and the exact topography.

R. Yulis: I am eager to collect all the evidence available pointing toward the existence of different cell populations in the SCO. Some cells might secrete into the bloodstream and others might project to the CSF. Might the use of different monoclonal antibodies help to differentiate these cellular elements?

F. Nürnberger: Has anybody so far investigated the pattern of the vascularity of the SCO distal to the hypendymal region? Do these vessels directly continue to the venous system or do they connect the SCO to other parts of the brain?

A. Oksche: In the SCO, there is no evidence of the existence of a portal system, at least according to my present knowledge. What about the possible occurrence of specific SCO glycoproteins in the general circulation? If such specific compounds of the SCO were released into the circulation, they should be detectable in the general circulation, at least if the venous vessels are punctured not far away from the site of release.

E. Rodríguez: I think we have to move step by step. Now, as we realize that there may exist a CSF-soluble secretory material, for which we can search in a big animal such as the bovine, and since we are aware of soluble messengers, which are independent of RF, we may be able to raise antibodies to such compounds, and hopefully, to establish an RIA. At the present stage we would be happy enough to have an enzyme immunoassay.

A. Oksche: The last point of our discussion is concerned with the seretoninergic input to the SCO. Would somebody like to make a comment on this innervation, which shows considerable species-dependent variety.

W. Severs: Two questions: (1) Is it a generalized fact that X-ray radiation damages the SCO and its serotoninergic system? Are there other serotonin systems, which are affected? (2) Substance P is generally related with serotoninergic systems. Are there certain substance P systems damaged by X-ray radiation?

N. Delhaye-Bouchaud: I am unable to answer your second question, since I do not have any experience concerning the destructive effects of X-rays on the substance P system. But with regard to the first question, X-rays obviously affect all serotoninergic systems of the brain stem, at least the fiber elements extending to the cerebellum and the spinal cord. I have never examined the serotonin input to the cortex, but I think that all raphe nuclei have been affected in our experiment. In our SCO system we have obtained results which are in marked contrast to the observations described for the cerebellum. During the early period of differentiation serotonin has been claimed to exert a signal essential for the differentiation of the target cells. It is rather curious that in our SCO system the ependymo-glial target conveys a signal to the serotonin-containing afferents. It may be that serotonin also acts as a differentiation signal on the SCO. This complex type of interaction is disturbed by X-ray irradiation. After the X-ray treatment the young animals are ataxic. A few months later they behave again quite normally and survive.

A. Oksche: I am impressed by the diversity of the effects concerning the basal and supraependymal serotonin systems. Do SCO cells possess serotonin receptors – apical, basal or both? This may be of importance also in connection with the pineal–SCO interrelationships. We know that in addition to melatonin its precusor serotonin is one of the important agents of the pineal organ. Serotonin could be released from the pineal organ into the CSF and then act on the SCO if there were apical receptors or binding sites for serotonin. We should not overlook that, in addition to the serotoninergic innervation, also in species where serotonin-containing afferents are lacking, the pineal organ may harbor a large amount of serotonin.

J.H.B. Diederen: When we injected radioactive serotonin into the CSF of frogs, we found much radioactive material on the SCO. We did not look particularly at RF; probably it did not bind serotonin.

A. Castañeyra Perdomo: Is it possible to obtain immunostainings before the SCO is fully developed in morphological terms?

K. Schoebitz: Yes, we can observe some positive immunohistochemical staining already before the SCO is completely developed in the rat.

A. Castañeyra Perdomo: At stage 38–40, I can see numerous characteristic cells of the SCO. Before that time point in the SCO area many cells resemble spongioblasts.

K. Schoebitz: In my opinion such "spongioblastic" elements are already capable of elaboration of immunoreactive material.

E. Rodríguez: I think this is a relevant point and I realize that there is a principal agreement between the individual reports. We learned from the publications and presentations by Drs. Schoebitz, Meiniel and Naumann that, in the chick embryo, the genes encoding for the secretory glycoproteins are expressed much earlier in ontogeny than the onset of the morphological differentiation of the SCO. With conventional staining such early immunoreactive cells cannot be distinguished from neighboring cellular elements. This holds particularly true for the chick embryo.

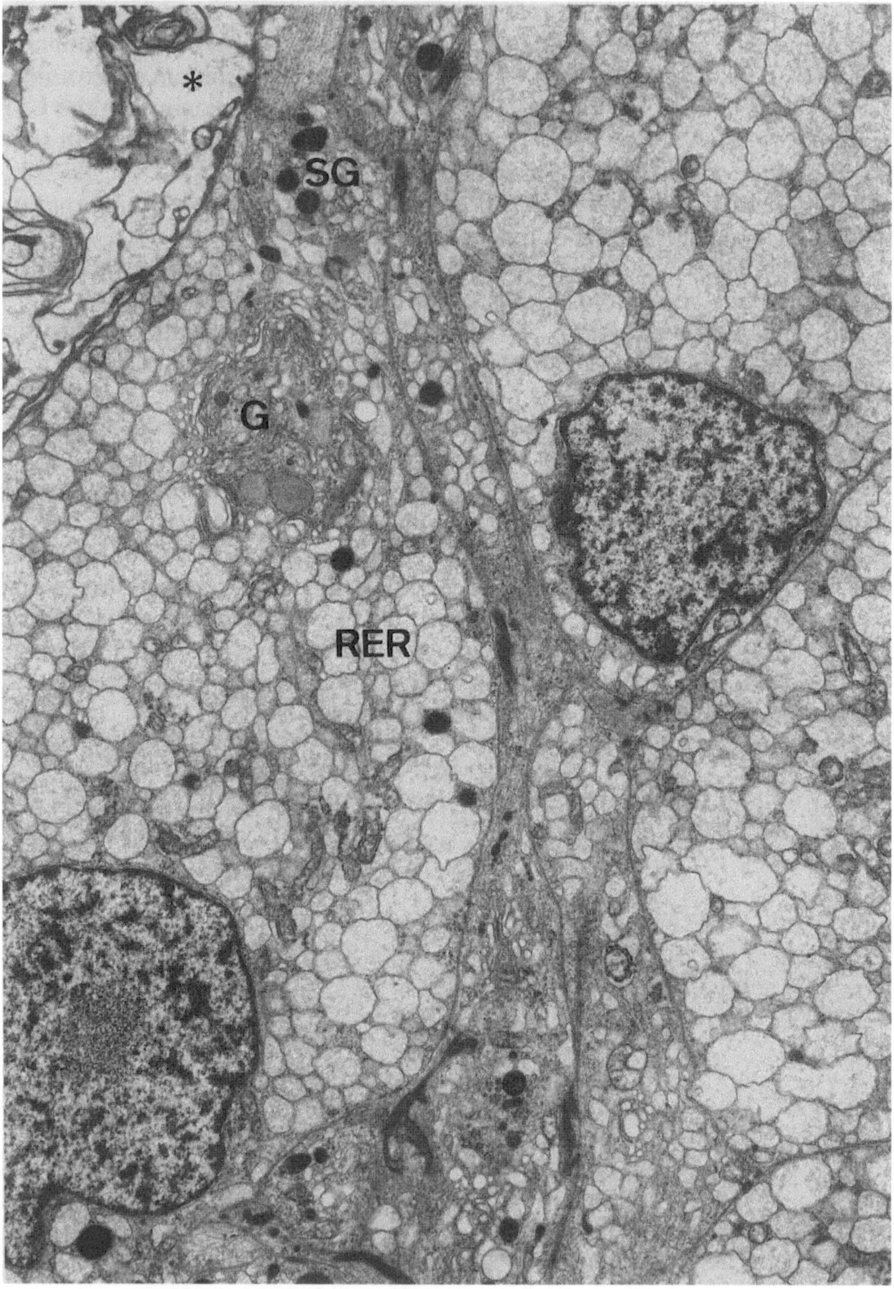

Fig. 1. Subcommissural organ, dog. Fixation by perfusion through the brain ventricles with a triple aldehyde mixture. Perikarya of secretory cells are filled with dilated cisternae of the rough endoplasmic reticulum (*RER*) and display a well-developed Golgi apparatus (*G*). *SG*, secretory granules; *asterisk*, wide cisternae of RER, ×8000. (E.M. Rodríguez, unpublished)

Discussion Session II

Cell Biology and Biochemistry

Chaired by E.M. Rodríguez

E. Rodríguez: For this session I propose the following outline in order to achieve a structured discussion: (1) Number of compounds in the secretion of the subcommissural organ (SCO). (2) Nature of the secretory products of the SCO. Glycoproteins: (a) molecular size; (b) type of glycosylation: O-linkage, N-linkage; proteoglycans; non-glycosylated proteins; peptides. (3) Basal pathways of secretion: vascular and subarachnoid route. (4) Two pools of secretory material? (5) Processing. Intracellular processing: (a) carbohydrate component; (b) proteinaceous component. Postrelease processing? (significance of the pre-RF stage?). (6) Technical considerations. Polyclonal/monoclonal antibodies: elaboration of a list of antibodies available; extraction procedures.

Number of Compounds in the Secretion of the SCO

A. Meiniel: To date we do not have sufficient information on whether one or more proteins occur in Reissner's fiber. Dr. Rodríguez has postulated the existence of two precursors in the bovine SCO proper, but how many components are present in RF?

E. Rodríguez: We have presented evidence indicating that bovine SCO secretes two types of compounds of high molecular weight, each of them capable of induction of antibodies with different immunoreactivities (F. Nualart et al., *Molecular Brain Research* 11: 227–238, 1991). In an experiment carried out in our laboratory by Nualart, mRNA purified from bovine SCO was translated into *Xenopus laevis* oocytes. Immunoblots of extracts of these oocytes exposed to an anti-RF serum showed three bands with a molecular mass corresponding to three of the six bands occurring in RF immunoblots. We interpret this finding as an indication that the mRNA of only one of the two putative precursors has been translated. We have to apply molecular tools, such as DNA probes, to find out whether the secretory material of the SCO is derived from one or several precursors.

Nature of the Secretory Products of the SCO

A. Meiniel: In my opinion the proper characterization of the antibodies reacting with the secretory material of the SCO is a fundamental point.

C. Hirschberg: For each antibody that is obtained it is very important to determine its epitope. For antibodies raised against glycoproteins it is fundamental to rule out that some of them recognize carbohydrates. An antiserum containing antibodies against carbohydrates would not be very specific. A relatively easy approach to solve this problem is to remove the bulk of sugar before using the glycoprotein for immunization. If after treatment of the glycoprotein with, for example, endo-H, the epitope is removed, then the result obtained is reliably interpretable; if after such a treatment the epitope is not removed, then the chances are 9 out of 10 that the epitope is located in the protein backbone of the glycoprotein. These are the antibodies worthwhile to be used for further studies, since they have the chance of being highly protein specific.

E. Rodríguez: I appreciate very much your suggestion. A similar suggestion was given to me by Prof. S. Stirm, Biochemistry Department of the University of Giessen, Germany. Thus, more than a year ago we started using endoglycosidase F to remove the carbohydrate moiety of the secretory glycoproteins of the bovine SCO for several purposes. Sialylated glycoproteins usually lead to difficulties, such as an "abnormal" migration, in electrophoresis. The removal of the carbohydrate component may help to solve this problem and also provide a more precise idea of the molecular mass of the protein backbone (see Hein et al., this volume). Another important aim is to raise antibodies against the protein backbone. Several of these experimental protocols are in progress in our laboratory. Furthermore, during the 13 years of our involvement in study of the SCO glycoproteins, we have had the permanent feeling that we are chasing a "ghost," facing one problem after the other. Thus, by removing the carbohydrate from these molecules we thought to improve our working situation. I would very much like to learn more about the experience of Drs. Sterba, A. Meiniel and Naumann with respect to the handling of these glycoproteins.

A. Meiniel: Yes, I agree with you; the SCO proteins do not belong to the "easy" molecules. We have also tried to deglycosylate RF material using endoglycosidase F and endoglycanase, but have not been successful, probably due to the high degree of packaging of RF material. The solubilization of this material is a particular problem. So far, we have not found a procedure which helps us to dissolve the RF material fully, followed by a cleavage of the carbohydrate component. I have a question to Dr. Hirschberg. If an antiserum against a glycoprotein contains antibodies against the carbohydrate chain, then in immunoblots this antiserum should produce a band pattern similar to that of concanavalin A or wheat-germ agglutinin binding. This seems, however, not to be the case for the antisera against SCO secretory glycoproteins so far investigated. Thus, I agree that in the case of these polyclonal antibodies a certain group of antibodies may react with the carbohydrate chain, but they would represent a minority when compared to those reacting with the protein component. Can you agree with this speculation?

C. Hirschberg: My guess would be that the worst scenario is the existence of mixed epitopes: partly protein, partly carbohydrate. In this case, one is indeed in trouble. It would be surprising that these proteins were so antigenic, and the sugars so extraordinarily nonantigenic that the problems I mentioned would not arise. With respect to the solubility of RF material I would suggest trying some chemical deglycosylation; even if it is not complete, it has the advantage that the probe should turn out to be more specific.

E. Rodríguez: I think that the SCO secretory proteins are extremely antigenic. The titers obtained when immunizing rabbits, rats, or mice are extremely high. For immunocytochemistry these antisera can be diluted more than 1 million times; furthermore, these high titers are attained at a relatively short time after immunization.

G. Sterba: We sent some milligrams of RF to Prof. Schauer of Kiel, who is a specialist for carbohydrate chemistry. He found that the RF glycoproteins exhibit two types of linkage: 90% of the protein is glycosylated through N-linkage, and 10% is O-linked. This result is similar to what Dr. Hädge and myself reported 20 years ago (D. Hädge and G. Sterba, *Acta Biologica et Medica Germanica* 30: 581–585, 1973).

E. Rodríguez: Thank you Prof. Sterba; I am very glad to learn from your experience. In fact, I included this point into the present discussion because there are very few references dealing with O-linked carbohydrates in the SCO glycoproteins. As far as I remember, there are two papers from your laboratory suggesting that the carbohydrates of RF glycoproteins are O-linked (see G. Wolf and G. Sterba, *Acta Zoologica (Stockholm),* 53: 147–154, 1972; and more recently a report by A. Meiniel et al., *Cell and Tissue Research* 253: 383–395, 1988). It was suggested that two types of linkage (O- and N-) may exist in SCO secretory proteins. Preliminary findings obtained in our laboratory by use of endoglycosidase F and SDS-PAGE suggest that one of the six compounds present in RF may possess O-linked carbohydrates. This might correspond to the 10% protein of RF displaying O-links as mentioned by Dr. Sterba.

C. Hirschberg: I want to raise a word of caution concerning this percentage. If in a protein you have one N-linkage and one O-linkage, and the N-linked chain has 15 sugars because it has the core and branches, and the O-linked chain has five sugars, then you would get a very low O-percentage as compared to the N-percentage: The percentage reflects the sugars in the branches but not how many O-links you may have.

E. Rodríguez: This applies to the case when you are quantifying the sugar residues.

C. Hirschberg: That is correct, and this is what is done very often.

G. Sterba: Prof. Schauer (University of Kiel) indicated to me that 90% of the protein of RF is N-linked; so I think this percentage is correct.

E. Rodríguez: I want to make a brief reference to a paper we published with H. Herrera (*Histochemistry* 93: 607–615, 1990). Some 60 h after injection of Tunicamycin into rat CSF, the SCO cells were completely devoid of ConA- and RCA-binding sites, indicating that N-linked sugars represent the bulk of carbohydrates of the SCO secretory glycoproteins, since Tunicamycin interferes with N-linking.

C. Hirschberg: The possibility that, in your experiment, Tunicamycin had interfered with protein synthesis has to be considered.

E. Rodríguez: We took care of that possibility by using Tunicamycin from three different sources; furthermore, the SCO of the treated rats which lacked ConA binding displayed immunoreactive secretory material, indicating that protein synthesis had not been severely disturbed.

E. Rodríguez: Now I invite the audience to move to the next point dealing with the nature of the secretory products: the presence or absence of *proteoglycans* in the SCO.

B. Krisch: Alcian blue, pH 1.0, is a simple method for screening acid, sulfated glycosaminoglycans. The perivascular space of all neurohemal regions of the circumventricular organs shows a positive reaction with the exception of the SCO.

C. Hirschberg: It is possible that the SCO cells do not secrete any proteoglycans, but I still think that in the absence of a direct sulfation measurement one should not exclude the presence of proteoglycans in the SCO.

E. Rodríguez: Many years ago Dr. Naumann (*Zeitschrift für Zellforschung und mikroskopische Anatomie* 87: 571–591, 1968) reported on a comprehensive histochemical study of the SCO. Dr. Naumann, did you ever gain evidence for the presence of proteoglycans in the SCO?

W. Naumann: I did not get any evidence suggesting that proteoglycans might be present in the SCO.

E. Rodríguez: The next point of the discussion stimulates us to analyze the possibility that the SCO might secrete *nonglycosylated proteins*. In a symposium held in Giessen in 1986, we reported on preliminary findings from immunoblot analyses of the bovine SCO. Although most of those findings have been confirmed (see Hein et al., this volume), some of them cannot be approved. In our early paper a group of immunoreactive compounds (60, 66, 68, and 74 kDa) were regarded as secretory products of the SCO. One of these bands did not bind ConA or WGA; thus, it was thought to correspond to a nonglycosylated secretory protein. Later we realized that these four bands were the result of a twofold technical artifact. First, they correspond to keratin polypeptides that had migrated into the gel from glassware contaminated with human keratin. But then, how to explain their immunoreactivity with the anti-RF serum? It took us a year to answer this question. When immunizing rabbits we use multiple intradermal injections, and these rabbits develop antibodies against both the injected antigen and the "implanted" keratin. When used in immunoblots this antiserum revealed both the secretory products of the SCO and the contaminating keratins. This technical artifact was overcome by using extremely clean glassware and by applying an anti-keratin serum as control for the immunoblots. Under these conditions all secretory compounds of the SCO were found to be glycosylated.

E. Rodríguez: I suggest continuing the discussion on the possibility that the SCO might secrete a peptide in response to Dr. R. Yulis. He, as well as Dr. S. Hein, have some evidence for the presence in acidic extracts of the bovine SCO of a secretory peptide (see this volume).

A. Meiniel: Is it a specific peptide of the SCO, or one of the known peptides?

E. Rodríguez: Apparently it is not related to any of the known neuropeptides; there is a possibility that it might be related to the somatostatin precursor.

Basal Pathways of Secretion: Vascular and Subarachnoid Route

H. Jones: I want to challenge the suggestions that the hypendymal cells may secrete directly into blood. The SCO has blood–brain-barrier capillaries and it seems unlikely that these capillaries would have transporting systems to transport SCO products into blood. I would suggest as a more likely possibility that the secretion of the SCO released into the perivascular space finally reaches the subarachnoid space. That the perivascular spaces do connect with the subarachnoid space has been shown elsewhere in the brain by use of tracers.

B. Ghiani-Uva: There is evidence that the tightness of tight junctions of the endothelial cells of brain capillaries can be modulated by neuromodulators, such as the atrial natriuretic peptide. Furthermore, we have evidence that the SCO possesses receptors for atrial natriuretic peptide.

H. Jones: There are, indeed, various substances that modulate permeability in blood–brain-barrier vessels, probably through the tight-junction pathway, but this would not be the case for the SCO vessels. In all other circumventricular organs that secrete into blood there are no blood–brain-barrier vessels.

E. Rodríguez: I would like to call attention upon the fact that the SCO cells are somehow in a unique position concerning barriers. Some species possess a very tight SCO-CSF barrier in addition to a functioning blood–brain barrier. Thus, here we have a small portion of the brain somehow "sequestered" (or protected?) by two barrier systems. This should be kept in mind when trying to decipher the unknown functional role of the SCO. There is convincing evidence by use of ultrastructural immunocytochemistry that secretory compounds of the SCO may be secreted into the perivascular space. Dr. Jones' suggestion that this material might be finally transported to the subarachnoid space, i.e., CSF, and not to the local bloodstream is a very good one, and I am inclined to think in the same direction. However, we must be aware that the blood–brain-barrier vessels of the SCO display special features in comparison to all other blood–brain-barrier vessels, such as the presence of a wide perivascular space and long-spacing collagen. The glycoproteins shifted by the SCO cells to the perivascular space might, by themselves, have some properties for making the barrier leaky. Transport of SCO secretory material into the local blood vessels can, at present, not be ruled out. I believe that the transport properties of the SCO capillaries are a crucial point to be further investigated.

W. Severs: I think you may be on the right line. Furthermore, the possibility of having blood flow control regulated locally without necessity of active compounds to be crossing from one side of the barrier to another has to be considered in SCO research.

F. Nürnberger: The SCO is not the only circumventricular organ endowed with two tight barriers. In several species, the pineal organ has two barrier systems as well, in analogy to the situation in the SCO, and in the pineal a vascular release of a well-known hormonal agent, melatonin, is fully established.

A. Meiniel: I would like to make a comparison with the situation concerning ependymins; these agents are synthesized in the meninges, released into the CSF, and reabsorbed in other parts of the brain. Is it possible that compounds secreted by the SCO reach the subarachnoid space and from here extend to other parts of the brain devoid of a blood–brain barrier?

E. Rodríguez: In the dog SCO, there are many secretory cells contacting the blood capillaries but not reacting with anti-RF sera. Here we have a secretion for which appropriate antibodies are not yet available. The secretion of these cells is apparently not related to RF material, and this product could be readily released into the perivascular space. The possibility of the existence in the SCO of an unknown compound with hormonal properties has to be considered as a working hypothesis.

R. Olsson: I think we have also to consider the evolutionary trend. Great differences may exist between the SCO of a fish and that of a mammal. I wish to refer to the situation of the endostyle, which is an organ located in the oral cavity of primitive animals and that later, in evolution, changes into the thyroid gland. This endostyle produces a mucous substance released into the oral cavity. Later in evolution, the gland discharges thyroid hormones from its basal aspect into the blood. A similar situation may hold true for the SCO.

A. Oksche: Since 1954, I have strongly supported the existence of a basal route of transport and release of SCO secretion (see. A. Oksche, *Zeitschrift für Zellforschung* 54: 549–612, 1961; Journal Neuro-Visceral Relations Suppl. IX: 111–139, 1969). Thus all comments in favor of a vascular and subarachnoid pathway give me great satisfaction. In a comparative context, attention should also be paid to the leptomeningeal organization in some lower vertebrates. In several of these animals a continuous subarachnoid space appears to be missing. Instead, local pools of extraventricular CSF may exist. The composition of the fluid in these compartments is open to discussion.

Pools of Secretory Materials and Their Processing

E. Rodríguez: When radiolabeled cysteine is injected into the rat CSF, the SCO labels strongly. It takes 1 h for labeled secretory compounds to appear as pre-RF material, and a further hour to become packed into RF. Thus, this is a rapid biosynthetic mechanism, since synthesis, processing, transport, and release occur within 1 h. However, 4 to 5 days after the injection of the labeled cysteine, labeled compounds are still present in the lumen of dilated cisternae of the RER. This material retained in the RER is apparently continuously released, since pre-RF material and the proximal end of RF continue to be labeled during the 5-day period after the injection of the radioactive cysteine. Which are the mechanisms underlying these two pathways, one that makes the secretory material readily releasable and enables rapid release into the ventricle, and the other one capable of storage of the material in the RER for days to be finally discharged into the ventricle?

C. Hirschberg: Is it the same secretory material following the two pathways? Obviously it may not be the same. If it is the same, then it is difficult to give an answer to your question. It is known that there is a group of proteins with the capacity to retain secretory proteins within the RER. This could also be the case for the SCO. The fascinating problem is the functional meaning of the long retention of secretory compounds within the RER of the SCO cells. With respect to what you call rapid release, 1 h time is average. The behavior you describe (quick and slow release) most likely represents the behavior of two different molecules.

J.H.B. Diederen: My view is that the huge amount of material stored in the RER represents unfinished material that will finally be released at the ventricular cell pole. I call it unfinished because glycosylation of this material has not yet been completed since it has not yet gained access to the Golgi apparatus.

Technical Considerations

A. Oksche: I would like to suggest discussing principal methodological aspects of SCO research, with the aim to establish an outline for further communication among the groups in order to promote the exchange of information and materials.

E. Rodríguez: We have prepared a list of the polyclonal and monoclonal antibodies raised in our laboratory against secretory products of the SCO. This list includes: (1) A clear-cut characterization of the antigen used, i.e., crude extract (type of extract), partially purified extract, purified compound, band from a gel; (2) the source of antigen, i.e., SCO, RF or cerebrospinal fluid; (3) the protocol of immunization; and (4) a characterization of the antibody by immunocytochemistry and/or immunoblotting. I put forward the following proposition to all colleagues raising antibodies against SCO secretory compounds: please prepare a similar list, so that we may finally have a sort of a "directory" of the antibodies available in this field. We could exchange this information continuously and eventually also exchange antibodies. This may help to bridge the gaps we have in the SCO research. Another source of discrepancies when reporting on biochemical findings is the extraction procedure used. I believe we are still at a rather initial, crude stage with respect to extraction procedures of the SCO secretions, and we lack the appropriate tests for evaluation of the extract itself. Taking into account these limitations, it would be of great benefit if investigators working in the field would present their bench protocols when publishing their findings. This could help to interpret discrepancies.

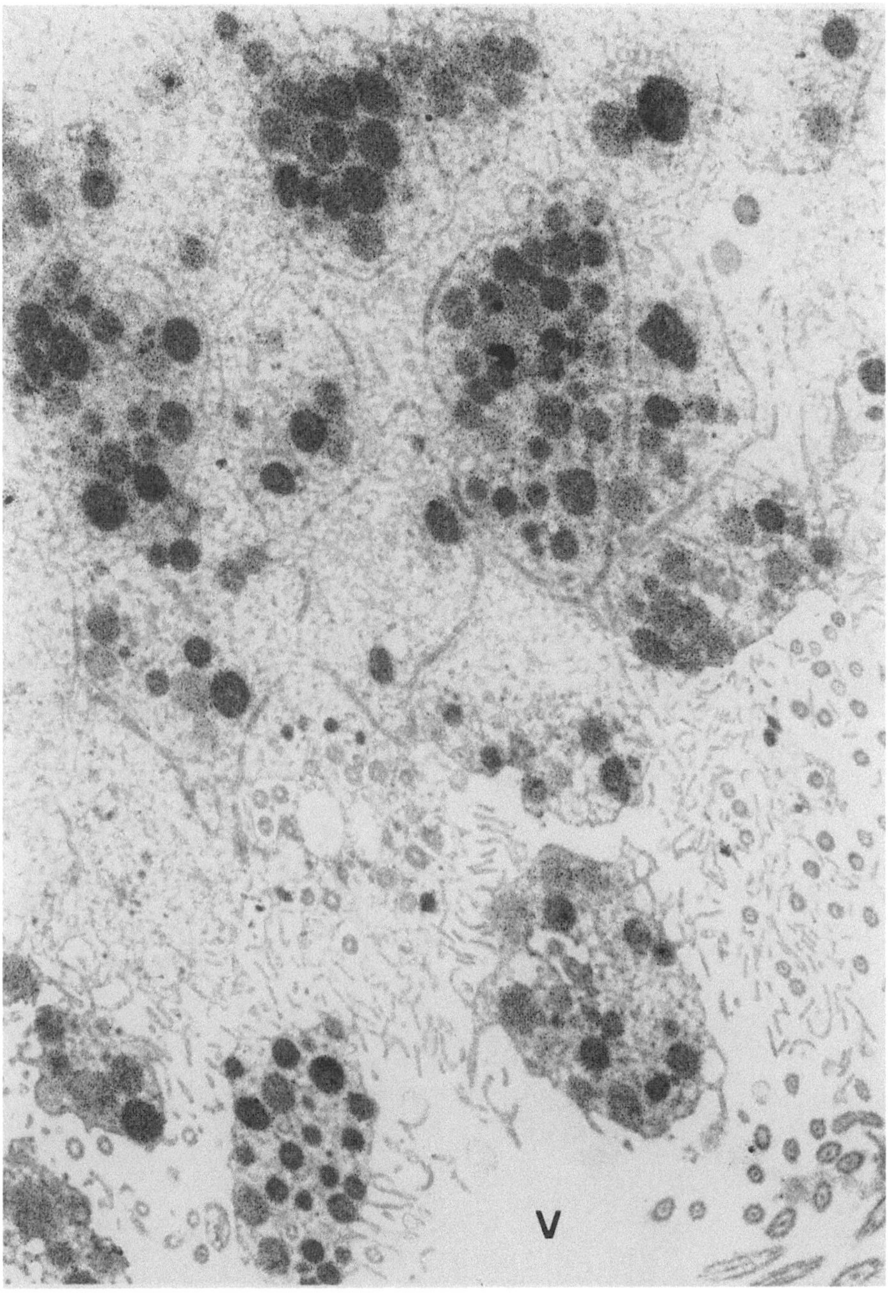

Fig. 2. Subcommissural organ, dog. Ventricular pole of secretory cells. Methacrylate embedding. Immunoperoxidase staining intensified with silver methenamine. An antiserum against glycoproteins of Reissner's fiber was used as primary antibody. Immunoreactive material is present in numerous granular structures of different sizes and densities. *V*, third ventricle, ×11300. (E.M. Rodríguez, unpublished)

Discussion Session III

Functional Aspects I

Chaired by R. Olsson

R. Olsson: During the 23 years since the first Reinhardsbrunn conference on the subcommissural organ (SCO) we have learned very much about this system, mainly concerning the fine-structural and biochemical mechanisms responsible for the elaboration of Reissner's fiber (RF) in the rat and some other mammals. However, it seems to me that we are not any closer to a basic understanding of the entire RF mechanism of chordates in its diversity than we were in 1968. However, several lectures presented at this meeting have shown that we are at least on the right path towards a better idea of these enigmatic structures and processes. Thus, let us now use the time available for an analysis of those problems that could not be discussed thoroughly after the respective lectures due to time limitation. In case enough time is left, I would also like to initiate an even wild and speculative discussion on possible functions of the RF system. The survey lectures by Drs. Leonhardt and Krisch dealt with the peculiar position and the neighborhood relationships of the SCO, and they gave the impression that the SCO is a rather strange member in this assembly. Could this be because it phylogenetically does not belong to that brain region, as I speculated in my lecture?

A. Oksche: A systematic approach is needed to classify the circumventricular organs under consideration of their median position, but one should not overestimate the singularities of their fine architecture and vascularization. Different species-specific and local specializations of the vascularity may exist; only in few species has the vascularization of the SCO been examined thoroughly. There are also other examples. The pineal body of the lower vertebrates serves primarily as a photoreceptive organ. Thus, one may assume that, in analogy to the retina, this extraocular photosensor – as a rule – possesses a tight blood–brain barrier. However, things turn out to be different. First, at least in some anurans, a tight blood–brain barrier has been shown to exist in the pineal organ. The latter is capable of melatonin production and, at the same time, it is also a functional photoreceptor. In contrast, in the rainbow trout, where the sensory function of the pineal organ has been verified by means of electrophysiology and its capacity to synthesize melatonin has been clearly proven, the blood–brain barrier is open to several conventional tracers. This situation shows that a greater number of species must be examined in order to establish definitive rules. All sorts of specializations and adaptations are possible, and so far we do not know what all this means in functional terms.

G. Sterba: Dr. Olsson, in your lecture you questioned whether the SCO indeed can be regarded as a paired organ and you provided two figures from Nicholls showing this. In my opinion these figures are hardly representative of the real situation, because the second figure is a selected image of the caudalmost paired portion of the SCO. However, in the same species, the frontal portion of the SCO is not paired. In very early embryonic stages of this group of animals one cannot find any sign that the SCO is paired. There is only one primordial placode in the midline of the embryonic brain.

R. Olsson: There are also other examples where only a part of the SCO is paired, e.g., the salamander *Ambystoma*, and I do not think that this situation is a better argument against than for a paired origin. I would, however, not call the SCO *anlage* a placode. Furthermore, I think that it could be interesting to analyze whether already the folds of the neural groove, which form this unpaired *anlage*, possess the immunocytochemical features being characteristic of the Reissner's material-producing floor-plate cells.

E.M. Rodríguez: Before I listened to Dr. Didier-Bazes providing evidence in favor of an inhibitory control of the SCO, I intended to challenge the inhibitory action of serotonin on this organ. I always thought that this assumption was based just on such weak morphological grounds as the amount of stainable secretory material or the appearance of more or less dilated cisternae of the rough endoplasmic reticulum. This is one of the cases starting with a mere suggestion; then the suggestion is repeated, and after a while the postulate is taken for granted. Although serotonin usually exerts an inhibitory influence on its targets, this analogy is not satisfactory to conclude that serotonin indeed plays an inhibitory role. If so, does it inhibit the synthesis or the release, or even both? Nobody has measured the release of the secretory material from the SCO when serotonin has been blocked or serotonin neurons have been destroyed. If serotonin exerts a permanent inhibition, then we never should find a Reissner's fiber in an animal with an intact serotoninergic input. Thus, it must be a partial inhibition, but is it partial in terms of time, i.e., sometimes off, sometimes on, or is it partial in the sense that it only partially blocks the release or the synthesis? These are some fundamental aspects of inhibition where we should be more critical, more specific.

And now a "wild speculation." Roberto Yulis described in the salmon a distal, caudal system of serotoninergic neurons beyond the urophysis. In this tissue surrounding the central canal that is virtually devoid of characteristic spinal-cord structures, a number of serotonin-containing neurons have been observed. Our "wild speculation" is that these CSF-contacting neurons at the caudalmost end of the central canal might be sensory. Bela and Ingeborg Vigh, among others, have displayed scanning electron-microscopic images showing RF in contact with the dendritic projections of such CSF-contacting neurons. We speculate that if a feedback mechanism for SCO regulation exists, a triggering mechanism must be involved to inform the serotoninergic neurons of the raphe nuclei when to turn on or off. This system of serotoninergic CSF-contacting neurons occurs at the ultimate end of the central canal where only RF can be regarded as a basic structural characteristic. It may be that these neurons are capable of sensing the RF material that arrives at the terminal point of the cerebrospinal fluid-containing ventricular system. These cells have processes projecting centrally, but we do not know where they end. To date, this is an entirely black area, but one may speculate that these axons project to an area somehow related – spatially or functionally – to the raphe nuclei, and thus a circuitry may exist that might be involved in a feedback mechanism.

A. Oksche: I would like to extend this line of thought concerning different sources of serotonin and place emphasis on serotonin receptors. The problem of serotonin receptors appears to be a most crucial one. Do different membrane receptors exist at the apical and the basal surface of the SCO cell? In this context, the pineal organ as a possible source of serotonin should not be overlooked. This serotonin could be conveyed via the blood circulation (?) or by the circulating CSF and then act on the SCO. The latter pathway is possible since the pineal lumen is in open communication with the third ventricle. The original idea of I. and B. Vigh that sensory pinealocytes of the lower vertebrates might be interpreted as CSF-contacting neurons may be conceptually integrated with the supposition of Drs. Yulis and Rodríguez concerning the role of the caudal serotoninergic CSF-contacting neurons. The serotonin-containing pinealocytes are spatially very close to the SCO.

It may be important to investigate when and where the serotonin innervation of the SCO has first appeared. Apparently, there is no serotoninergic innervation of the SCO in the poikilothermic vertebrates, at least in several species of anurans and teleosts examined in our laboratory. This innervation may have appeared in sauropsids, but nobody has examined this aspect in a representative, comparative way. In parallel, there is no noradrenergic input to the pineal organ of the anurans or fish; it appears for the first time in sauropsids. Some lizards show an abundant noradrenergic input to the pineal proper and in mammals this autonomic, sympathetic innervation has become a very important afferent pathway to the pineal organ. Is there already some evidence concerning the earliest appearance of the serotoninergic innervation of the SCO in phylogeny?

G. Sterba: In my opinion this question cannot be answered as long as appropriate comparative studies are lacking. An interesting problem in this context is the flow of the cerebrospinal fluid within the central canal. As shown by P. Fernández-Llebrez, the velocity of the flow may be influenced by the RF. This observation induces several questions: (1) Are the structures in the ependymal wall of the central canal capable of measuring the velocity of the CSF? (2) Are these structures the CSF-contacting neurons? (3) Is the velocity of the flow different in resting and in moving animals? (4) If we speculate that some substances (e.g., metabolites that are toxic by concentration in loco) are released from the surrounding tissues into the CSF, then the velocity of the CSF-flow may play an inevitable role in the process of clearance (variation of the Olsson hypothesis). In this connection the results of Hauser and Rühle may be recalled. They showed alterations of the vertebral column in animals in which the RF was experimentally prevented from entering the central canal.

R. Yulis: There is a special system of CSF-contacting neurons located exclusively in the midline of the ventral wall of the central canal; these cells produce a urotensin II-like peptide. Somatostatinergic, CSF-contacting neurons in the wall of the central canal are located at other sites. The population of the urotensin II neurons is situated in all species so far studied below the Reissner's fiber, suggesting a relationship to this fiber. A mechanoreceptor theory may be considered, and a functional relation of the fiber to these receptors might be suspected. More evidence in this context can be derived from flatfish which actually swim sideways. In these animals the peptidergic neurons of the spinal cord still occur in a ventral position, but they are displaced laterad. In these teleost species the amount of peptides in the neuronal perikarya and axons is very low.

A. Oksche: I have always been rather concerned about the interpretation of experiments involving mechanical procedures of ablation of the SCO. In my opinion, in all these cases damage to the posterior commissure is unavoidable and nobody can guarantee the specificity of the changes in behavior and of other effects resulting from this type of mechanical damage.

G. Sterba: I forgot to mention that Hauser and Rühle did not destroy the posterior commissure in their experiments. They injected chemically neutral granular material (beads of plastic) into the ventricles, and by this procedure the connection between the fourth ventricle and the central canal was blocked.

A. Oksche: I am aware of other experiments based on a mechanic ablation or thermic lesion of the SCO. In these cases there was damage in the vicinity of the SCO and I had the very strong impression that the posterior commissure had been lesioned, and this may have affected the specificity of symptoms and subsequently led to functional misinterpretations. Furthermore, an artificial opening in the ventricular wall may lead to changes in CSF pressure with unknown generalized effects on neuroendocrine and neural functions.

W.B. Severs: I am attracted by the concept that pineal and SCO form some sort of joint unity as far as function and activity goes. A suggestion to those who are using SCO antibodies, and perhaps a way to avoid the problem of damage, is that it is rather easy to pinealectomize the rat cleanly without bleeding. Have any of you tried to use pinealectomized animals, treating them with appropriate antisera to see what kind of biological changes emerge in the SCO.

E. Rodríguez: As a young student I pinealectomized a group of rats with the aim to look at the SCO and I have kept many unstained sections from these experiments. As soon as specific antibodies against the SCO secretion became available, I immunostained brain sections of these pinealectomized rats, which belonged to different groups, namely, 1 day, 5 days, and up to 1 month after pinealectomy. Simply by judging the amount of immunoreactive material in the SCO I did not see any changes after pinealectomy.

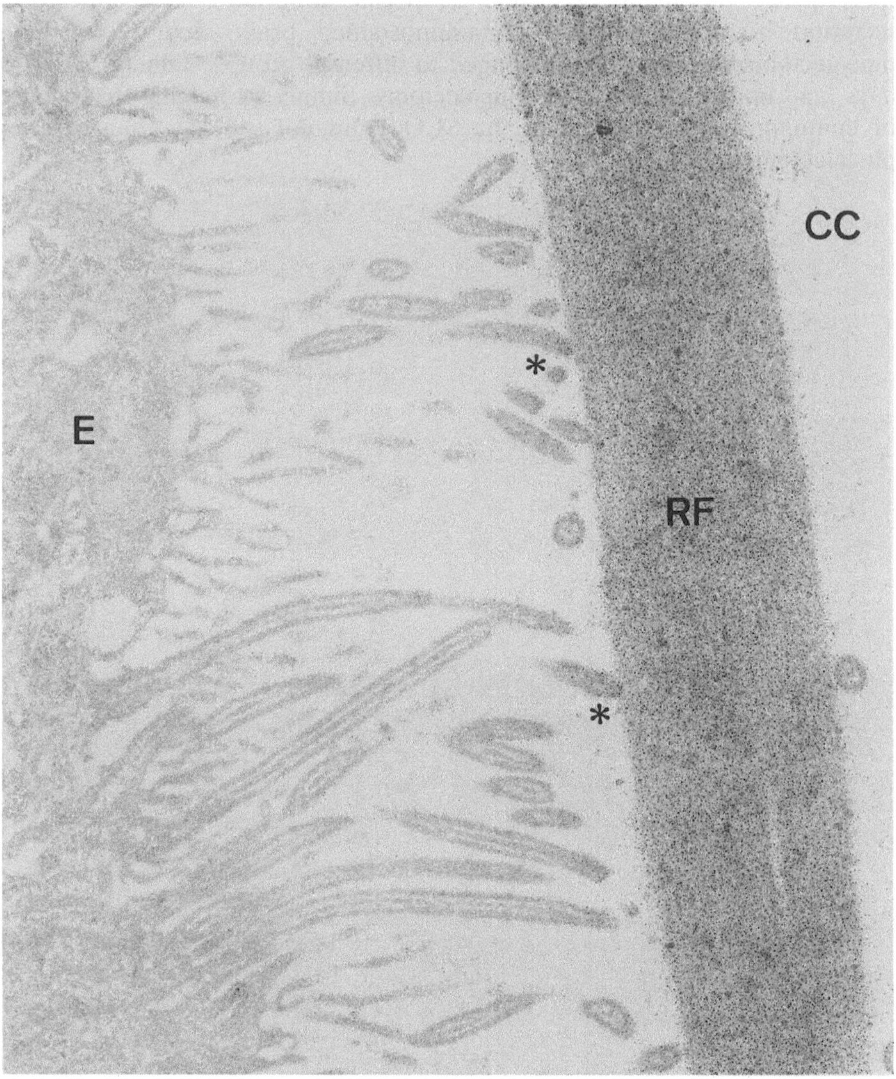

Fig. 3. Spinal cord of the lamprey, *Geotria australis*. Immunoperoxidase staining intensified with silver methenamine after the use of an antiserum against Reissner's fiber material. Reissner's fiber (*RF*) is strongly labeled. *CC*, central canal; *E*, ependyma. Tips of cilia (*asterisks*) touching RF also display attached immunoreactive material, ×18 800. Modified after Peruzzo B, Rodríguez S, Delannoy L, Hein S, Rodríguez EM, Oksche A (1987) Ultrastructural immunocytochemical study of the massa caudalis of the subcommissural organ–Reissner's fiber complex in lamprey larvae (*Geotria australis*): evidence for a terminal vascular route of secretory material. Cell Tissue Res 247, 367–376. For a scanning electron micrograph of an isolated Reissner's fiber, see Fig. 20 in Rodríguez EM, Oksche A, Hein S, Yulis CR (1992) Cell biology of the subcommissural organ. Int Rev Cytol 135:39–121

Discussion Session IV

Functional Aspects II

Chaired by W.B. SEVERS

W. Severs: It is fair to say that comparatively little is known about the functional role of the SCO, especially as it relates to day-to-day life. Although many possibilities have been put forth, the experimental record does not now provide strong support for any proposed function. An incomplete list of some possible functions include a "volume," "stretch," or "osmotic" (ion?) sensing mechanism that might make adjustments in cardiovascular and salt/water balance parameters. A "detoxification" function has been discussed, wherein the SCO removes excess endogenous or exogenous chemicals entering ventricular cerebrospinal fluid. Other possible functions of the SCO that have emerged are a role in the physiology of sleep and reproduction. Data presented at this symposium emphasize a new aspect of SCO function: namely, does it affect intracranial salt/water blance and CSF circulation? Implicit in these areas of inquiry is that the SCO could be involved in the pathophysiology of several models of hydrocephalus.

H. Jones: Neurophysiologists know that transections removing the forebrain are compatible with life, but transections that damage hindbrain control of cardiovascular, respiratory, and motor functions are not. The location of the SCO at the entrance to the aqueduct would permit it to act as a "caretaker" or "guardian" of these sensitive hindbrain areas by removing biologically active substances from CSF flowing towards the hindbrain. The area postrema is in a particular position, as this circumventricular organ in the hindbrain is intimately involved with cardiovascular and respiratory centers and transduces bloodborne signals.

F. Nürnberger: None of the classical brain transections permit animals to survive beyond a short period without extensive postoperative care. Even lower vertebrates need much of the forebrain for survival. Thus, the sense of a "caretaker" role based on the SCO position does not seem to offer an evolutionary advantage.

E. Rodríguez: I usually think that sometimes we may overlook very simple facts, such as the location of the SCO: why it's there! We could similarly ask why the adrenal medulla is placed where it is. There is always an explanation and with good reason. Here, the location of the SCO at the entrance of the funnel through which virtually all the ventricular fluid has to pass is telling us something. We just cannot identify what. Our experiments with central injections of an antibody against RF provide some perspective about the SCO location and possible protection of vital brain-stem centers.

At least for 6h after injection, not only is RF missing, but we assume that all kinds of secretory material released from the SCO into the ventricle would be blocked by the antibody. However, the animals seem to do fine in the acute period, even though we did not measure cardiovascular and respiratory effects. It is possible that a "short" RF is functional after disruption by the antibody and this may be adequate to serve a "clearance" function; I don't like the word "detoxication." Animals seem to survive well over a longer recovery period (about a month) when RF grows back near the entrance to the central canal, but does not enter it. Along these lines, the SCO may provide some protection to vital brain-stem structures, as Dr. Jones suggests, although acute disruption of RF by central antibody injections is not, by itself, lethal.

G. Sterba: There are animals that lead a normal life without a brain in the strict sense of the word. Maybe a spinal cord is satisfactory for basal functions of the organism. The SCO should be looked at from this point of view too.

R. Olsson: This situation appears to exist in amphioxus. Furthermore, the apendicularian central nervous system as far as we know develops in the same way as that of the chordates. There is a neural groove, and this groove closes its cavity all the way, so that the central nervous system is compact up to the trunk region. But in the tail the neural cord is hollow and there, far away from the brain, a single cell produces an intraluminal fiber. It is clear that embryonic cells from the appendicularian neural tube may have the capacity to produce RF substance wherever an RF is needed, and that in areas where a fiber is not required the CNS can become compact.

W. Lehmann: We have heard that the SCO and RF may be related to some morphogenetic aspect of nervous system development, and also to pathological conditions including hydrocephalus. However, little is known about the number of secretory proteins and how they function. Perhaps at least some of RF material exhibits carbohydrate (CHD)-binding properties, similar to lectins. Some such soluble CHD-binding substance could be a component of the extracellular matrix of the brain or be involved in a cell recognition processes. This may be important in morphogenesis and embryogenesis.

W. Severs: The circumventricular organs are often referred to as secretory structures. We know the SCO secretes RF material into CSF, and we are informed about the secretion release into the circulation in the neurohypophysis. How about other circumventricular organs? The subfornical organ (SFO), for example, was long considered to be secretory; it may contain structures similar to Herring bodies of the posterior hypophysial lobe. Has any progress been made related to secretory activity of circumventricular organs other than the SCO?

A. Oksche: The SCO secretion aggregating in RF and the pineal synthesis of melatonin are reasonably well known. The SFO possesses receptors for angiotensin II and other receptors. It may gather information via the bloodstream or by afferent neuroendocrine pathways and its action may be primarily neural, i.e., maintained by nerve cells sending specific signals to other brain areas. The possibility of secretory substances produced in the SFO, however, cannot be excluded.

W. Severs: Serotonin and GABA influences on the SCO are quite prominent. However, low amounts of tyrosine hydroxylase have been detected, indicating also a catecholaminergic representation. Several neuropeptides have been shown by immunohistochemical techniques, albeit at low densities. Thus, the SCO is neurochemically "rich," and the potential role of substances whose concentration or fiber density is low should not be overlooked. We do not know what signals these "low-level" neurochemicals transmit, nor how much amplification those signals may undergo.

F. Nürnberger: The paraventricular nucleus (PVN) provides an example supporting this idea of potentiation. The lateral septal nucleus receives many afferent vasopressin-immunoreactive projections, and tracer studies suggest that they originate from a very small portion, maybe 10–15 neurons, of the PVN. It is possible that small amounts of neuropeptides in the SCO may exert in a paracrine fashion important functions.

H. Jones: Why does RF have to be continuously regenerated? A possibility is that if it acts like a separation column, or ion-exchange column to take up active substances, it would become saturated if it were not continuously regenerated by new secretions.

W. Severs: Can some associations be made between a neurochemical present in the SCO and its biological effects when elevated in CSF during disease or under defined experimental conditions? For example, somatostatin is clearly elevated in CSF of animals after certain types of experimentally induced seizures, and one paper reported that the peptide is elevated in CSF from patients with benign intracranial hypertension. Could a somatostatin–SCO linkage be contributory to seizure-related events or CSF pressure?

B. Krisch: Apparently there are no somatostatin receptors on SCO cells.

W. Severs: Is it possible that something like somatostatin 1–28 is processed somewhere in the SCO to another peptide for which there would be a receptor?

J.H.B. Diederen: Morphologically based binding studies may not be sufficient. Biochemical studies may be more sensitive than morphological receptor studies. To my knowledge, no biochemical receptor studies have been performed along this line.

R. Yulis: At a certain stage in the development of the rat SCO in its basal region droplets are present which cross-react with anti-somatostatin. Thus, there appears to be some sort of functional relationship. The peptide may be used by the SCO itself.

W. Lehmann: Does the immunological blockade of RF also imply inactivation of all broken fragments of RF?

P. Fernández-Llebrez: The injection into the CSF of an anti-RF serum leads to an invasion of the CSF by immunocompetent cells. They appear to migrate from blood vessels of the choroid plexuses. These are the cells that break down RF; it is an immunological reaction.

W. Lehmann: After the injection into the CSF of the anti-RF serum, RF is broken down, the antibody is cleared from the CSF after 1 day, and the SCO remains secretory active as shown by Rodríguez et al. (this volume). Is it possible that the SCO secretes a soluble material that cannot be inactivated by the injected antibody?

P. Fernández-Llebrez: I do not know. I assume that the injected antibody would recognize the antigens present in the CSF, either soluble or in the form of RF.

R. Yulis: It seems possible that not all sites of the secretory molecule are blocked by the injected antiserum. The biological activity of these molecules would be blocked in the case the antibody binds to the active site of these molecules.

E. Rodríguez: I have the impression that something remains definitively wrong with the SCO after a single injection into the CSF of an anti-RF serum. Firstly, it takes a month to the newly formed RF to reach the entrance of the central canal of the spinal cord; this is a very low growth rate. Secondly, the newly formed RF, although it reaches the entrance of the central canal, cannot proceed beyond this point despite the fact that the central canal is wide open. Thirdly, and this may be relevant to Dr. Lehmann's question, in the rabbits injected with an anti-RF serum into the CSF, the amount of soluble RF material, assayed by ELISA, increased enormously.

A. Oksche (final remark, added in proof): Although it is possible that the SCO produces a hormonelike agent involved in the control of homeostatic mechanisms, there are good reasons to follow the line of thought linking the secretory product of the SCO with the extracellular matrix, cell-surface molecules, trophic (growth) factors and embryogenesis. At the present time, potent in vitro and in situ systems are available which could serve in the experimental exploration of these questions. Within this conceptual framework even a combination of endocrine and trophic (or communicative) aspects appears debatable. For example, recent regeneration studies in ascidians [Bollner et al. (1993), Cell Tissue Res 272:545–552] have shown that certain neuropeptides (e.g., substance P, CCK) are relevant to both normal development and regeneration. In addition to their potential as growth-promoting factors these peptides act as neuroendocrine agents in the normal physiology of these animals. Also the close molecular relationship between certain neurohormones of invertebrates (insects, mollusks) and insulinlike growth factors should be kept in mind (see Joosse 1986; in: Ralph CL (ed) Comparative endocrinology: developments and directions. Liss, New York). With respect to the SCO and its secretory products having access to the inner and the outer CSF-space, it appears worthwhile to consider these possibilities and to design adequate experiments.

Subject index

Springer-Verlag and the Environment

MIX
Papier aus verantwortungsvollen Quellen
Paper from responsible sources
FSC® C105338

If you have any concerns about our products,
you can contact us on
ProductSafety@springernature.com

In case Publisher is established outside the EU,
the EU authorized representative is:
Springer Nature Customer Service Center GmbH
Europaplatz 3, 69115 Heidelberg, Germany

Printed by Libri Plureos GmbH
in Hamburg, Germany